SURFACE AND COLLOID SCIENCE

Volume 16

SURFACE AND COLLOID SCIENCE

Volume 16

Edited by

EGON MATIJEVIĆ

Center for Advanced Materials Processing
Clarkson University
Potsdam, New York

Kluwer Academic / Plenum Publishers
New York, Boston, Dordrecht, London, Moscow

ISBN 0-306-46456-X

©2001 Kluwer Academic / Plenum Publishers, New York
233 Spring Street, New York, New York 10013

http://www.wkap.nl

10 9 8 7 6 5 4 3 2 1

A C.I.P. record for this book is available from the Library of Congress

Printed in the United States of America

Preface to the Series

A need for a comprehensive treatise on *surface and colloid science* has been felt for a long time. This series endeavors to fill this need. Its format has been shaped by the features of this widely roaming science. Since the subjects to be discussed represent such a broad spectrum, no single person could write a critical review on more than a very limited number of topics. Thus, the volumes will consist of chapters written by specialists. We expect the series to represent a treatise that will describe theories, systems, and processes in a comprehensive manner, and indicate solved problems and problems which still require further research. Purely descriptive colloid chemistry will be limited to a minimum. Qualitative observations of poorly defined systems, which in the past have been so much in evidence, will be avoided. Thus, the chapters are neither supposed to possess the character of *advances,* nor to represent reviews of authors' own studies. Instead, it is hoped that each contribution will treat a subject critically, giving the historic development as well as a digest of the newest results. Every effort will be made to include chapters on novel systems and phenomena.

It is impossible to publish a work of this magnitude with all chapters in a logical order. Rather, the contributions will appear as they arrive, as soon as the editor receives sufficient material for a volume. A certain amount of overlap is unavoidable and uniform treatment and style cannot be expected in a publication that represents the effort of so many. Notwithstanding these anticipated difficulties, the series as described appears to be the only practical way to accomplish the task of a high-level and modern treatise on surface and colloid science.

Some general remarks may be in order. In modern times, few disciplines have fluctuated in "popularity" as much as colloid and surface science. However, it seems that these sporadic declines in interest in the science of "neglected dimensions" have been only apparent. In reality, there has been a steady increase in research through the years, especially in industrial laboratories. The fluctuations have been most noticeable in academic institutions especially with regard to teaching of specialized courses. It is natural that university professors with surface and colloid science as their abiding interest have expressed frequent concern for and have repeatedly warned of the need for better and more intensive education, especially on the graduate level.

There are several reasons for the discrepancy between the need of industrial and academic research laboratories for well-trained surface and colloid scientists and the efforts of the academic institutions to provide specialization in these disciplines. Many instructors believe that a good background in basic principles of chemistry, physics, and mathematics will enable a professional person to engage in research in surface and colloid science. This may be true, but only after much additional professional growth. Indeed, many people active in this area are self-educated. Furthermore, this science deals with an unusually wide range of systems and principles, this makes a uniform treatment of problems in surface and colloid science not only challenging but also a very difficult task. As a matter of fact, certain branches of colloid science have grown into separate, independent disciplines which only in a broad sense are now considered a part of the "parent" science. Finally, there is often a stigma associated with the name "colloids." To many, the term symbolizes empirically and poorly described, irreproducible, etc., systems to which exact science cannot as yet be applied. The latter is in part based on the fact that a considerable number of papers were and are published that leave much to be desired with regard to the rigorousness of the approach.

Yet, during the first half of the last century some of the most acclaimed scientists have occupied themselves with colloid and surface science problems. One needs to mention only a few, such as Einstein, von Smoluchowski, Debye, Perrin, Loeb, Freundlich, Zsigmondy, Pauli, Langmuir, McBain, Harkins, Donnan, Kruyt, Svedberg, Tiselius, Frumkin, Adam, and Rideal, who have made substantial contributions to the classical foundations of colloid and surface science. This work has led to many fundamental theoretical advances and to a tremendous number of practical applications in a variety of systems such as natural and synthetic polymers, proteins and nucleic acids, ceramics, textiles, coatings, detergents, lubricants, paints, catalysts, fuels, foams, emulsions, membranes, pharmaceuticals, ores, composites, soils, air and water pollutants, and many others.

It is therefore our hope that this treatise will be of value to scientists of all descriptions, and that it will provide a stimulating reference work for those who do not need to be convinced of the importance of colloid and surface science in nature and in application.

EGON MATIJEVIĆ

Preface to Volume 16

The preceding "Preface to the Series," written in 1969, contains statements which are even more appropriate now in view of the present status of the science of colloids and interfaces. Therefore the continuation of this publication is fully justified. Indeed, this series is the only surviving such endeavor in the literature.

As with previous works in this series, Volume 16 contains contributions on three diverse topics of interest to readers in both industry and academia, with extensive literature citations, including the most recent publications on the relevant subjects.

Although there has been some delay in the appearance of the present volume, it is expected that future books in this series will be issued more frequently as a part of the Kluwer Academic/Plenum Publishers program.

EGON MATIJEVIĆ

Contents

2. Ionization Processes and Proton Binding in Polyprotic Systems: Small
   Molecules, Proteins, Interfaces, and Polyelectrolytes
   Michal Borkovec, Bo Jönsson, and Ger J. M. Koper

3. Combined Application of Radiochemical and Electrochemical
   Methods for the Investigation of Solid/Liquid Interfaces
   Kálmán Varga, Gábor Hirschberg, Pál Baradlai, and
   Melinda Nagy

Physical Chemistry of Cetyl Alcohol: Occurrence and Function of Liquid Crystals in O/W Creams

Shoji Fukushima and Michihiro Yamaguchi

Introduction

Cetyl alcohol is one of the important components in various preparations, such as cosmetic creams or lotions, or pharmaceutical hydrophilic ointments. Sometime ago a curious phenomenon was observed, namely, the cream was unstable when prepared with 1-hexadecanol instead of cetyl alcohol. This finding triggered studies on the difference between 1-hexadecanol and cetyl alcohol by many workers, including the authors of this review.

Much could be written on these studies, and the explanations proposed here about the difference between 1-hexadecanol and cetyl alcohol are based on a survey of previous achievements. The results of other investigators are presented, since the explanation of the effects offered here are not easily understood and have been reported fragmentarily in different publications.

The leading aspect of the effects observed is the crystal transformation of the higher alcohol, 1-hexadecanol, the transition point of which decreases if it includes 1-octadecanol. Why does such a phenomenon occur? This problem of molecular interaction could be solved by the molecular orbital method, but it has not been done so far. Nevertheless, there is good reason to describe the problems and effects in detail, because they will certainly be useful to many researchers who are working on the formulation of emulsion-type cosmetics or pharmaceuticals.

There are some aspects that are useful in relation to a live body as well as to fundamental research. The former refer to the study of the physiological influence of higher alcohols. The weight of a rat brain is about 10 g and it contains 1 mg of higher alcohols, 1-hexadecanol and 1-octadecanol being the main ones. What kind of role do these higher alcohols play in the brain?

Shoji Fukushima and *Michihiro Yamaguchi* • Basic Research Laboratory, Shiseido Research Center, Yokohama, Japan.

Another example applies to cell membranes, the main component of which comprises phospholipids. The aliphatic part of the latter consists of esterified higher acids. Since the transition of the lipophilic layer affects the nature of the membrane, the study is essential for knowledge about living cells. These transitions are similar to those of higher alcohols. Thus, the studies described here may help us to understand processes in the latter.

1. Cetyl Alcohol

1.1. Description of Cetyl Alcohol

There are numerous alcohol molecules which differ in the number and position of OH groups they bind. Among these, cetyl alcohol is said to be 1-hexadecanol, which is a primary straight-chain compound having 16 carbons. This alcohol is an important raw material in the production of surfactants, along with lauryl and oleyl alcohols. The most characteristic property of cetyl alcohol is its ability to increase the viscosity and stability of O/W emulsions. As such, it is an essential constituent in making O/W cream-type cosmetics or pharmaceutical hydrophilic ointments. It is no exaggeration to state that nobody can produce cosmetic creams without cetyl alcohol. For this reason many colloid chemists have been challenged to establish the mechanism of underlying reactions that make this ingredient indispensable in the cosmetics industry.

Some studies have focused on the interfacial film viscosity due to interactions with a surfactant, while others have examined the properties from the standpoint of liquid crystal formation. Since no other higher alcohols have exhibited similar properties, no investigations have been carried out to consider the possible influence of mixing such alcohols. In experiments, it is customary to use chemicals of high purity. Thus, many workers have used "pure" 1-hexadecanol instead of "impure" cetyl alcohol, which is considered to be synonymous with the former, but few have noted that 1-hexadecanol does not exhibit emulsion-stabilizing or viscosity-increasing effects at ordinary room temperatures.

Essentially, cetyl alcohol is not 1-hexadecanol but a higher alcohol obtained by hydrolyzing spermaceti, a wax component of the sperm whale head oil, 1-hexadecanol being one of its main components. However, investigators working on the role of higher alcohols in cosmetic creams realized that the commercial cetyl alcohol contained much 1-octadecanol and, therefore, there are many nomenclatures applied to this material, as summarized in Table 1.

1.2. Short History of Cetyl Alcohol

To retrace the history of cetyl alcohol, one may go back to 1813.[1] Chevreul found that a new material was obtained by saponifying spermaceti, which is a wax component of the sperm whale head.[2] He called this component "et al." In 1836,

Table 1.   *Various Names of Cetyl Alcohol and Analogous Compounds or Materials*

Name	Definition
Cetanol	The higher alcohol adopted in the Japanese Pharmacopoeia. It consists mainly of 1-hexadecanol
Cetostearyl Alcohol	The higher alcohol adopted in the British Pharmacopoeia. It consists mainly of 1-hexadecanol and 1-octadecanol
cetyl alcohol	A higher alcohol obtained by hydrolysis of spermaceti, but frequently identified with 1-hexadecanol
Cetyl Alcohol	The higher alcohol adopted in the United States' Pharmacopoeia. It consists mainly of 1-hexadecanol
Cetylstearylalkohol	The higher alcohol adopted in the German Pharmacopoeia
1-hexadecanol	A primary straight-chain saturated higher alcohol with 16 carbons
1-octadecanol	A primary straight-chain saturated higher alcohol with 18 carbons
stearyl alcohol	A higher alcohol obtained by hydrolysis and reduction of beef tallow, but frequently identified with 1-octadecanol
Stearyl Alcohol	The higher alcohol adopted in the United State's Pharmacopoeia. It consists mainly of 1-octadecanol

Dumas and Perigot found the material to have the qualities of alcohol,[3] and in 1854 Heintz established it to be an impure compound.[3] Later on, Krafft proved this alcohol to consist mainly of 1-hexadecanol and 1-octadecanol.[4] Andre and François also reported that small amounts of 1-tetradecanol and 1-octadecanol are contained in cetyl alcohol obtained by fractionation, saponification, and recrystallization after esterification of the higher alcohol of a sperm whale.[3]

Thus, the term "cetyl alcohol" originally did not imply the composition of 1-hexadecanol, a saturated higher alcohol with 16 carbons, but described a higher alcohol obtained by hydrolyzing spermaceti. Nowadays, it is usual to use synonymously the words "cetyl alcohol" and "1-hexadecanol," but caution is needed, since there are cases where the former is used for the product of the original meaning.

In 1878, Ludwig reported that cetyl alcohol was found in the fat of dermoid cysts,[5] which was also studied by von Zeynek.[6] However, Ameseder identified the compound as eicosyl alcohol, $C_{20}H_{42}O$, and it is unclear whether cetyl alcohol is contained in this fat at all.[7]

There are also studies on the physiological nature of cetyl alcohol. For example, Izumi *et al.* studied the effect of higher alcohols like the sexual hormone.[8] Stewart, and Vines and Meakins investigated the use of cetyl alcohol for the purpose of retarding evaporation of water from pools.[9,10]

The most important application of cetyl alcohol is in the preparation of O/W emulsion-type cosmetics or pharmaceuticals, which appear to have been originally developed in the 1930's. Before that time, spermaceti and borax had been used to prepare the so-called cold creams. Since spermaceti is saponified by borax to produce the fatty acid soap and cetyl alcohol, it can be said that the specific functions

Table 2. Formulation of Acid Cream According to Kalish[11]

Components	wt%
Cetyl alcohol	15.0
Spermaceti	5.0
Sodium laurylsulfate	2.0
Lactic acid (85%)	1.2
Water	71.8
Glycerol	5.0

of this alcohol to increase the viscosity and to stabilize the emulsion were utilized without the realization of its true role. The same combination of ingredients was used for a long time after cetyl alcohol became available as a commercial raw material.

Kalish introduced several formulations with cetyl alcohol, as exemplified in Tables 2 and 3.[11]

In the early days cetyl alcohol contained considerable amounts of 1-octadecanol.[12] Afterwards stearyl alcohol, derived by reduction of tallow fatty acid, increased in consumption. In the United Kingdom other higher alcohols were used, designated as "alcohol C14 mixture" or "alcohol C16-18 mixture," consisting of linear saturated chains with carbon numbers indicated by respective numerals.

1.3. Definitions in Official Books

The definition of cetyl alcohol and its analogues varies not only with nation, but also with time, as enumerated below in chronological order.

Year Nomenclature and Definition

1932 Cetanol (Japanese Pharmacopoeia V and VI)
      Cetanol is the one obtained by saponification of spermaceti or on high pressure contact reduction of sperm oil. It consists mainly of cetyl alcohol ($C_{16}H_{33}OH$).

Table 3. Formulation of Soft Cream According to Kalish[11]

Components	wt%
Cetyl Alcohol	15.0
Sodium laurylsulfate	1.5
Lanolin	1.5
Lactic acid (85%)	1.0
Water	81.0

Year Nomenclature and Definition

1947 Stearyl Alcohol (United States Pharmacopoeia XIII to XV)
Stearyl Alcohol is a mixture of solid alcohols consisting chiefly of
stearyl alcohol [$CH_3(CH_2)_{16}CH_2OH$].

1950 Cetyl Alcohol (National Formulary IX to XIV)
Cetyl Alcohol is a mixture of solid alcohols consisting chiefly of
cetyl alcohol.

1951 Stearyl Alcohol (Japanese Pharmacopoeia VI)
Stearyl Alcohol is a mixture of solid alcohols consisting chiefly of
stearyl alcohol [$CH_3(CH_2)_{16}CH_2OH$].

1958 Cetostearyl Alcohol (British Pharmacopoeia 1958 to 1973)
Cetostearyl Alcohol is a mixture of solid aliphatic alcohols consisting chiefly of stearyl and cetyl alcohols. It may be obtained by
the reduction of the appropriate fatty acids or from sperm oils.

1960 Stearyl Alcohol (United States Pharmacopoeia XVI to XVII)
Stearyl Alcohol is a mixture of solid alcohols, consisting chiefly
of stearyl alcohol ($C_{18}H_{38}O$).

1968 Cetostearyl Alcohol (British Pharmaceutical Codex 1968)
Cetostearyl Alcohol is a mixture of solid aliphatic alcohols and
consists chiefly of stearyl alcohol, $CH_3(CH_2)_{16}CH_2OH$, and cetyl
alcohol, $CH_3(CH_2)_{14}CH_2OH$, with minor amounts of other alcohols, mainly myristyl alcohol $CH_3(CH_2)_{12}CH_2OH$. The proportion
of stearyl alcohol to cetyl alcohol varies considerably, but cetostearyl alcohol usually consists of about 50 to 70 per cent of stearyl
alcohol and about 20 to 35 per cent of cetyl alcohol. Material which
is known in commerce as "cetyl alcohol" unless specified as pure,
is a cetostearyl alcohol usually containing about 60 to 70 per
cent of cetyl alcohol and 20 to 30 per cent of stearyl alcohol.
Pure cetyl alcohol, containing not less than 98 per cent of
$CH_3(CH_2)_{14}CH_2OH$, is also available. Cetostearyl alcohol occurs
as a white or cream coloured unctuous mass, or almost flakes or
granules, with a faint characteristic odour and a bland taste. It has
a melting point lower than 43 °C, and melts to a clear colourless
or pale yellow liquid which is free from cloudiness or suspended
matter.

1968 Cetylstearylalkohol (Deutsches Arzneibuch VII)
Gemisch aus Cetylalkohol ($C_{16}H_{34}O$:242.4) und Stearylalkohol
($C_{18}H_{38}O$:270.5).

1970 Stearyl Alcohol (United States Pharmacopoeia XVIII)
Stearyl Alcohol contains not less than 90.0 per cent of stearyl
alcohol ($C_{18}H_{38}O$), the remainder consisting chiefly of cetyl alcohol ($C_{16}H_{34}O$).

Year Nomenclature and Definition

1971 Cetanol (Japanese Pharmacopoeia VIII)
      This material is a mixture of solid alcohols, consisting chiefly of
      cetanol ($C_{16}H_{34}O$).
1971 Stearyl Alcohol (Japanese Pharmacopoeia VIII)
      This material is a mixture of solid alcohols, consisting chiefly of
      stearyl alcohol ($C_{18}H_{38}O$).
1973 Cetostearyl Alcohol (British Pharmaceutical Codex 1973)
      Cetostearyl Alcohol is a mixture of solid aliphatic alcohols and
      consists chiefly of stearyl alcohol, $CH_3(CH_2)_{16}CH_2OH$, and cetyl
      alcohol, $CH_3(CH_2)_{14}CH_2OH$, with small amounts of other alco-
      hols, mainly myristyl alcohol $CH_3(CH_2)_{12}CH_2OH$. Cetostearyl
      alcohol usually consists of about 50 to 70 per cent of stearyl
      alcohol and about 20 to 35 per cent of cetyl alcohol, but propor-
      tions may vary considerably. Material which is known in com-
      merce as "cetyl alcohol" unless specified as pure, is a cetostearyl
      alcohol usually containing about 60 to 70 per cent of cetyl alcohol
      and 20 to 30 per cent of stearyl alcohol. Pure cetyl alcohol,
      containing not less than 98 per cent of $CH_3(CH_2)_{14}CH_2OH$, is also
      available.
1975 Cetyl Alcohol (National Formulary XIV)
      Cetyl alcohol is a mixture of solid alcohols consisting chiefly of
      cetyl alcohol.
1979 Cetostearyl Alcohol (British Pharmaceutical Codex 1979)
      A mixture of solid aliphatic alcohols and consists chiefly of
      stearyl alcohol, $CH_3(CH_2)_{16}CH_2OH$, and cetyl alcohol,
      $CH_3(CH_2)_{14}CH_2OH$, with small amounts of other alcohols, mainly
      myristyl alcohol, $CH_3(CH_2)_{12}CH_2OH$.
1980 Cetostearyl Alcohol (British Pharmacopoeia 1980 to 1990)
      Cetostearyl Alcohol is a mixture of solid aliphatic alcohols con-
      sisting chiefly of stearyl and cetyl alcohols. It may be obtained by
      the reduction of the appropriate fatty acids.
1980 Cetyl Alcohol (National Formulary XV)
      Cetyl Alcohol contains not less than 90.0 per cent of cetyl alcohol
      ($C_{16}H_{34}O$), the remainder consisting chiefly of stearyl alcohol
      ($C_{18}H_{38}O$).
1980 Stearyl Alcohol (National Formulary XV)
      Stearyl Alcohol contains not less than 90.0 per cent of stearyl
      alcohol ($C_{18}H_{38}O$), the remainder consisting chiefly of cetyl alco-
      hol ($C_{16}H_{34}O$).

2. Physical Properties of Cetyl Alcohols

2.1. Polymorphism of Higher Alcohols

Higher alcohols have specific boiling and melting points, as do other organic compounds. Below its melting point, each alcohol has a specific temperature at which it changes from a crystalline structure to another form, a phenomenon referred to as transition, and the temperature being called the transition temperature or transition point. When 1-hexadecanol is cooled from a temperature above its melting point (49 °C), it solidifies at that temperature, adopting a hexagonal structure, and at 43 °C the structure changes to monoclinic. This transition point is easily determined with a differential scanning calorimeter.

The hexagonal crystal form, in which 1-hexadecanol appears at temperatures between 49 and 43 °C, is referred to as the high temperature stable type,[13–15] or α-type. In this case the hydrocarbon chains are oriented vertically to the *c* plane as shown in Fig. 1(A). Below 43 °C the crystal forms of 1-hexadecanol are designated as β- and γ-types. Both are monoclinic, with the angle of tilt of the hydrocarbon in β-type crystals equal to 91–92° and that of γ-type 122–123° [Fig. 1(B) and (C)]. The crystal parameters of 1-hexadecanol, 1-heptadecanol, and 1-octadecanol are listed in Table 4.

The names of the three (although some reports in the literature give four or six) are associated with different researchers, as shown in Table 5. The designation by Wilson and Ott is used in this text.

The X-ray powder diffraction (XRD) pattern of α-type higher alcohols is characterized by 4 to 5 diffraction peaks which appear at regular intervals in the region below 10° of Bragg's angle and one strong single peak which appears at 21.4°. The structural peculiarity of this type is as follows. The lipophilic chains,

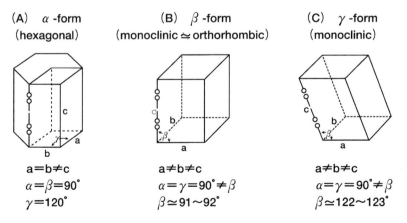

(A) α-form (hexagonal)	(B) β-form (monoclinic ≃ orthorhombic)	(C) γ-form (monoclinic)
$a = b \neq c$	$a \neq b \neq c$	$a \neq b \neq c$
$\alpha = \beta = 90°$	$\alpha = \gamma = 90° \neq \beta$	$\alpha = \gamma = 90° \neq \beta$
$\gamma = 120°$	$\beta \approx 91 \sim 92°$	$\beta \approx 122 \sim 123°$

Figure 1.   Schematic illustration of crystal structures of n-higher alcohol.

Table 4.   Crystal Parameters of β- and γ-Higher Alcohols

Alcohol	Crystal type	Space group	Lattice parameters (nm)			β-angle^a	Ref.
			a (nm)	b (nm)	c (nm)		
1-Hexadecanol	γ	$C_{2h}^6 - A2/a$	0.895	0.493	8.81	122° 23′	16
1-Heptadecanol	β	$C_{2h}^5 - P2_1/c$	0.503	0.740	9.46	91° 18′	17
1-Octadecanol	γ	$C_{2h}^6 - A2/a$	0.896	0.493	9.97	123° 03′	17

^{a}Angle of tilt of the c-axis with respect to the c-plane.

which are hexagonally packed as "logs," are piled up and oriented rectangularly against the plane of the OH hydrogen bond networks. Oxygen atoms of the OH groups are placed at the three apexes of the base and the center of the tetrahedron, the hydrogen atoms forming hydrogen bonds with the adjacent oxygens to produce a network. These structures are shown in Fig. 2.

In the α-structure, the hydrocarbon chains are said to rotate around their long axis. The possibility of molecular rotation within the crystal has been pointed out by Pauling.[28] Bernal interpreted the results with 1-dodecanol, obtained using a microscope and XRD, in terms of molecules rotating around the long axis at a few degrees below the melting point.[29,30]

Table 5.   Various Designations for Crystal Forms of Higher Alcohols

	Crystal type			Reference
Smith	—	$β_1$	$β_2$	19
Wilson and Ott	α	β	γ	18
Hoffman and Smyth	α	$β_1$	$β_2$	20
Kolp and Lutton	α	Sub-α	β′	21
Chapman	α	$β_1$	$β_2$	22
Tanaka *et al.*	$α_0$	β	$γ_1$	13
	$α_1$	—	$γ_2$	
	$α_2$	—	—	
Vines and Meakins	α	Sub-α	—	23
Abrahamsson *et al.*	α	β	γ	16
Stewart	α	Sub-α	—	24
Benton	α	α′	β	25
	—	Sub-α	—	
Brooks	α	Sub-α	β	26
Fukushima *et al.*	α	β	γ	27

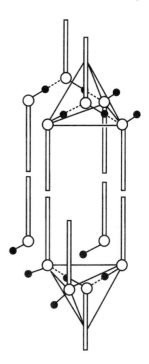

Figure 2.   Structure of a-type 1-hexadecanol. Open and full circles
represent oxygen and carbon atoms, respectively, and hollow bars
show hydrocarbon chains. Solid and dotted lines represent covalent
bonds and hydrogen bonds, respectively. Tetrahedrons are drawn to
show positional relations of oxygen atoms.[27]

Frosch carried out the dielectric constant measurements and observed a strong
frequency dispersion, suggesting molecular rotation, but the data were not reproducible.[31] Ott could not detect rotation in the XRD experiments.[32] The dielectric
constant measurements by Hoffman and Smyth indicated the rotation of higher
alcohols over a certain temperature range below the solidification point (24 °C).[20]
Figure 3 show the results on 1-dodecanol. The dielectric constants measured at 0.5,
5.0, and 50 kHz with temperatures below 32.1 °C show a sudden decrease at around
23.5 °C.

Analogous data for 1-tetradecanol (Fig. 4) show a characteristic feature, i.e.,
the dielectric constant curve at 0.5 kHz suddenly rises on cooling at 36 °C and falls
again at 34.5 °C. In the measurements at other frequencies, the rise did not occur
and only the fall was observed, which took place as at 0.5 kHz. Figure 5 displays
the results at 0.25 kHz with measurements carried out by the heating method,
yielding a transition point of 35.7 °C.

Similar dependencies of the dielectric constant on temperature were observed
with other higher alcohols. The temperature at which the dielectric constants
suddenly vary differs somewhat, which is based on the time requirement to change
the structure, i.e., to complete the crystal transformation. Table 6 shows the
temperatures at which the higher alcohols suddenly exhibited a change in the
dielectric constant.

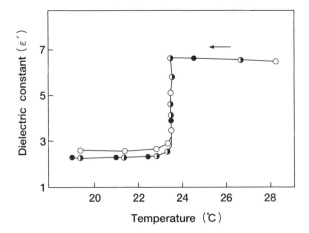

Figure 3. Temperature dependence of the dielectric constant of 1-dodecanol. Open circles represent values at 0.5 kHz, half-filled circles those at 5.0 kHz, and full circles those at 50 kHz.[20]

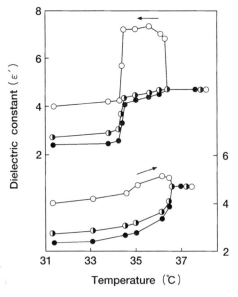

Figure 4. Temperature dependence of the dielectric constant of 1-tetradecanol. Open circles represent values at 0.5 kHz, half-filled circles that at 5.0 kHz, and full circles those at 50 kHz. The upper curve scale is on left, while the lower curve scale is on right. Arrows indicate thermal direction.[20]

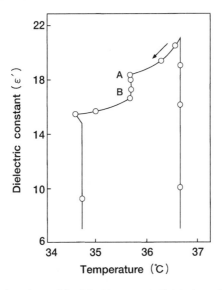

Figure 5.   Temperature dependence of the dielectric constant of 1-tetradecanol. The sample was cooled in the usual way to 35.7 °C, where it was held for one hour. The dielectric constant decreased from A to B. Normal cooling was then allowed to resume.[20]

Table 6.   Temperatures at which the Dielectric Constant of Higher Alcohols Abruptly Changes[a]

Alcohol	Frequency (kHz)				Transition point[b]
	0.25	0.5	5.0	50	
Cooling method		(°C)			(°C)
1-Dodecanol	—	23.5	23.5	23.5	23.8
1-Tetradecanol	35.7	—	—	—	33.5–34.9
	34.7	34.4	34.4	34.4	
1-Octadecanol	55.2	—	—	—	52–54
	53.4	53.4	53.4	53.4	
1-Docosanol	—	64.5	64.5	64.5	63.5
Heating method		(°C)			(°C)
1-Tetradecanol	36.7	36.3	—	—	35.2
1-Octadecanol	57.7	57.7	57.7	57.7	54–54.5
1-Dodecanol	—	69.6	69.6	69.6	67.6

[a] The table is based on Hoffman and Smyth's data.[20]

[b] See Table 8.

The temperatures at which the dielectric constants rise are in good agreement with the corresponding melting points and those at which they fall are consistent with the transition points. From these measurements, it is inferred that the hydrocarbon chains of higher alcohols rotate around the long axis of their molecules in the α-phase region; i.e., in the region between the solidification and transition points.

It should be emphasized here that, although the α-structure is hexagonal in the solid state, as lyotropic liquid crystals it appears lamellar, with a repetitive overlap of half molten lipophilic and OH layers. In contrast, the solid state is maintained by the hydrogen bond networks of the OH groups.

2.2. Crystal Structure of Higher Alcohols

The hexagonal structure of higher alcohols changes to monoclinic at a lower temperature. The transition temperature differs depending on the nature of the alcohol, being 43 °C in the case of 1-hexadecanol. Although there are two types of solids in the monoclinic forms (β and γ), so far it is not clear under which conditions the α-form will transform into either the β- or γ-structure. XRD patterns of β- and γ-alcohols closely resemble each other. They are characterized by 4 to 5 diffraction peaks, which appear at regular intervals in the region below 10° Bragg's angle and by twin peaks at 21.4° (strong) and 24.0° (weak). The two forms can, however, be clearly distinguished by accurately calculating interplanar spacings. The structures are shown in Figs. 6 and 7.[16,20]

The most important difference between the β- and γ-forms is that the latter is composed of *trans* molecules only, while the former consists of equal numbers of *trans* and *gauche* molecules.[15] In the former all skeletal carbon and oxygen atoms are coplanar, while in the latter oxygen atoms are out of the skeletal plane. The difference in the infrared spectra of the *b*- and *g*-forms is interpreted in terms of this rotational isomerism.

Tanaka *et al.* discussed the relative stability of β- and γ-structures at higher temperatures. They observed by the heating method that β–α transition takes place at 42 °C and γ–α transition at 4 °C. Neither β–γ transition was noted in the heating process nor α–γ transition in the cooling process.[14] In the present authors' experience, the γ-form showed a tendency to appear at a higher temperature than the β-form, while at an intermediate temperature there was a case in which both forms were detected.[33] Table 7 summarizes the structural parameters of higher alcohols collected from the literature.

2.3. Melting Point and Transition Point of Higher Alcohols

There are several ways to determine the melting point and transition point of higher alcohols: (1) to visually observe the external appearance of the sample while heating or cooling it at a constant rate, (2) to follow the change in the sample

$C_{17}H_{35}OH$ (β-form)

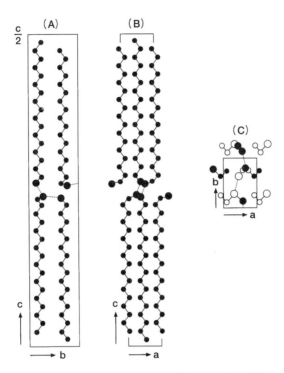

Figure 6.   Molecular arrangement of 1-heptadecanol in the β-form as viewed along the *a*-axis (A), the *b*-axis (B), and the axis of the molecule (C). Large circles represent oxygen atoms, and full and hollow circles in (C) represent upper and lower layers, respectively.[17]

temperature with time while heating or cooling it at a constant rate, (3) to measure the thermal absorption or release while heating or cooling at a constant rate, (4) to measure the volume change with temperature using a dilatometer, (5) to measure the dielectric constant change with temperature, and (6) to investigate the change in the crystal structure with temperature (e.g., with an XRD).

Methods (2) and (3) indicate that thermal variations take place, while they do not offer information about structural changes. Dilatometry (4) yields only the volume increase. The dielectric constant method (5) reveals any variation in the dielectric property, i.e., the state of dipoles. With the XRD method (6), one obtains directly the structural variations. This method, however, is inconvenient and consumes much time in order to accurately control the temperature of the specimen in an air bath.

In all cases, the information is obtained on physical variations taking place at a specific temperature. Some small difference is noted between the melting and

C₁₈H₃₇OH (γ -form)

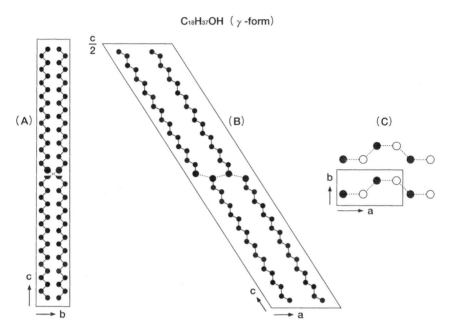

Figure 7.   Molecular arrangement of 1-octadecanol in the γ-forms as viewed along the *a*-axis (A), the *b*-axis (B), and the axis of the molecule (C). Large circles represent oxygen atoms, and full and open circles in (C) represent upper and lower layers, respectively.[17]

solidification points, and between the transition points, as dependent on the thermal direction of measurements (i.e., raising or lowering the temperature), since a certain length of time is required for cylindrical and large molecules, such as of higher alcohols, to initiate the molecular transformation. In these cases, the transition point is higher when measured by the temperature raising rather than lowering method. The melting and transition points data are summarized in Table 8.

Higasi and Kubo employed the transition point measurements using a dilatometer and found that the transition of higher alcohols from the β- to the α-form was accompanied by a volume increase of about 4%.[47] The extent of this expansion reflects a reasonable volume change due to rotational transition, characteristic for dipoles in the α-modification.

2.4.   Transition Point and Infrared Absorption

The polymorphism of higher alcohols may be investigated by infrared (IR) absorption. According to Chapman,[22] the IR spectrum of a long-chain primary alcohol (such as cetyl alcohol) obtained in the liquid state develops different changes when the alcohol is gradually cooled down to room temperature. The observed shifts in the OH stretching frequency indicate increased hydrogen bond-

Table 7. Long Interplanar Spacings and b-Angles of Higher Alcohols

Carbon number	α-Form LIPS[a] (nm)	β-Form LIPS (nm)	β-Angle[b]	γ-Form LIPS (nm)	β-Angle	Ref.
11	2.88	3.22	91° 27′	—	—	14
12	3.41	3.49	91° 25′	—	—	14
13	3.61	3.73	91° 24′	—	—	14
	3.69	—	—	—	—	19
14	3.89	3.96	91° 23′	—	—	14
	3.97	—	—	—	—	19
	3.98	—	—	—	—	34
15	4.13	4.24	91° 22′	—	—	14
	4.21	—	—	—	—	19
16	4.38	4.49	91° 21′	3.73	52° 31′	14
	4.46	—	—	—	—	19
	4.49	3.74	—	—	—	34
17	4.63	4.75	91 20′	—	—	14
	4.72	—	—	—	—	19
	—	$a = 0.503$	91° 18′	—	—	29
	—	$b = 0.740$	—	—	—	
	—	$c = 9.46$	—	—	—	
18	4.89	5.03	91° 19′	4.16	52° 48′	14
	—	—	—	$a = 0.895$	122° 23′	16
	—	—	—	$b = 0.493$	—	
	—	—	—	$c = 8.81$	—	
	—	—	—	$d_{001} = 7.44$	—	
	—	—	—	$a = 0.896$	123° 03′	29
	—	—	—	$b = 0.493$	—	
	—	—	—	$c = 9.97$	—	
	—	—	—	$d_{001} = 4.28$	—	
	4.954	4.10	—	—	—	19
	5.02	4.14	—	—	—	34
19	5.11	5.28	91° 18′	—	—	14
20	5.31	—	—	4.57	53° 02′	14
21	5.53	5.80	91° 18′	—	—	14
22	5.77	—	—	4.99	53° 13′	14
23	5.98	6.31	91° 16′	—	—	14
24	6.19	—	—	5.42	53° 23′	14
25	6.40	6.81	91° 15′	—	—	14
26	6.61	—	—	5.80	53° 31′	14
27	6.83	7.33	91° 15′	—	—	14
28	7.07	—	—	6.26	53° 39′	14

Table 7.   Continued

Carbon number	α-Form LIPS[a] (nm)	β-Form LIPS (nm)	β-Angle[b]	γ-Form LIPS (nm)	β-Angle	Ref.
29	7.35	7.84	91° 14′	—	—	14
30	(7.80)	(8.11)	—	—	—	14
31	7.72	—	—	—	—	35
32	7.94	—	—	7.10	—	35
33	8.12	—	—	—	—	35
34	8.28	—	—	7.51	—	35
35	8.66	—	—	—	—	35
37	9.09	—	—	—	—	35

[a]Long interplanar spacing.
[b]Angle of tilt of the c-axis with respect to the c-plane.

ing in the transition sequence: liquid → α → β states, increased resolution of the skeletal bands near 1060 cm^{-1}, and a prominent change in the 720 cm^{-1} band, i.e., the band assigned to the CH_2 rocking vibration. The latter band is single in the spectrum of the liquid, but on solidification of the alcohol at the freezing point into the α-form, it narrows and increases in intensity. On further cooling to the lower transition point to the β-form, two components of this band appear at approximately 720–730 cm^{-1}.

Tasumi et al.,[15] however, did not observe such a change in the absorption band at 720 cm^{-1}. The difference may have been due to the way the specimen was prepared. Chapman placed the sample between rock salt plates,[22] while Tasumi et al. incorporated it in a solid film state, after it was melted with an IR beam on a rock salt plate.

The latter authors[15] also used IR spectroscopy to examine higher alcohols having 11 to 37 carbons, except those with 30 and 36 carbons, and classified them into two types, A and B, according to the spectral specificity.

The A type exhibited (1) an absorption signal at 1408 cm^{-1}, (2) complex absorption signals in the 1400–1150 cm^{-1} region, (3) a strong absorption at 1125 cm^{-1}, and (4) strong absorptions in the 950–850 cm^{-1} region. All higher alcohols of 12 to 16 carbons, and alcohols of odd carbon number between 17 to 29 belonged to this category.

The B type showed (1) absorption at 1427 cm^{-1}, (2) monotonous increase in absorptions in the 1400–1150 cm^{-1} region, (3) a weak absorption at 1125 cm^{-1}, and (4) weak absorptions in the 950–850 cm^{-1} region. This behavior was characteristic of alcohols of even carbon numbers between 18 to 30 and of all higher alcohols from 31 to 37. Alcohols of carbon numbers 27, 29, 30, 31, and 35 displayed

Table 8. Melting Point and Transition Point of Higher Alcohols

Carbon Number	Melting or freezing point (°C)	Transition point (°C)	Method	Direction of Measurement	Ref.
10	5.99	—	—	—	36
	6.4	—	—	—	37
	—	5.3	T-T[a]	Cool	38
11	14.0	—	—	—	39
	15.85	—	—	—	16
	16.3	—	—	—	17
	—	14.25	T-T	Cool	38
12	21.45	23.6	T-T	Cool	38
	23	—	—	—	34
	23.5	23.5	DC[b]	Cool	20
	23.8	—	—	—	37
	23.87	—	—	—	36
	24	—	—	—	39,41,50
	24	16	T-T	Cool	29
13	30.0	—	—	—	39
	30.63	—	—	—	36
	—	25.0	T-T	Cool	35
14	36.3	34.4	DC	Cool	20
	36.4	—	—	—	20
	36.7	—	—	—	20,42
	37.2	34.9	T-T	Cool	38
	45–46	—	—	—	27
	45.5	—	—	—	39
	—	41.5	T-T	Heat	35
	—	36.0	T-T	Cool	35
16	47.8	—	—	—	43
	49.0	—	—	—	23,41
	49	—	—	—	34,44
	49.1	—	—	—	45
	49.1	40	T-T	Cool	46
	49.1–2	—	—	—	47
	49.25	—	—	—	48
	49.27	—	—	—	27,36,40,49
	49.5	45.7	T-T	Heat	39
	49.5	44.6	T-T	Cool	26
	49.5	—	—	—	50,51
	50	43	DSC[c]	Cool	27
	—	43	T-T	—	47

Table 8. *(Continued)*

Carbon Number	Melting or freezing point (°C)	Transition point (°C)	Method	Direction of Measurement	Ref.
	—	41.8	T-T	Cool	47
	—	45	Dil[d]	Heat	47
	—	44	Dil	Cool	47
	—	45	DC	Heat	47
	—	43	DC	Cool	47
	—	43.0	T-T	Cool	39
	—	43.8	T-T	Cool	21
17	53.31	—	—	—	36
	54	—	—	—	52,53
	54.5	47.5	T-T	Heat	35
	42.0	—	T-T	Cool	35
18	57.7	55.2	DC	Cool	20
	57.85	—	—	—	36
	57.9	52.2	T-T	Cool	26
	57.95	—	—	—	27,48,49
	58.5	—	—	—	41,51,52,53
	58.5	54.5	T-T	Cool	39
	58.5–9	—	—	—	54
	59	—	—	—	34
	59	51.3	T-T	Cool	44
	—	52	DSC	Cool	27
	52.5	T-T		Cool	39
	—	53.4	DC	Cool	20
	—	54	T-T	Heat	47
	—	54	Dil	Cool	47
	—	54.7	T-T	Cool	21
	—	55	Dil	Heat	47
19	62	58.0	T-T	Heat	35
	—	53.0	T-T	Cool	35
20	64.5	—	—	—	51
	65.0	63.5	T-T	Heat	35
	71.0	—	—	—	41
	—	58.2	T-T	Cool	35
21	68.5	60.5	T-T	Heat	35
	—	54.5	T-T	Cool	35
	70.8	—	—	—	55
22	69.6	64.5	DC	Cool	20
	70.5	67.6	T-T	Heat	39

Table 8. (Continued)

Carbon Number	Melting or freezing point (°C)	Transition point (°C)	Method	Direction of Measurement	Ref.
	—	63.5	T-T	Cool	39
	70.8	43			
23	73.2	—	—	—	39
	—	66.5	T-T	Heat	35
	—	60.5	T-T	Cool	35
24	75.5	72.8	T-T	Heat	39
	—	71.5	T-T	Cool	39
25	77.5	73.0	T-T	Heat	35
	—	67.5	T-T	Cool	35
26	79.3	78.1	T-T	Heat	40
	—.	73.5	T-T	Cool	39
27	80.5	76.5	T-T	Heat	35
	—	72.0	T-T	Cool	35
28	82.5	77.5	T-T	Heat	39
29	83.3	81.0	T-T	Heat	35
	—	74.0	T-T	Cool	35
31	86.5	82.0	T-T	Heat	35
	—	77.0	T-T	Cool	35
32	88.5	86.2	T-T	Cool	39
33	88.5	85.0	T-T	Heat	35
	—	81.2	T-T	Cool	35
34	91.0	86.2	T-T	Cool	39
35	91.0	85.0	T-T	Heat	35
	—	82.5	T-T	Cool	35
37	91.2	87.5	T-T	Heat	35
	—	86.0	T-T	Cool	35

[a]Temperature-time method.
[b]Dielectric constant method.
[c]Dilatometry.
[d]Differential scanning calorimetry.

different spectra, which were assumed to be due to impurities. Such differences in the IR spectral patterns were related to the rotational isomerism as mentioned above. Crystals of the β-form show the A-type pattern, while γ-form crystals show the B-type pattern. Al-Mamun also described the relationship between the polymorphism of higher alcohols and the IR spectrum patterns.[56] The spectral characteristics of higher alcohols are summarized in Table 9.

Table 9. Relationship of Crystal Types and IR Spectra of Higher Alcohols[56]

Crystal type	Specific character
α	(1) Singlet at 6.75 and 13.75 mm arising from CH_2 rocking vibrations
	(2) Simplification and reduction in intensities of most of the band between 7–3.5 μm
	(3) Intensity increase, simplification, and lower wavelength shift of OH stretching vibration band near 3 μm
β	(1) Noticeable intense OH bending (in-plane) band at ~7.1 μm
	(2) Complexity of the absorption between 7.14–8.7 μm
	(3) Reasonable strong band at ~8.9 μm
	(4) Occurrence of OH bending (out-of-plane) band at ~14.6 μm
	(5) OH stretching vibration band occurring with low frequency shoulder
	(6) Strong band in the region 10.5–11.8 μm
γ	(1) OH bending (in-plane) band occurring at ~7.03 μm with extremely low intensity
	(2) Simplicity of the band in the region 7.14–8.7 μm
	(3) Weak intensity of the band at ~8.9 μm
	(4) Occurrence of the OH bending (out-of-plane) band at ~16.3 μm
	(5) OH stretching vibration band occurring as a doublet
	(6) Weak intensity of the band in the region 10.5–11.8 μm

3. Specific Interaction between 1-Hexadecanol and 1-Octadecanol

The transition point of 1-hexadecanol decreases substantially if it contains 1-octadecanol, but not as much in the presence of other higher alcohols. This specific interaction between 1-hexadecanol and 1-octadecanol characterizes the interfacial chemical properties of commercial cetostearyl alcohol, which consists mainly of these two constituents. In this section the variation in the transition point of 1-hexadecanol/1-octadecanol mixtures is described.

3.1. Composition of Commercially Available Cetyl Alcohol

Analytical data for several commercially available cetyl alcohols given in Table 10[57–59] show that the compositions differ; some products contain more and less 1-octadecanol with varying amounts of other higher alcohols, although the latter are less abundant.

In Japan, materials sold as cetanol or cetyl alcohol often contain a considerable amount of 1-octadecanol or equivalents to 1-hexadecanol. Hilditch and Lovern reported higher alcohols in the head and blubber oils of the sperm whale, as given in Table 11.[60]

3.2. Transition Point of 1-Hexadecanol

As indicated above, cetyl alcohol, which had been used earlier, often contained a considerable amount of 1-hexadecanol. Stewart collected 18 commercial cetyl

Table 10.   Compositions and Transition Points of Various Commercially
Available Cetyl Alcohols

Alcohol	1-Tetradecanol (%)	1-Hexadecanol (%)	1-Octadecanol (%)	Others (%)	Transition point (°C)	Ref.
Product A	—	41	45	14	—	57
Product B	—	53	40	7	—	57
Sample 2	5	88	2	5	—	58
Sample 3	5	81	8	6	—	58
Product I	4.5	88.0	2.5	5.4	31.8	59
Product II	5.0	81.0	8.0	6.3	21.8	59

alcohols originating in the United Kingdom, in Europe, and in the United States, and determined the respective transition points. The latter varied from 19.5 to 34.9 °C, and were in all cases appreciably below the transition point of 1-hexadecanol (43.8 °C).[59] He further examined the influence of other higher alcohols, such as lauryl, myristyl, stearyl, or oleyl alcohol, on the transition point of 1-hexadecanol, and found that stearyl alcohol depressed it most strongly,[59] exhibiting a minimum at 20 °C in the presence of 30% stearyl alcohol (Fig. 8).[61]

Benton employed XRD to examine the crystal structure variations of 1-hexadecanol when mixed with 1-octadecanol[62] and reported the transition point of 1-hexadecanol to increase from 38 to 40 °C with the phase change from the a- to the g-form, while that of samples to which 10% 1-octadecanol was admixed ranged from 30 to 32 °C. Twin XRD peaks corresponding to short interplanar spacings of 0.184 to 0.399 nm appeared after the transition occurred, which Benton called the α-phase, while he designated the β-phase formed at 24 °C as the sub-α-phase. No transition from the β- to the γ-phase could be observed; instead of still higher temperature, transformation to the α-phase at 32 to 37 °C occurred, as summarized in Table 12.

Table 11.   Higher Alcohol Composition of a Sperm Whale
Head Oil Alcohol[60]

Alcohol	Content (%)
Tetradecyl alcohol	13
Cetyl alcohol[a]	44
Octadecyl alcohol	2
Hexadecenyl alcohol	4
Oleyl alcohol[b]	27
Eicosenyl alcohol	10

[a]Hexadecyl alcohol?
[b]*cis*-9-Octadecen-1-ol?

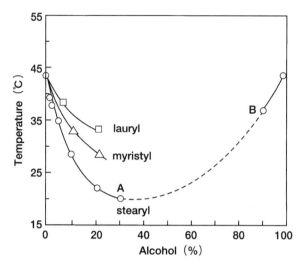

Figure 8.   Influence of lauryl, myristyl, and stearyl alcohol on the transition point of cetyl alcohol. The upper left dotted line represents the melting point curve of cetyl alcohol–stearyl alcohol mixtures.[60]

Al-Mamun produced phase diagrams from time–temperature curves of binary systems 1-hexadecanol/1-octadecanol, 1-pentadecanol/1-hexadecanol, 1-tride-canol/1-pentadecanol, and 1-dodecanol/1-octadecanol.[63] The transition point of the 1-hexadecanol/1-octadecanol system reached a minimum at about 17 °C.

The present authors determined the melting points and transition points of 1-hexadecanol, 1-octadecanol, and of their mixtures in various ratios with a differential scanning calorimeter, and found that the transition point reached the minimum of about 25 °C at a 3:2 weight ratio as shown in Fig. 9.[64]

The difference between Stewart's and the authors' data is considered to be due to the purity of materials rather than the method of measurements. The alcohols used by the present authors were of such high purity as to give only one peak in the

Table 12.   Transition Point of Hexadecanol, Octadecanol, and Their Mixtures[62]

Alcohol	Transition point (°C)	Transition direction
Hexadecanol	38–40	$\alpha \rightarrow \beta$
90% Hexadecanol + 10% Octadecanol	30–32	$\alpha \rightarrow \gamma$
	24	$\gamma \rightarrow \beta$
	32–37	$\gamma \rightarrow \beta$
50% Hexadecanol + 10% Octadecanol	32	$\alpha \rightarrow \beta$
	20–15	$\alpha \rightarrow \beta$
Octadecanol	47	$\alpha \rightarrow \beta$

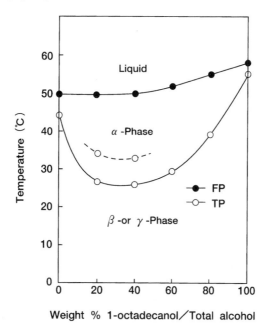

Figure 9. Freezing points (●) and transition points (○) of the 1-hexadecanol/1-octadecanol binary system.[64]

gas chromatography analysis. Stewart claimed the cetyl alcohol he used was more than 99.8% pure, but offered no information as to the purity of stearyl alcohol. The transition point of the latter, estimated from the figure, is about 43–44 °C, which is considerably lower than that given in Table 8 (51–55 °C). Apparently, the stearyl alcohol he used must have contained a certain amount of other alcohols, as well as 1-hexadecanol, since the minimum value of 20 °C at 30% stearyl alcohol is still lower than the authors' minimum value of 25 °C at 40% 1-octadecanol (Fig. 9).

While data are available on transition points of binary mixtures of 1-hexadecanol with other alcohols, no such information can be found for ternary systems. Table 10 shows the transition points that Stewart determined on two commercial cetyl alcohols.[59] Product II, containing 8% of 1-octadecanol, has the transition point of 21.8 °C, which is lower than that (about 25 °C) of a sample containing 10% 1-octadecanol (Fig. 9). It is highly possible that inclusion of 1-tetradecanol or other higher alcohols, other than 1-octadecanol, is the cause of this decrease. The question arises as to the reason for the lower transition point of 1-hexadecanol when 1-octadecanol is present. As mentioned earlier, the hydrocarbon chain layer in the α-form is in a half molten state, while that in the β- or γ-form is in the crystalline state. Thus, the hydrocarbon layer may become more difficult to crystallize if constituents of different length coexist in a mixture. It remains to be explained why the combination of 1-hexadecanol and 1-octadecanol is so specific for the lowering of the melting point of the hydrocarbon layer.

4. Interaction between Higher Alcohols and Water

If a higher alcohol coexists with water, the transition point from the α-form to the β- or γ-form is greatly lowered, while the melting or solidification points rise. The present section considers this interesting behavior.

4.1. Experimental Facts

Higasi and Kubo were the first to note that the transition point of a higher alcohol is lowered in the presence of water.[65] They used a dilatometer to determine the transition point of 1-octadecanol and noticed a decrease of about 10 °C when water was used instead of mercury in the experiment. The melting point of the same system was not much affected under the same condition.

Trapeznikov observed the same effect by measuring the change of the equilibrium spreading pressure (ESP) with temperature using 1-tetradecanol, 1-hexadecanol, and 1-octadecanol.[66] Stewart found that a higher alcohol mixture containing 54% 1-octadecanol formed the *a*-phase after a few days storage in water at 25 °C.[67] One intense diffraction peak appeared at the angle corresponding to 0.417 nm interplanar spacing. The sample was again restored to the α-form after a few days, if left in air. These phenomena suggest that the α–β transition point of higher alcohols is lowered in the presence of water, since the α-form is the stable modification at higher temperatures.

Brooks investigated the change in the transition points of 1-tetradecanol and 1-hexadecanol with the inclusion of water and found that a minimum is reached at a molar ratio of alcohol and water of 2:1.[68]

Brooks and Alexander studied the interaction of higher alcohols with water by measuring the ESP of higher alcohols on the water surface as a function of temperature.[69] Points B and C in the schematic Fig. 10 corresponded to the transition point of the water-containing higher alcohols from the α- to the β-form and to the melting point, respectively. The system resembles a liquid crystal-like hydrate formed in the region from points C to C′.

The ESP data were also used to evaluate the heat of fusion of 1-hexadecanol, which amounted to 25.5 kJ/mol, while a direct calorimetric determination at its melting point gave a value of 35.4 kJ/mol. They suggested that this difference was due to the formation of a hydrate of the α-phase of solid 1-hexadecanol, assuming that the liquid above the point C in the ESP experiment was the same as the liquid in the heat of fusion experiment. Therefore, the value for the heat of hydration of the solid α-phase is calculated to be 9.9 (= 35.4–25.5) kJ/mol, which is reasonable for the reaction

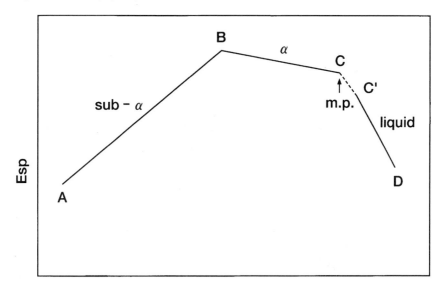

Temperature

Figure 10.   Schematic presentation of the variation of the equilibrium spreading pressure (ESP) with temperature for a typical higher alcohol.[69]

$$\text{alcohol} + H_2O \rightarrow \text{alcohol hydrate}$$

based on the hydrogen bond energies. Incidentally, values such as 17.79, 20.93, 23.69, 26.79, 28.05, and 32.23 kJ/mol have been obtained as the hydrogen bond energy of water, through various calculation methods.[70]

Lawrence *et al.* determined the temperature at which higher alcohols in contact with water lost their optical anisotropy or melted, as observed in a microscope with crossed nicols.[71] Although the temperature at which optical anisotropy disappears is considered to correspond to the *b*- to *a*-transition, values obtained are appreciably higher than those reported by other workers (Table 13), which could have been due to the insufficient time for the water molecule to penetrate into the alcohol crystal.

In order to form hydrates by association with water, higher alcohols require a certain length of time. Therefore, if Lawrence and co-workers had carried out their experiments after a sufficiently long time following dipping of the alcohol into water, lower values would have been obtained corresponding to the transition point in the presence of excess water. They also examined the change in the α-freezing point and transition point of 1-decanol, 1-dodecanol, and 1-tetradecanol when water was included,[72] and these results are shown in Fig. 11. It is somewhat doubtful to have transition points higher than the freezing points of the *a*-phase in case of 1-dodecanol over the region of lower water content.

Table 13.   *Melting/Solidifying Points and Transition Points for Anhydrous and Hydrated Higher Alcohols*

Alcohol	Melting/Solidifying Point (°C)		Transition Point °C		Method	Ref.
	Anhydrous	Hydrated	Anhydrous	Hydrated		
1-Decanol	—	0.75	5.3	1.9	T-T^a(C^b)	72
1-Undecanol	—	11.75	14.25	8.1	T-T (C)	72
1-Dodecanol	21.45	24.04	23.65	15.4	T-T (C)	72
	23	25	—	18	Mic^c(H^d)	71
1-Tetradecanol	38.0	40.1	—	—	GC^e(H)	68
	—	—	—	27.7	T-T (H)	68
	—	—	33.5	23.0	T-T (C)	68
	—	—	35	25	ESP^f	66
	37	39	—	27	Mic(H)	71
	37.2	40.2	34.9	27.4	T-T (C)	72
1-Hexadecanol	49.5	52.1	—	—	GC(H)	68
	—	—	—	39.5	T-T (H)	68
	—	—	44.6	35.5	T-T (C)	68
	—	—	44	30	ESP	66
	47	51	—	38	Mic(H)	71
	50.0	51.0	43.0	31.0	DSC^g (C)	73
	49.5	50.8	39.5	31.7	DTA^h (C)	65
1-Octadecanol	—	—	52.5	46	ESP	66
	58	62	—	47	Mic(H)	71
	58	62	—	47	Mic(H)	71
	59.0	59.5	52.0	40	DSC(C)	73
	58.5	59.5	53.5	41.0	DTA(C)	75
Hexaoctadecanol (3:2)	47.5	54.0	25.0	10.0	DSC(C)	73
1-Docosanol	64.5	61.0	56.5	30.0	DTA(C)	75

Methods: ^{a}Temperature-time, ^{b}Cooling, ^{c}Microscopic, ^{d}Heating, ^{e}Glass capillary, ^{f}Equilibrium spreading pressure, ^{g}Differential scanning calorimetry, ^{h}Differential thermal analysis.

    The present authors used differential scanning calorimetry (DSC) to follow the changes in the freezing and transition points of 1-hexadecanol, 1-octadecanol, and their 3:2 mixture [hexaoctadecanol (3:2)] with increased amounts of included water.[73,74] The results in Fig. 12 show also that the freezing point rose slightly, but the transition point decreased considerably with the water content. The latter effect was especially significant with hexaoctadecanol (3:2). In Section 3, it was shown that the transition point of 1-hexadecanol was lowered when 1-octadecanol was incorporated. The addition of water enhanced this effect.

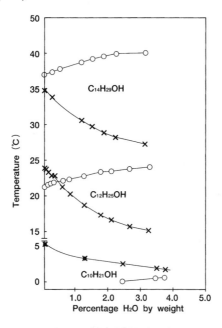

Figure 11. Effect of water on the freezing points of the *a*-phase (○), transition points of decanol, 1-dodecanol, and 1-tetradecanol (×), and freezing point direct to the opaque phase (●).[72]

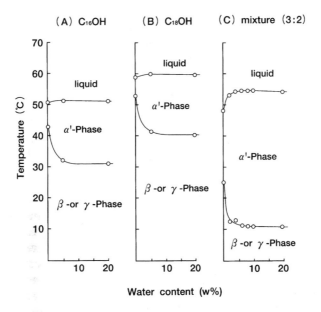

Figure 12. Effect of water on the freezing and transition points of 1-hexadecanol, 1-octadecanol, and the 3:2 mixture of both.[73]

Figure 13 displays the change in the freezing and transition points as a function of the 1-hexadecanol/1-octadecanol ratio in the presence of excess water, based on Fig. 12, while Table 13 summarizes analogous data of alcohols in the absence (dry) and presence (wet) of water.

4.2.   Formation of Hemihydrate

What is the reason for the melting or freezing points of higher alcohols to rise and the transition points to fall when water is contained? As previously mentioned, Trapeznikov proposed the possibility of the existence of hydrates,[66] while Brooks, Lawrence *et al.*, and Kuchhal *et al.* suggested the formation of hemihydrates.[68,72,75] Pachler and Stackelberg found that the central feature of a stoichiometric alcohol + water compound (1,1,2,2-tetramethyl-1-propanol hemihydrate, TMP hemihydrate) was a continuous hydrogen-bonded chain in which the water molecules were tetrahedrally coordinated as in ice.[76] The distance between neighboring oxygen atoms was 0.38 nm. Lawrence *et al.* referred to such a structure by considering a water molecule placed at the center of a tetrahedral grouping of four long-chain alcohol molecules in a layered lattice. Assuming a similar dimen-

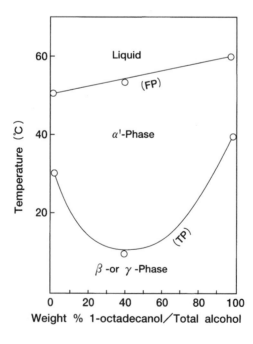

Figure 13.   Freezing point (FP) and transition point (TP) of 1-hexadecanol, 1-octadecanol, and the 3:2 mixture in the presence of excess water.[74]

sion to that in the TMP hemihydrate lattice, a side spacing of 0.38 nm could be calculated with an interlayer vertical spacing of 0.22 nm between alcohol oxygen atoms. Thus, it might be relatively easy for water molecules to fit into a hydrogen-bonding system of the alcohol α-phase.

Lawrence found that the unit cell lengths of anhydrous (β-form) and hydrated (α′-form) 1-dodecanol at 20 °C was 3.54 and 3.51 nm, respectively.[72] The fact that both lateral and vertical spacings might be reduced by the presence of water agrees with the observations of Trapeznikov and Lawrence *et al.*,[66,72] who have both noted the contraction in volume when water penetrated the α-phase.

The present authors compared the lattice parameters of anhydrous and hydrated 1-hexadecanol, 1-octadecanol, and Hexaoctadecanol (3:2) at the same temperatures at which the former appeared in α and the latter in the α′-structure.[73,74] The results are summarized in Table 14, the interpretation of which is suggested as follows. The structure of anhydrous and hydrated alcohols is such that the long hydrocarbon chains are oriented vertically with respect to the plane of the oxygen atoms, while OH groups build a hydrogen-bond network with water as shown in Fig. 14 (A and B). Molecules of anhydrous alcohol are attached to each other only by the hydrogen bond between the OH groups, whereas the hydrated alcohol forms bonds between the OH group and the water molecule in the ratio 2:1.

In the structure (A), three oxygen atoms from OH groups are placed at the vertex of the base plane of a regular tetrahedron, and a fourth is at the center. Assuming that hydrogens are connected to neighboring oxygens by a hydrogen bond, one can visualize a network as illustrated in Fig. 14(A), which is characterized by a closed hydrogen-bonded ring. No other closed ring can exist, providing one ring is formed within one OH cluster. Hydrocarbon chains are thought to be oriented normal to the plane of paper, yielding a hexagonal structure. In the latter, hydrogen

Table 14. *Comparison of Interplanar Spacings Between Anhydrous and Hydrated Higher Alcohols*[73]

		Anhydrous alcohol		Hydrated alcohol	
Alcohol	Measured at (°C)	IPS^a (nm)	Miller index^b	IPS (nm)	Miller index^c
1-Hexadeacanol	47 ± 1	0.42	($2\bar{2}00$)	0.42	(100)
		1.52	(0006)	1.56	(006)
1-Octadecanol	56 ± 1	0.421	($2\bar{2}00$)	0.417	(100)
		1.62	(0006)	1.65	(006)
Hexaoctadecanol (3:2)	43 ± 1	0.414	($2\bar{2}00$)	0.414	(100)
		1.57	(0006)	1.61	(006)

^{a}Interplanar spacing.

^{b}Assumed hexagonal.

^{c}Assumed tetragonal.

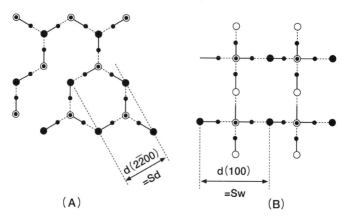

Figure 14. Possible hydrogen-bond network built with (A) hydroxyl groups of anhydrous *a*-alcohol, and (B) hydroxyl groups of alcohol and water in the hydrated *a'*-phase. Full, double, and open large circles represent oxygen atoms below, in, and above the plane of the paper, respectively. Small full circles show hydrogen atoms. Solid and dotted lines between circles represent the covalent and hydrogen bond, respectively. Double circles in (B) are oxygen atoms of water.[73]

atoms are not necessarily located between all nearest oxygen atoms because of the insufficient number of hydrogen atoms necessary to achieve a complete hydrogen-bonded network. The coulombic repulsion between two oxygen atoms not connected by a hydrogen bond will cause the network of anhydrous alcohol to be mechanically weak. The overall structures including hydrocarbon chains are illustrated in Fig. 15(A).

In the case of hydrated alcohols, oxygen atoms of OH groups of alcohol are placed at the four vertex positions of a regular tetrahedron, and that of water at the center. Assuming the hydrogen bond connecting the alcohol hydrogens to the neighboring water oxygens and the water hydrogens to the neighboring alcohol oxygens, one can produce a network as illustrated in Fig. 14(B). The molar ratio of the OH groups to water, required to complete such a structure, is 2:1. Hydrocarbon chains are also considered to orient normal to the plane of the paper, resulting in a tetragonal configuration. The overall structure including hydrocarbon chains is illustrated in Fig. 15(B). This network, characterized by perfect hydrogen bonds with no missing connections, is mechanically stronger than that of anhydrous alcohol. Incidentally, the theoretical water content of the hemihydrate of 1-hexadecanol and 1-octadecanol is 3.58 and 3.22 wt%, respectively, which is in reasonable agreement with the minimum water content at which the transition points reach the saturated values, as shown in Fig. 12.

The height of a regular triangle d is $\sqrt{3}/2$ times the side a. Since oxygen atoms form triangles, the interplanar spacing SD of the (2200) plane for the anhydrous species can be written as

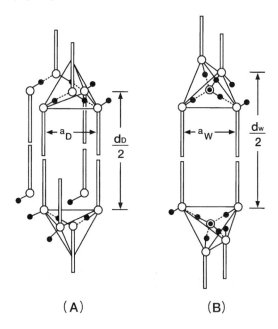

Figure 15. Possible molecular arrangements of (A) anhydrous a-alcohol and (B) hydrated α'-alcohol. Open and double circles represent oxygen atoms from alcohol and water, respectively, and full circles are hydrogen atoms. Solid and dotted lines connecting the circles stand for the covalent and hydrogen bond, respectively. Tetrahedrons are drawn to show positional relations of the oxygen atoms.[73]

$$S_D = \frac{\sqrt{3}}{2} a_D \qquad (1)$$

by taking the length between the nearest oxygens as a_D.

For hydrated species, the interplanar spacing S_W of the (100) plane yields the length of the side of a tetrahedron made by the OH oxygen atoms, a_W (Fig. 15). Therefore, one can write

$$a_W = S_W \qquad (2)$$

The distances between oxygen atoms obtained using these expressions are listed in Table 15, which shows that a shrinkage of the order of 0.06 to 0.07 nm takes place on hydration.

4.3. Structure of Hydrated Alcohols

In order to verify the structure of hydrated higher alcohols, the authors have extended the considerations regarding the results of Table 14.[74] Assuming that no

Shoji Fukushima and Michihiro Yamaguchi

Table 15.   Comparison between Nearest Interoxygen Distances of Anhydrous and Hydrated Alcohols

Higher alcohol	Measured at (°C)	Anhydrous a_D (nm)	Hydrated a_w (nm)	Difference $a_D - a_w$ (nm)
1-Hexadecanol	47 ± 1	0.485	0.420	0.065
1-Octadecaol	56 ± 1	0.486	0.417	0.069
Hexaoctadecanol (3:2)	43 ± 1	0.478	0.414	0.064

[a]Eq. (1).
[b]Eq. (2).

change occurs in the hydrocarbon chain lengths on inclusion of water, the long interplanar spacings can be calculated by

$$\frac{d_D}{2} = \frac{a_D}{2\sqrt{6}} + l \tag{3}$$

and

$$\frac{d_w}{2} = \frac{a_w}{\sqrt{2}} + l \tag{4}$$

respectively, where d_D and d_W are long interplanar spacings of the (0001) plane of the anhydrous and (001) plane of the hydrated alcohol, respectively, and l is the length of the hydrocarbon chain.

The length of d_D, d_W, S_D, and S_W can be determined experimentally by XRD. Therefore, it is possible to estimate d_W using Eq. (5) below, even though one may not know the value of l.

$$d_w = d_D + \sqrt{2}\,S_w - \frac{\sqrt{2}}{3} S_D \tag{5}$$

The values of d_W calculated using the quantities given in Table 14 and listed in Table 16 are in good agreement with the experimentally determined data. Thus, the interesting observation that the melting point rises and the transition point falls by inclusion of water in higher alcohols can be explained by taking into consideration the above structures. As stated before, the lipophilic layer of an α-alcohol is in a half molten state, and that of β- or γ-alcohols is in the crystalline state. Therefore, the transition from the α-form to the β- or γ-form represents a rearrangement of the lipophilic layer. Any factor inhibiting the latter will cause a lowering of the transition point and raising of the melting point. One may consider as one of the inhibiting factors the build-up of hydrogen-bond networks on hydration of alcohols. Such networks are less readily distorted as compared to those of anhydrous alcohols, since sites lacking in hydrogen bonds are compensated by interposition of water molecules.

Table 16.   Comparison between Measured and Calculated Values of Long Interplanar Spacings in Hydrated Higher Alcohols[73]

Higher Alcohol	Measured at at (°C)	Measured value (nm)	Calculated value (nm)
1-Hexadecanol	47 ± 1	9.36	9.32
1-Octadecanol	56 ± 1	9.90	9.92
1-Hexaoctadecanol (3:2)	43 ± 1	9.66	9.61

In order for the lipophilic layer to crystallize, it is necessary to change the angle of the hydrocarbon chain with respect to the network plane, which requires a force to break the hydrogen-bonded network. The stronger the resistance against the force to break up the latter, the more easily will the lipophilic layer melt.

Assuming this mechanism, the hydrated alcohol should clearly be in a different phase than the anhydrous one, although the XRD pattern may be nearly the same. For this reason, the hydrated phase has been described as a' in order to draw a clear distinction between the two structures (Figs. 14 and 15).

4.4.   Phase Diagram of the 1-Hexadecanol/1-Octadecanol/Water Ternary System

Based on data shown in Figs. 9 and 13, it is possible to construct a phase diagram of the 1-hexadecanol/1-octadecanol/water ternary system, which is given in Fig. 16.[74]

The following features are noted: (A) above 60 °C, which corresponds to the melting point of 1-octadecanol, two phases consisting of an aqueous saturated alcohol solution and a liquid solution of 1-octadecanol and 1-hexadecanol cover the entire area. (B) At 60 °C, at which the hemihydrate of 1-octadecanol is formed, conditions along the W–O line can yield either the hemihydrate and water or hemihydrate and a crystal coexisting phase, represented by pure hemihydrate (HH). (C) Below 60 °C and somewhat above 50 °C, the area of hemihydrate extends toward the 1-hexadecanol as the temperature is lowered. (D) At 50 °C, water and the β- or γ-phase of the mixed alcohols (rich in 1-octadecanol) appear in the vicinity of 1-octadecanol. (E) Between 50 and 40 °C, the range of water + anhydrous alcohol extends. (F) At 40 °C, another β- or γ-phase of the mixed alcohols (rich in 1-hexadecanol) appears near the area of water/1-hexadecanol. (G)–(J) Below 40 °C, areas of both phases extend with decreasing temperature. (K) At 10 °C, the hexaoctadecanol, which forms a hemihydrate, consists only of the 3:1 mixture. (L) Finally, below 10 °C, the entire phase diagram consists of two phases, i.e., of a β- or γ-mixed crystalline phase and water.

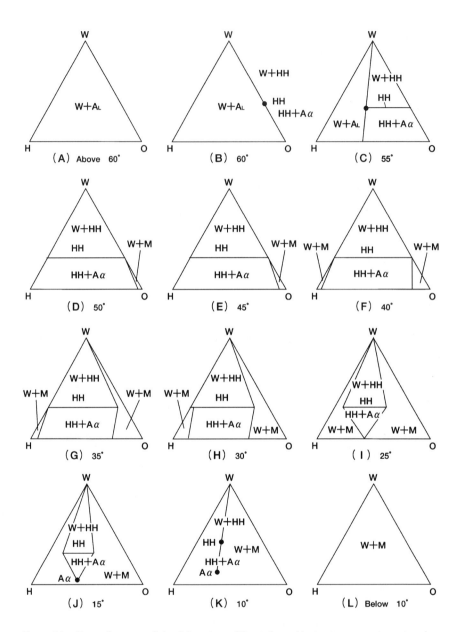

Figure 16. Phase diagrams of the 1-hexadecanol/1-octadecanol/water ternary system at various temperatures: (A) above 60 °C, (B) at 60 °C, (C) at 55 °C, (D) at 50 °C, (E) at 45 °C, (F) at 40 °C, (G) at 35 °C, (H) at 30 °C, (I) at 25 °C, (J) at 15 °C, (K) at 10 °C, and (L) below 10 °C. W, water; H, 1-hexadecanol; O, 1-octadecanol; AL, molten alcohol; Aα, α-alcohol; HH, hemihydrate (or α'-alcohol); M, anhydrous alcohol (β- or γ-phase).[74]

5. Studies on Higher Alcohol/Surfactant/Water Systems

This section describes studies on ternary systems consisting of a higher alcohol, a surfactant, and water. Table 17 shows that many ternary systems in various combinations have been reported.[77-105] Some investigations produced detailed phase diagrams while others yielded only one combination. Since the purpose of this writing is to address the specific nature of cetostearyl alcohol, and especially its effect on the increase in viscosity of O/W emulsions, the latter aspect is mainly considered.

5.1. The 1-Decanol/Sodium Caprylate/Water System

The ternary phase diagram of 1-decanol/sodium caprylate/water (Fig. 17) as completed by Ekwall *et al.* is the most detailed of such reported systems consisting of a higher alcohol, a surfactant, and water.[84] Regions L1 and L2 represent homogeneous isotropic solutions in water and in 1-decanol, respectively. Five regions designated as B, C, D, E, and F represent the areas of homogeneous mesomorphous phases. Dotted-lined regions are for two-phase systems, while regions connected with solid lines and those marked with numerals are for three phases. The five mesomorphous phases consist of a liquid layered structure (B), a structure in which cylindrical micelles are tetragonally packed (C), another layered structure (D), a structure in which cylindrical O/W micelles are hexagonally packed (E), and a structure in which cylindrical W/O micelles are hexagonally packed (F).

Ekwall *et al.* established many three-phase diagrams using 1-pentanol, 1-hexanol, 1-octanol, and 1-nonanol instead of 1-decanol and cetyltrimethylammonium bromide, potassium caprate, potassium oleate, sodium alkylsulfate, sodium cholate, sodium desoxycholate, sodium octylsulfate, sodium octylammonium chloride, and Triton X-100 instead of sodium caprylate, which were essentially similar to that reproduced in Fig. 17.[78]

5.2. The 1-Hexadecanol/OTAC/Water System

Yamaguchi and Noda studied the ternary system 1-hexadecanol/octadecyltrimethylammonium chloride (OTAC)/water,[92,95] and found an association complex at 3:1 molar ratio of 1-hexadecanol to OTAC, and other complexes in ratios from 0 to 1.5.

5.3. Rheology of Ternary Systems Containing 1-Hexadecanol or a Homologous Alcohol

Barry *et al.* studied the effect of increasing emulsion viscosity in ternary system.[96-101,106,107] Cationic, anionic, or nonionic surfactants, such as alkyltrimethylammonium bromide, sodium dodecylsulfate (SDS), or Cetomacrogol

Table 17.   Literature Survey of Alcohol/Surfactant/Water Ternary System

Alcohols	Surfactants	Ref.
1-Butanol	Na caprylatel	77,78
1-Pentanol	Na caprylate	77,78
	Teepol[a]	79
2-Methyl butanol-2	Teepol	79
3-Methyl butanol-1	Teepol	79
3-Pentanol	Teepol	79
1-Hexanol	Cetyltrimethylammonium bromide	77,80
	Na caprylate	77
1-Heptanol	Na caprylate	77
Benzyl alcohol	Na laurate	79
1-Octanol	Cetyltrimethylammonium bromide	77,81
	K caprate	78
	Na caprylate	80
	Na dodecylsulfate	82
	Hexaethoxylaurylether	83
1-Nonanol	Na caprylate	78
1-Decanol	K caprylate	78
	K oleate	78
	Na caprylate	77,84
	Na cholate	78
	Na desoxycholate	78
	Na octylsulfate	85
	Cetrimide[b]	81
	Octylammonium chloride	85
	Triton X-100[c]	78
1-Dodecanol	Na laurylsulfate	86–89
	Na myristylsulfate	87,91
	Cetrimide	88,90
	Cetomacrogol 1000[d]	88,90
	Sorbester Q 12[e]	88,90
1-Tetradecanol	Na myristylsulfate	86
	Octadecyltrimethylammonium chloride	92
1-Hexadecanol	Na dodecylsulfate	93,94
	Octadecyltrimethylammonium chloride	92,95
Oleyl alcohol	Na laurylsulfate	88
	Cetrimide	88
	Cetomacrogol 1000	88
	Sorbester Q 12	88
Cetostearyl Alcohol	Na laurate	90
	Na laurylsulfate	90

Table 17. *Continued*

Alcohols	Surfactants	Ref.
	Cetrimide	90,96,97
	Dodecyltrimethylammonium chloride	98,99
	Tetradecyltrimethylammonium chloride	98,99
	Hexadecyltrimethylammonium chloride	98,99
	Octadecyltrimethylammonium chloride	98,99
	Cetomacrogol 1000	90,100,101
	Sorbester Q 12	90
Hexaoctadecanol (3:2)	Dodecaethoxydodecylether	103
	Pentadecaethoxyoleylether	102–105

[a]Branched chain sodium alkylsulfate.

[b]Cetyltrimethylammonium bromide.

[c]Alkylarylpolyether alcohol.

[d]Polyoxyethyleneglycol 1000 monocetylether.

[e]Polyoxyethylenesorbitan monolaurate.

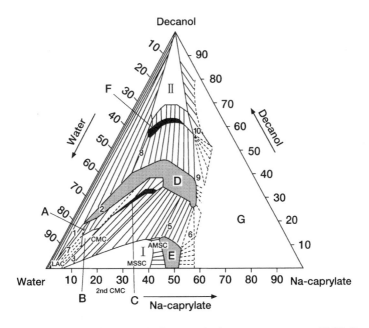

Figure 17.   Phase diagrams of 1-decanol/sodium caprylate/water ternary system at 20 °C. Concentrations are given in wt%. I, homogeneous, isotropic solutions in water; II, homogeneous, isotropic solutions in 1-decanol; B, C, D, E, F, homogeneous mesomorphous phases; G, solid crystalline sodium caprylate and hydrated sodium caprylate with fiber structure; 1–10, three-phase triangles; LAC, Limiting association concentration; AMSC, Alcohol-middle soap corner; MSSC, Middle soap separation concentration.[84]

1000, were used. Talman *et al.* also investigated the same effect on similar systems and surfactants.[81,88,90]

5.3.1. Influence of the Amount of 1-Hexadecanol

Barry and Shotton[107] examined rheological properties of ternary systems prepared with 4 g SDS, 400 g water, and varying amounts of 1-hexadecanol to give molar ratios to SDS (R_n), ranging from 1 to 10. Systems R_1 to R_3 (subscript refers to the molar ratio) were fluid, turbid, off-white liquid, the consistency and viscoelasticity of which increased in the same order. System R_5 was off-white, shiny, and translucent with noticeable elastic recoil, yet flowed slowly under its own weight. As the value of n in R_n became larger consistency increased steadily, with R_{10} being a white, glossy, and smooth soft solid.

Microscopic observations showed in all systems the existence of dispersed particles with optical anisotropy. Some of these anisotropic particles melted at ~50 °C, corresponding to the melting point of 1-hexadecanol, to form isotropic globules. Some particles exhibited optical anisotropy up to 65 °C, which Barry and Shotton suggested to be "frozen liquid crystals." To determine the rheological properties of these systems, they used a Ferranty–Shirley cone-and-plate viscometer, which was programmed to continuously increase in the rotating velocity of the cone from zero to 100 min^{-1}, and then decrease to zero.

Figure 18 shows a typical plot of the average of three runs for each system at 25 ± 0.1 °C. System R_1 gave deflections too small to be measured with reasonable accuracy. All experimental plots showed a common hysteresis effect in that the "down" curve lay to the left of "up" curve.[107] The nonlinear behavior or hysteresis loops indicated that all samples were non-Newtonian. A cream-like material generally gives the latter type of flow curves.

The area enclosed by the loop became larger with increase in the molar ratio. From R_4 onward there was a "spur" on the "up" curve which virtually disappeared with R_{10}. Areas enclosed by the hysteresis loops, the values of the shear stress, the shear rate, and the shear strain at the spur point for different ternary systems described above are given in Table 18.

Barry and Shotton also examined the effect of temperature on the rheological properties of various systems,[107] and showed that at >60 °C, all compositions gave very low readings. System R_8, stored overnight at temperatures ranging between 25 and 60 °C, had a maximum apparent relative viscosity at about 42.5 °C. On aging, all composite materials underwent a change in appearance, becoming more fluid, and forming silvery crystalline deposits. In systems with a low alcohol content such deposits were noticeable, as they formed a pearly layer adhering to the inner surface of the container. Compositions R_4 to R_{10} progressively lost their rigidity and became sufficiently liquid to flow under their own weight. They claimed that the pearly crystalline deposits, which melted at 28 °C, were 2:1 adducts of SDS with 1-hexadecanol, but its crystal structure was not established.

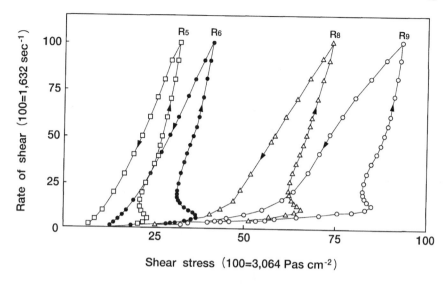

Figure 18. Flow curves for mixed systems of 1-hexadecanol, sodium dodecylsulfate, and water in different molar ratios of the alcohol to surfactant (R_n): R_5 (□), R_6 (●), R_8 (△), and R_9 (○). A Ferranty–Shirley viscometer was used in the automatic mode at 25 °C.[107]

Table 18. Areas of Hysteresis Loops and Values of Shear Stress, Shear Rate, and Shear Strain at Spur Point for Tenary Systems, 1-Hexadecanol, Na dodecylsulfate, and Water at Different Ratios of the Alcohol to Surfactant[107]

System	Area of hysteresis loop $(cm^2)^a$	Shear stress at spur point (Pa)	Shear rate at spur point (s^{-1})	Shear strain at spur point
R_2	2.6	—	—	—
R_3	5.6	—	—	—
R_4	21.4	36	98	1760
R_5	44.6	70	98	1760
R_6	55.3	112	114	2400
R_7	57.7	145	147	3970
R_8	66.6	202	163	4900
R_9	101.9	259	196	7060
R_{10}	167.8	$*^b$	$*$	$*$

^{a}1 cm^2 = 6.40 × 10^4 Pa/s.

^{b}The "spur" point disappeared in the "up" curve.

5.3.2. *Influence of the Nature of Surfactant and of Higher Alcohol*

Talman *et al.* studied the rheological behavior, i.e., apparent viscosity and static yield value with a cone-and-plate viscometer, of the cetostearyl alcohol, surfactant and water ternary systems using Cetomacrogol 1000, Sorbester Q 12, Cetrimide, and SDS as surfactants.[88] Table 19 shows the composition of the samples used. They also investigated the rheological behavior of quaternary creams containing 50% liquid paraffin, and found that the surfactant molar concentration giving maximum rheological parameters is the same, irrespective of the nature of the surfactant at a given cetostearyl alcohol concentration.[88]

Based on this hypothesis, the present authors recalculated their data for the ternary systems, the molar ratios of alcohol to surfactant, and compared them with the viscosity and yield values.[108] Table 20 is the results and it shows that the latter have reached a maximum at the alcohol/surfactant ratio of about 10. This ratio is quite close to that of the M phase referred to in Section 7.

Talman *et al.* also examined the effect of increasing viscoelasticity using oleyl or lauryl alcohols instead of cetostearyl alcohol. Creams prepared with oleyl alcohol showed neither an increase in viscosity nor in the static yield value. In contrast, creams prepared with lauryl alcohol had enhanced both parameters, although not as effectively as cetostearyl alcohol.[88]

Barry and Saunders studied the rheological properties of cetostearyl alcohol/cetyltrimethyl- ammonium bromide (Cetrimide)/water ternary systems, formulated as given in Table 21.[96] The ternary system T_2 was very fluid, T_5 was a mobile semisolid, and T_{10} was a rigid semisolid which flowed only slowly under its own weight. All systems were white and glossy, which on microscopical examination displayed a dispersion of deformed globules, some of which containing small spherical inclusions near the centers. Between crossed nicols these anisotropic structures often appeared as a distorted Maltese cross.

It was also found that the viscosity of T_{10} increased with temperature, reaching a maximum at 43 °C.[97] The temperature at the maximum consistency was close to the transition temperature from the frozen smectic to liquid crystalline phase, and it was lower than the penetration temperature of 1% Cetrimide solution into cetostearyl alcohol. Above 43 °C, the consistency decreased as the network became a true smectic phase, which was still viscoelastic, i.e., shoulders were noted on the

Table 19.   Composition of Ternary Systems Used by Talman et al.[88]

Components	Weight ratio (%)				
Cetostearyl Alcohol	10.0	10.0	10.0	10.0	10.0
Surfactant*ᵃ*	0.5	1.0	2.0	4.0	10.0
Water	89.5	89.0	88.0	86.0	80.0

*ᵃ*Cetomacrogol 1000, Sorbester Q 12, Cetrimide, or Na dodecylsulfate.

Table 20.   *Relationship between Cetostearyl Alcohol to Surfactant Molar Ratio and the Viscosity or the Static Yield Value (SYV) in Ternary Systems Listed in Table 19*[88]

Surfactant		Concentration of surfactant (wt%)				
		0.5	1.0	2.0	4.0	10
Cetomacrogol 1000	Molar ratio	92.1	46.0	23.0	11.5	4.60
	Viscosity (mPa · s)	50	102	162	261^a	55
	SYV (Pa)	132	302	373	470^a	33
Sorbester Q 12	Molar ratio	39.5	19.8	9.8	4.94	1.98
	Viscosity (mPa · s)	34	165	220^a	152	29
	SYV (Pa)	94	119	192^a	109	94
Cetrimide	Molar ratio	28.8	14.4	7.21	3.60	1.44
	Viscosity (mPa · s)	118	144	177^a	72	44
	SYV (Pa)	192	236	246^a	172	38
Na dodecylsulfate	Molar ratio	22.7	11.4	5.68	2.84	1.14
	Viscosity (mPa · s)	107	167	175^a	60	30
	SYV (Pa)	197	292^a	259	121	15

^{a}The maximum value in the series.

continuous shear and creep curves between approximately 50 and 57 °C. Above 57 °C, the smectic phase dissolved to form an isotropic solution.

5.3.3.   Influence of Alkyl Chain Length of Surfactant

Barry and Saunders also investigated the influence of alkyl chain lengths (dodecyl, tetradecyl, hexadecyl, and octadecyl) and of the concentration of alkyl-trimethylammonium bromide (ATAB) on the "self-bodying" action. The composition of the latter systems are listed in Table 22.[98] Consistencies of the ternary systems increased markedly as the concentration of the mixed emulsifier (alcohol

Table 21.   *Composition of Ternary Systems Used by Barry and Saunders*[96]

Components	Systems		
	T_2^a	T_5^a	T_{10}^a
Cetostearyl Alcohol	7.2	18.0	36.0
Cetrimide^b	0.8	2.0	4.0
Water	360	360	360

^{a}Molar ratio of cetosteary1 alcohol to Cetrimide is 13:1 in these formulations.
^{b}Cetyltrimethylammonium bromide.

Table 22.   Composition of Ternary Systems Used by Barry and Saunders[98,99]

| Components | T Series[a] | | | TC Series[b] | | | |
| | T_2 | T_6 | T_{10} | TC_{12} | TC_{14} | TC_{16} | TC_{18} |
	(g)				(g)		
Cetostearyl Alcohol	7.2	21.6	36.0	57.6	36.0	36.0	36.0
ATAB[c]	0.8	2.4	4.0	6.4	4.0	4.0	4.0
Water	360	360	360	360	360	360	360

[a]In T series, dodecyl-, tetradecyl-, hexadecyl-, or octadecyltrimethylammonium bromide was used as ATAB.
[b]In TC series, subscripts indicate carbon number of alkyl group in ATAB used.
[c]Alkyltrimethylammonium bromide.

plus surfactant) increased and were greater at the same concentrations of surfactants as their chain length became longer.

Furthermore, it was established microscopically[99] that the temperature for each ternary system yielding maximum consistency (T_{max}) was close to or agreed with the penetration temperature (T_{pen}) of a 1% aqueous solution of the respective ATAB into cetostearyl alcohol (Table 23).

These findings were explained by assuming that during mixing of a ternary system, aqueous quaternary solution penetrates the molten cetostearyl alcohol to form a smectic LC consisting of spherulites, platelets, and filaments. When the ternary system cools below the penetration temperature, the interaction is diminished, and the system precipitates to form a three-dimensional viscoelastic gel network of "frozen" smectic phase.

Finally, the ternary system cetostearyl alcohol/polyoxyethyleneglycol 1000 monocetylether (Cetomacrogol 1000)/water system with a constant ratio of the first two constituents (Table 24) was also studied at 9:1 by weight ratio (or ~44:1 by mole). The maximum apparent relative viscosity was reached at 50 °C. It was not possible, however, to establish the relation between T_{max} and T_{pen} as described above, since the latter could not be determined.[100] The consistency of ternary systems increased with the mixed emulsifier concentration from T_1 to T_3, and in each case it became more pronounced during the first few hours. T_1 and T_2 were

Table 23.   Relationship between T_{max} and T_{pen} for the Ternary System Given by Table 22[99]

	TC_{12}	T_{14}	TC_{16}	TC_{18}
T_{max} (°C)	—	45	50	52
T_p (°C)	39	45	50	54

Table 24.   Composition of Ternary Systems Used by Barry and Saunders[100]

Components	T_{10}	Systems T_{12} (g)	T_{14}
Cetostearyl Alcohol	36.0	43.2	57.6
Cetomacrogol 1000	4.0	4.8	6.4
Water	360	360	360

mobile liquids when first prepared, but turned semisolid within 24 h, as did the emulsions of low mixed emulsifier concentrations. System T_3 was a semisolid immediately after preparation and increased in consistency rapidly over the first few days. It appears that at the same temperature and the length of time over which they form, the networks of nonionic mixed emulsifiers differ from those formed by anionic and cationic mixed emulsifiers.[100,101]

5.3.4.   Conclusion

(i) Ternary systems behave in a similar way as creams in which oil phase is emulsified.

(ii) The consistency of ternary systems varies with temperature and reaches a maximum at approximately the same temperature as that at which an aqueous solution of the surfactant begins to penetrate into the crystals of alcohol.

(iii) The consistency of a ternary system reaches the maximum at about 10:1 molar ratio of cetostearyl alcohol to surfactant, at a constant concentration of the alcohol.

(iv) These findings do not depend on the nature of surfactants. These facts also suggest that the ternary systems may yield increased viscosity by building an internal middle phase.

6.   Nature of Ternary Systems Prepared with Hexaoctadecanols

In Section 4 it was shown that 1-hexadecanol, 1-octadecanol, and the 3:2 mixture, i.e., Hexaoctadecanol(3:2), can form hemihydrate at temperatures between 31 and 51 °C, 40 and 60 °C, and 10 and 53 °C, respectively. In other words, this means that neither 1-hexadecanol nor 1-octadecanol can appear as hemihydrate at lower temperatures, i.e., between 10 and 31 °C. Furthermore, the hemihydrate can form a liquid crystalline (LC) phase by association with a surfactant and water. Thus, in order to achieve the stable LC phase at room temperature, it is advantageous to mix higher alcohols in an appropriate proportion rather than to use them pure.

In doing so the hemihydrate-forming and, therefore, the LC-forming temperature is lowered to near-room temperature. Such an effect is not only of scientific interest, but of practical importance, since it is a key to the preparation of stable O/W creams. Temperatures of 10 to 50 °C are just normal for the environment in which many people spend their daily lives. Present authors use the expression "Hexaoctadecanol (3:2)" to denote a mixture of 1-hexadecanol with 1-octadecanol. The numbers in parentheses indicate the ratio by weight of the components in the mixture.

6.1. Stability and Rheological Properties of Ternary Systems as a Function of Temperature

As shown in Section 5, it is possible to obtain creamy matter, even if oily components, such as Vaseline or liquid paraffin, are not incorporated in the formulation of an ordinary O/W emulsion-type cream. The external appearance or rheological behavior of the creamy matter not containing such oils will be hardly distinguishable. So, such ternary systems will be referred to hereafter as ternary creams.

Crystalline phases which show some structural regularities are contained in ordinary O/W creams made up with higher alcohols. In order to investigate the internal structure or thermal properties of the crystalline phases, the use of a ternary cream as a test sample has advantages over ordinary creams, since in doing so one can increase the concentration of crystalline materials per unit volume and, consequently, enhance the detectability. The present authors studied the formation of the crystalline phase and the variation of viscosity using ternary creams, the formulations of which are listed in Table 25.[109]

6.1.1. Variations in External Appearance

All ternary creams were quite similar to ordinary ones in softness to the touch and in the external appearance immediately after their preparation. However, when stored at 5 °C, they became pearly lustrous and mobile after one month storage.

Table 25. Composition of Ternary Systems Used in the Study of Ternary Creams[109]

	Systems					
	T_1	T_2	T_3	T_4	T_5	T_6
Components			(g)			
1-Hexadecanol	150	120	90	60	30	0
1-Octadecanol	0	30	60	90	120	150
OE-15[a]	25	25	25	25	25	25
Water	350	350	350	350	350	350

[a]Polyoxyetylene (15 mol) oleylether, commercial grade.

Ternary creams T_1 and T_6 (Table 25) changed similarly when stored at 25 °C, and T_6 when kept at 40 °C for the same period of time. Table 26 summarizes the stability of all those creams by indicating which ones changed in appearance and consistency as unstable and which did not show such change as stable.

6.1.2. Variations in Viscosity

Viscosity variations of three typical ternary creams (Fig. 19) show that those which were regarded as stable maintained high viscosities (above 2,000 Pa.s), while those regarded as unstable had lower viscosity (below 1000 Pa.s). The viscosity of unstable creams decreased faster and reached lower constant values with increasing storage temperature. For example, the viscosity of the cream coded T_1, which eventually broke down when stored at 5 and 25 °C, respectively (Fig. 19), was initially as high as that of the same sample aged at 40 °C, which remained stable.

6.1.3. Microscopic Observation

Immediately after preparation, all ternary creams showed many spherical particles dispersed uniformly over the field of vision in a microscope. Under polarized light, the particles showed optical anisotropy (Fig. 20), similar to observations of Barry et al.[110-114] On long-term storage, the ternary creams maintained at temperatures at which they remained stable continued to show such an image. However, at lower temperatures, these creams appeared as rhombic particles as shown in Fig. 21.

Similar particles were first observed by Mapstone,[115] who described in detail the viscosity variation and the change in microscopic appearance, etc., as a function of temperature, as will be described in Section 9.

Table 26. *Stability of the Ternary Creams Determined Based on the Change in the Appearance and Consistency After One Month Storage*[109]

System	Mixing ratio of Alcohols[a]	Storage temperature		
		5 °C	25 °C	40 °C
T_1	5:0	Unstable[b]	Unstable	Stable[c]
T_2	4:1	Unstable	Stable	Stable
T_3	3:2	Unstable	Stable	Stable
T_4	2:3	Unstable	Stable	Stable
T_5	1:4	Unstable	Stable	Stable
T_6	0:5	Unstable	Unstable	Unstable

[a]1-Hexadecanol:1-Octadecanol by weight.
[b]Became pearly lustrous and decreased in viscosity.
[c]Did not show marked change in appearance.

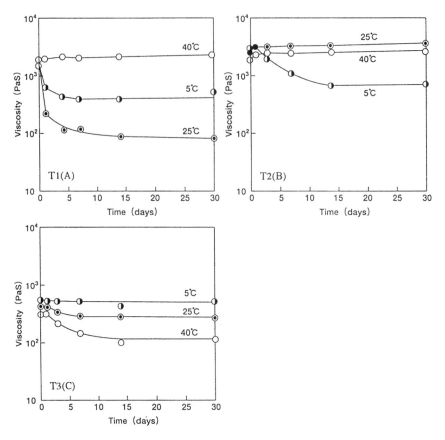

Figure 19. Change in viscosity of the ternary creams on aging described in Table 25. Creams: T1 (A),
T3 (B), and T6 (C).[109]

6.1.4. X-Ray Diffraction Analysis

Results of the X-ray diffraction analyses, illustrated schematically in Figure
22, were classifiable into two types: one which showed only one peak at 21° and
the other which had two or more peaks. It was not possible to identify the crystal
form when only one peak was given. Crystals of hydrated higher alcohol (α'-phase)
should yield several peaks corresponding to a longer interplanar spacing at Bragg's
angles near 10°, which were not observed. Instead, an LC phase may have formed,
which would shift the peaks into a region of lower angles. Although the structures
of two kinds of LC phases in the creams which gave only one peak are described
in detail later, the phase existing in the cream is referred to as the λ-phase at present.

The analysis of diffractograms, which gave two or more peaks, revealed β- or
γ-crystals, as indicated in Fig. 22. A comparison of these results with data in Table

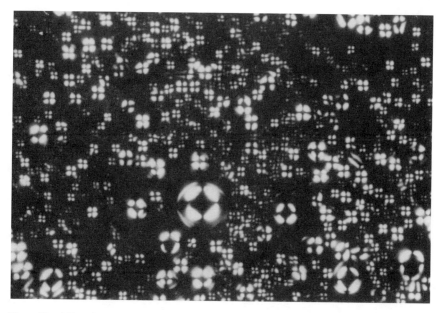

Figure 20. Microphotograph of the ternary cream T3 observed between crossed nicols, immediately after preparation.[109]

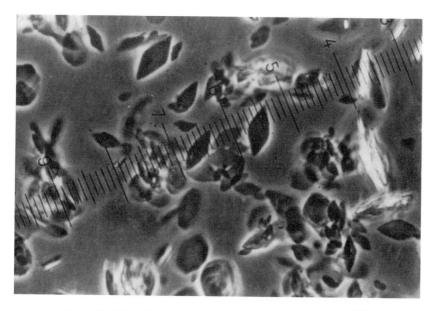

Figure 21. Microphotograph of T1 after one month storage at 25 °C.[109]

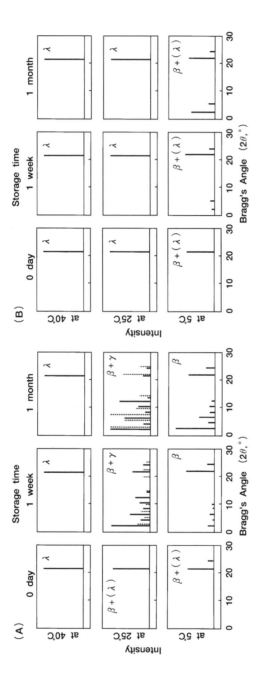

Figure 22.   Change in the XRD patterns of the ternary creams on aging at various temperatures. Greek letters written in each section denote the identified crystal forms. (A) T1, (B) T3, and (C) T6.[109]

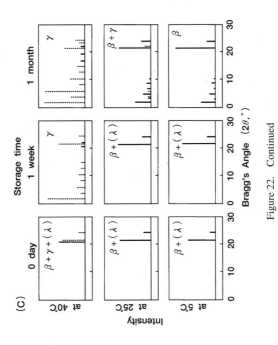

Figure 22. Continued

26 leads to the conclusion that β- or γ-crystals may develop in ternary creams which deteriorate on storage. The rhombic particles seen in Fig. 21 are indeed those of either or both crystals. The interplanar spacing of crystals obtained in creams formulated by mixing 1-hexadecanol and 1-octadecanol assumed intermediate values of crystal parameters characteristic of β- and γ-phases.

Thus, the change in the X-ray diffraction powder pattern on aging is correlated with the appearance of pearly lustrous sheen, with the viscosity decrease, or with the growth of their rhombic crystals.

6.1.5. Low-Angle X-Ray Diffraction Analysis

Figure 23 shows the low-angle XRD diagrams for the ternary cream T_1 immediately after preparation (upper) and after storage for one month at 25 °C (lower).[109] The freshly obtained sample gave two peaks of 5.89 and 11.8 nm, with ratio 1:2, which indicates the existence of lamellar liquid crystals (LCs). After one month, the original two peaks disappeared and different ones appeared, characteristic of (002) planes of crystals of β- and γ-alcohols. These data suggest that the

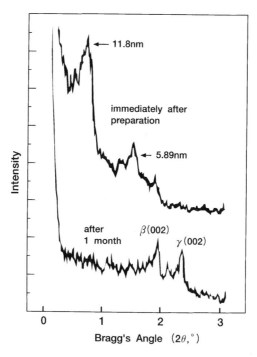

Figure 23.   Low-angle XRD patterns for the ternary cream T_1, immediately after preparation (upper), and after one month storage at 25 °C (below).[109]

degradation of lamellar LCs is closely related to crystallization of the alcohol. The degradation of ternary creams should also be dependent on the same processes.

6.1.6.   *Thermal Property*

The differential scanning calorimetry (DSC) curves obtained with ternary creams stored at lower temperature at which alcohol crystals appeared are reproduced in Fig. 24. Each curve displays essentially two endothermic peaks, which may be due to the transition point and melting point of the alcohol crystals which grew in the cream. For each curve one can establish two inflection parts, indicated by curves for the sample T_1. The first of these points designates the transition from

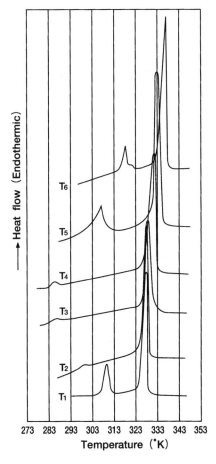

Figure 24.   Differential scanning calorimetry heating curves for ternary creams described in Table 25.[109]

the β- or the γ-phase to the λ-phase, while the second is characteristic of the transition from the λ-phase to liquid.

Figure 25 summarizes these points for various mixtures of 1-hexadecanol and 1-octadecanol. These plots are very similar to those of the transition and freezing point of alcohols in the presence of excess water, shown in Figure 13.

6.2.   *Polymorphism of Hexaoctadecanol (3:2) and Stability of a Ternary Cream*

Experimental results described above suggest that the stability of a ternary cream is closely related to the crystalline state of alcohols when they coexist with water. Such ternary creams are stable at temperatures at which the alcohol is in the α'-form in the presence of excess water, but unstable at temperatures at which the alcohol cannot appear as this phase. On addition of nonionic surfactants, these may form LCs in combination with the α'-phase.

This rule may appear invalid, since as in the case of T_1, the LC phase formed at 25 °C immediately after the preparation of the cream, although the temperature is below the limit at which α'-phase should exist. However, one needs to consider the cream preparation process. The mixture of all components is first heated to temperatures above the melting point of added alcohols (70 °C in the case of this study), and then cooled to room temperature. In the course of the cooling process,

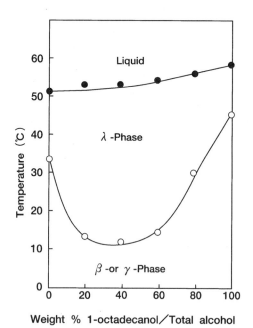

Figure 25.   Plot of the temperatures at inflection points as indicated in Fig. 24 as a function of the weight ratios of 1-hexadecanol to the sum of 1-hexadecanol and 1-octadecanol in ternary creams.[109]

the mixture passes through a temperature range (from 51 to 31 °C) in which added alcohol assumes the α'-form, yielding LC phases.

The so-produced LCs do not break instantaneously, even if the temperature reaches the range at which they are unstable (below 31 °C in case of 1-hexadecanol). Since molecules of both the higher alcohol and the surfactant are large and extended, it will take a certain amount of time for the LC phase to disintegrate, and for the alcohol molecules to crystallize into the β- or γ-form.

The suggested process is supported by data of viscosity lowering as a function of time (Fig. 19), and from the change in the XRD patterns (Figs. 22 and 23). Approximately two weeks were required for the viscosity to reach a limiting value, even when the ternary creams were stored at breakdown temperatures.

Yamaguchi *et al.* showed that the rate of breakdowns depended on the amount of LCs present in the cream.[116] They compared the viscosity lowering rates of two ternary creams, formulated in the same way as shown in Table 25, but using different cooling rates, one fast (in iced water) and the other slow (in air). Figure 26 indicates that the viscosity lowering of the ternary cream made by the fast cooling process proceeds more rapidly than by the slow cooling process. The amount of the LC phase contained in the latter case should be larger than that contained in rapidly cooled creams, since the time period during which the α'-phase exists is longer.

Over the temperature range where a higher alcohol hemihydrate is formed, the reaction proceeds to form the LC phase, according to:

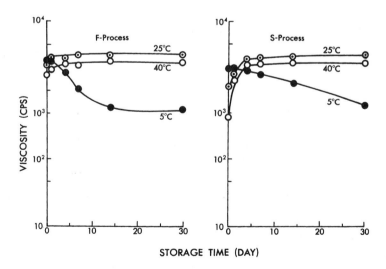

Figure 26. The change in the viscosity of the ternary cream T_1 (Table 25) as a function of time at three different temperatures. Left-side data refer to the fast cooling process using an ice/water bath and right-side data to slow cooling in air. In both cases the creams were stirred with a homogenizer.[116]

Alcohol hemihydrate + Surfactant + Water $\rightarrow$ liquid crystals
(α'-phase)   (λ-phase)

and, therefore, a close relationship exists between the λ-phase and the stabilization of ternary creams. This process also explains phenomena often encountered in the actual cream production, so far considered as unaccountable. Creams, which appeared excellent when produced but lost smoothness and became fluid on storage, must have been made with cetyl alcohol of high purity.

Sometimes, the viscosity of a cream does not reach a high enough value during tens of hours after production, even though an appropriate higher alcohol is used. This effect may be possibly due to the fact that the LC phase did not fully form during the cooling time, because of too fast a cooling rate. In such cases, no trouble will occur even if the cream is bottled as it is. If in doubt, however, the creams should be left in a tank until the viscosity equilibrates. Such a process is referred to as maturation, and experience shows it to be effective and it is applied in practice.

7.   Liquid Crystalline Phases in Hexaoctadecanol (3:2)/Surfactant/ Water Ternary Systems

7.1.   Phase Diagram

The present authors investigated the region of liquid crystalline phases in the Hexaoctadecanol (3:2)/pentadecaethoxyoleylether(OE-15)/water ternary system. The higher alcohol was chosen, since it forms the α'-phase, i.e., tetragonal structure at room temperature. The experimental method was as follows. Three components were weighed into glass test tubes and their top was sealed by fusing. The systems were vigorously shaken at 70 °C, then cooled in a water bath at 25 °C, and stored at the same temperature.

Since all samples were highly viscous (and turbid) over nearly the entire region of constituent concentrations, the phases could not be separated even in a high speed centrifuge. Thus, they were examined as prepared by the XRD analysis, the result of which is shown in Fig. 27.[117] The boundary lines of the regions D_2 and M, indicated by dotted lines as they were not clearly defined, have been drawn for conditions at which only the XRD peaks designated to these phases could be observed. Weak XRD peaks on the outside of these regions suggested the coexistence of one or two other phases. Numerals from 75 to 105 along the D_2 phase region in Fig. 27 indicate the long interplanar spacings (in Å) of the D_2 phases, while the lines projecting from the periphery of the region designate the coexistence of the latter with the other phases. The interplanar spacings appear to lengthen with increase in water content.

The I_w phase is an optically isotropic transparent liquid. It could be an aqueous solution of OE-15 in which a small amount of Hexaoctadecanol (3:2) is dissolved.

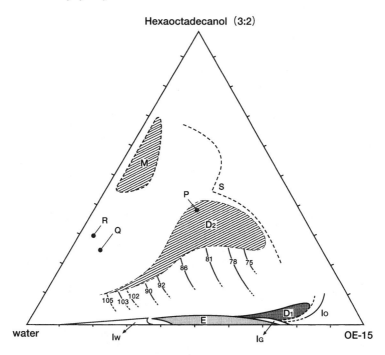

Figure 27.   Phase diagram of the Hexaoctadecanol (3:2)/OE-15/water ternary system: IW, isotropic aqueous solution; IG, transparent gelatinous phase; E, gelatinous phase having optical anisotropy; IO, isotropic oil solutions; S, semi-solid phase; D_1, lamellar liquid crystals; D_2, lamellar liquid crystals; M, region of many closely packed spherulite particles. Lines drawn at the periphery of the D_2 phase indicate coexisting phases and the added numerals refer to interplanar spacings of the D_2 phase in Å. Points P, Q, and R show the position of samples described in Sections 7.3, 7.5, and 10.1.[117]

Phase E is a translucent gel showing optical anisotropy, with the XRD indicating hexagonal structure. The latter may consist of a small amount of Hexaoctadecanol (3:2) included into the optically anisotropic, translucent, and gelatinous phase, termed the AG phase in the binary system of OE-15/water.[117]

The I_G phase is an optically isotropic, translucent gel. Since the XRD pattern is of a face-centered cubic structure, the same as the $I_{G'}$ phase in the OE-15/water binary system, spherical OE-15 micelles dissolving small amount of Hexaoctadecanol (3:2) may be situated at the lattice point of the structure in the phase.[117] The region D_1 in contact with E and I_G phases consists of a translucent gel which shows optical anisotropy. The low-angle XRD analysis gave peaks corresponding to an interplanar spacing ratio of 2:1 at ~1.5, and 3.0° Bragg's angles (corresponding to interplanar spacings of about 6.0 and 3.0 nm, respectively, varying with constituent ratios) showed this phase to be a typical lamellar LC phase. Phases denoted as D_2 and M are explained in detail in the following sections.

7.2. D₂ Region

The systems in the region designated D_2 which covers the concentration of Hexaoctadecanol (3:2) higher than that in the D_1 phase exhibit an XRD pattern characteristic of lamellar structures. The peaks in the XRD pattern at 21.4° (Fig. 28) for Hexaoctadecanol (3:2):water:OE-15 = 4:3:3 (corresponding to the composition at P in Fig. 27) is generally observed in the anhydrous *a*-crystal or in the hydrated α′-crystal of an *n*-higher alcohol. This peak is specific for the α-form of normal paraffins when the hydrocarbon chains are aligned hexagonally.[118] Thus, the hydrocarbon layer of the D_2 phase is considered to have the same hexagonal structure as an α-alcohol. Another characteristic aspect of the XRD pattern in Fig. 28 is the appearance of five or more peaks at regular intervals at angles less than 10°. The longest corresponding interplanar spacings divided by the shorter ones give integer values (Table 27), suggesting the D_2 phase to be lamellar LC.

The electron micrograph of a sample characteristic of the D_2 phase (Fig. 29) clearly displays the lamellar structure, which coincides with the G phase, explained in Section 8. However, the interplanar spacing of the G phase is 11.8 nm, meaning that it includes much water. Figure 30 represents a partial phase diagram, which was redrawn in terms of mole fractions, based on the phase diagram of Fig. 27, suggesting that the D_2 phase appears along the 3:1 line.

It has been known that mixtures of two surfactants exhibit various specific properties at their 3:1 molar ratio.[119–122] For instance, mixed monolayers of

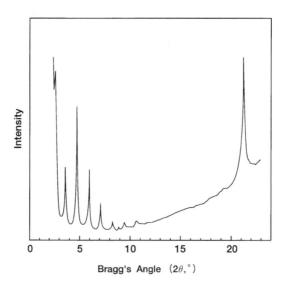

Figure 28.   XRD pattern for the formulation corresponding to the D_2 phase.[117]

Table 27. Long Interplanar Spacings and Their Ratios with Respect to the Longest One for the D_2 Phase of the Composition Designated by P in Fig. 27[117]

Bragg's angle (°)	Interplanar spacing d (nm)	Ratio 7.48[a]/d
1.18[b]	7.48	1.00
2.38[b]	3.71	2.02
3.58	2.46	3.04
4.73	1.87	4.00
5.95	1.23	6.08
8.30	1.06	7.06

[a]The longest interplanar spacing measured.
[b]Determined by the low-angle X-ray diffraction.

cholesterol–egg lecithin and stearyl alcohol–stearic acid exhibited minimum area per molecule at the 3:1 molar ratio.[123] The rate of evaporation of water through mixed monolayers of stearyl alcohol–stearic acid reached a minimum at this ratio, and so did the rate of drainage of foams of decanol–decanoic acid systems. Under these condition the foam stability was at its maximum.[123] Optimum solubilization of water in microemulsions occurred when the molar ratio of hexanol:potassium

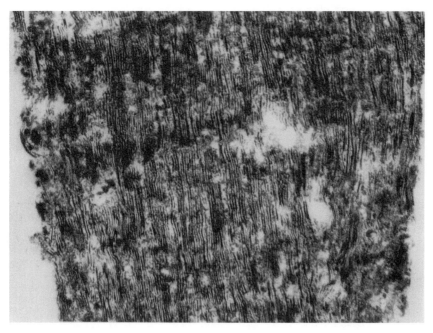

Figure 29. Electron micrograph of a sample designated by P in the D_2-phase region in Fig. 27.[117]

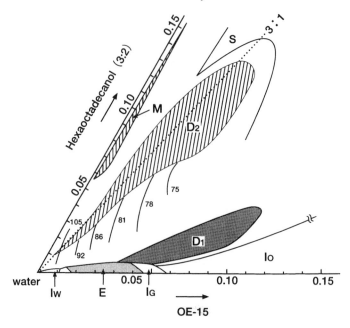

Figure 30. Partial phase diagram of the hexaoctadecanol (3:2)/OE-15/water ternary system in terms of mole fractions based on results shown in Fig. 27, using the same abbreviations.[117]

oleate was close to 3:1[123] and also the extraction of oil from emulsions stabilized by hexadecylpyridinium chloride and sodium dodecylsulfate was most efficient at the 3:1 surfactant composition. [123]

Based on these observations, Shah suggested that such specific changes in the nature of mixed surfactant systems at the 3:1 molecular ratio were due to two-dimensional hexagonal packing of molecules, resulting in a closer molecular arrangement and a greater stability of mixed films at the interface, as shown in Fig. 31.[123]

In the D_2 phase, the same arrangement of Hexaoctadecanol (3:2) and OE-15 could be constructed, in which the hydrocarbon groups are oriented vertically with respect to the paper plane, as shown in a three-dimensional array in Fig. 32.[124] From such empirical facts and considerations, the hydrocarbon chains in the D_2 phase are thought to maintain a hexagonal structure as α-higher alcohols, although the interplanar spacings of hydrophilic layers may be longer by inclusion of water.

The reaction equation in this case can be written as

alcohol hemihydrate + 1/3 surfactant + n_2 water $\rightarrow$ D_2 phase
(α'-phase)   (lamellar)

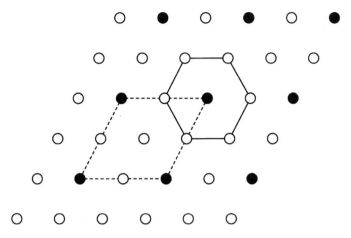

Figure 31. Proposed two-dimensional hexagonal arrangement of molecules at the 1:3 molecular ratio in mixed surfactant systems. Molecules of one type occupy the corners and those of the other type occupy the center of hexagons.[123]

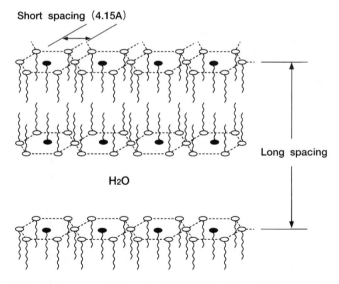

Figure 32. Molecular arrangement of a system in the D_2 phase. Open circles represent the hydroxide groups of 1-hexadecanol or 1-octadecanol, and full circles represent ethyleneoxide groups of OE-15.[124]

in which n_2 takes various large numbers.

When either 1-hexadecanol or 1-octadecanol is used instead of Hexaoctadecanol (3:2), the D_2 phase will form over the temperature range of 31 to 51 °C or 40 to 60 °C, respectively.

7.3. M Region

Another phase was found in a region of much higher alcohol concentrations, which consisted of a translucent and a very rigid gel in which spherical and optically anisotropic particles were closely packed. Each particle appeared as a tetrapetalous flower under a polarized light as shown in Fig. 33.[117] In this case the molar ratio of alcohol to surfactant is ~10, corresponding to that which gave the maximum static yield value, as shown in Table 21.

Barry and Shotton first observed such particles and described them as "frozen LC showing a Maltese cross,"[125] the latter being the reason for choosing the description "M" phase. We may, however, consider this phase to be heterogeneous with spherulites or "M" particles dispersed in very small amounts of the D_2 phase.

The low-angle XRD patterns of the M phase had only one peak at 21.4° (corresponding to interplanar spacing of 0.415 nm), which is specific of alcohols

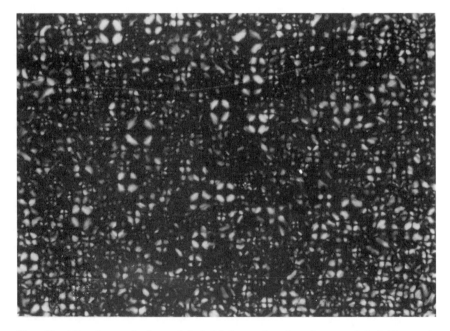

Figure 33.   Microphotograph of a sample in the M-phase region taken using crossed nicols. The sample contained hexadecanol (3:2), surfactant EO-15, and water in weight ratios 10:1:9.[117]

in both α- (anhydrous) and α'- (hydrated) forms. The scanning electron microscopy showed the particle to be concentric spheres with multilamellar layers. When the molecular arrangement is concentric, the angles of X-rays reflected from the upper and lower spherical surfaces are not the same, and consequently the X-rays do not interfere and no diffraction peaks corresponding to the long interplanar spacings are observed.

The schematic illustration of M particles in Fig. 34 shows that the multilamellar layers overlap with repeated orientation of every two hydrophilic surface facing each other, and vice versa. The M particles may form in the temperature range between approximately 31 and 51 °C, and 40 and 60 °C, respectively, if 1-hexadecanol and 1-octadecanol are used instead of Hexaoctadecanol (3:2). Thus, at temperatures for which the hydrated alcohol appears in a'-form, reaction of the type

Alcohol hemihydrate + ~1/10 surfactant + n_3 water → M particle
(α'-phase)   (spherulites)

may occur, in which n_3 takes various large numbers.

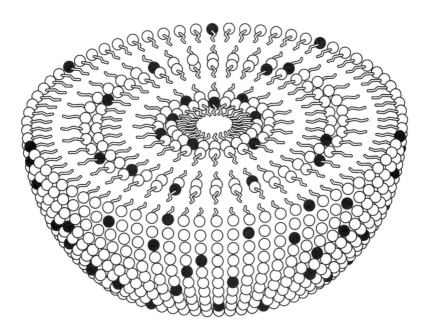

Figure 34.   Schematic molecular arrangement of a spherulite particle dispersing in the M region. (Illustrated by S. Sakai, Shiseido Research Center.)

When χ_α and χ_s grams of the alcohol and the surfactant are formulated in the condition that enough water is present, molar fraction p of the surfactant utilized to form the D_2 phase is expressed as

$$p = 10/7 - 1/7 \; (\chi_\alpha/m_\alpha/\chi_s/m_s) \qquad (6)$$

where m_α and m_s are the molecular weight of the alcohol and surfactant, respectively. If the molar ratio of the alcohol to surfactant $(\chi_\alpha/m_\alpha/\chi_s/m_s)$ is 3, all surfactant molecules are used to form the D_2 phase. If it is 10, all terms are used to form the M phase.

7.4. The Location and State of the D₂ Phase and M Particles in a Ternary Cream

The question arises as to the presence and location of the D_2 phase and M particles in a ternary cream. For this purpose, Yamaguchi *et al.* used a cryo-scanning electron microscope (cryo-SEM) equipped with a defrosting device. If the

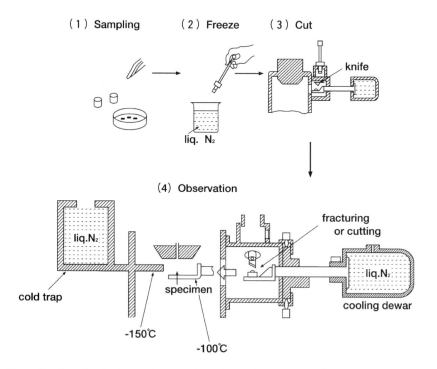

Figure 35. Procedure for the sample preparation to observe inner structure of creams by cryo-scanning electron microscopy.[126]

specimen contains water, the latter freezes at the surface of the specimen, which can be avoided by the installation of a cold trap. The actual sample preparations reproduced in Fig. 35: (1) about 50 mg of a specimen is mounted on an aluminum holder, (2) the sample is dipped into liquid nitrogen for 30 minutes, (3) it is then mounted onto a sample holder and the sample room is cooled to -100 °C, (4) the frozen sample is cut with a cooled keen-edged glass-knife, and (5) finally observed in the cryo-SEM.

The cold trap is cooled to -150 °C, i.e., the temperature 50 °C lower than that of the sample stage of cryo-SEM. In doing so, water which vaporizes from the specimen is trapped and the frost formation on the specimen surface is thus prevented.

Figure 36 displays a micrograph of a cut surface for a material of the D_2 phase corresponding to point P in Fig. 27 [Hexaoctadecanol (3:2):OE-15:water = 30:40:30], as observed by the cryo-SEM. Many very thin (200 to 300 nm) sheets appear to overlap, corresponding in thickness to several tens times LC lamellae

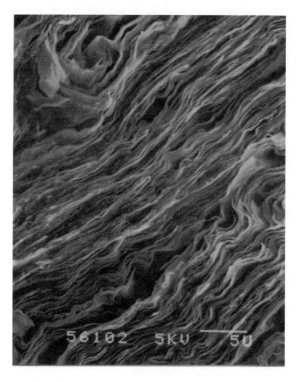

Figure 36.   Cross-section of the D_2 phase of the composition designated by P in Fig. 28 as observed with a cryo-SEM. (Taken by K. Yoshida, Shiseido Research Center.)

(7.48 nm), as evaluated by the XRD. The micrograph (Fig. 37) of a cut surface of the ternary cream corresponding to point R in Fig. 27 [Hexaoctadecanol (3:2):OE-15:water = 30:5:65] displays spherical M particles surrounded by irregularly shaped matter, all dispersed in the aqueous matrix. The hollow-like crater in the upper right of this photo is formed by a spherical particle that has been removed.

On the left side a particle is seen from which half of the shell is peeled away, exposing the core. These spheres are assumed to be the M particles. According to the phase diagram, the D_2 phase and M particles ought to coexist in the samples between M and D_2 areas. In other words, the ternary cream is a dispersion of M particles in a gel made of thin D_2 phase sheets, dispersed in the aqueous phase. A cross-sectional cryo-SEM photograph of a single M particle of the composition R in Fig. 27 is shown in Fig. 38.

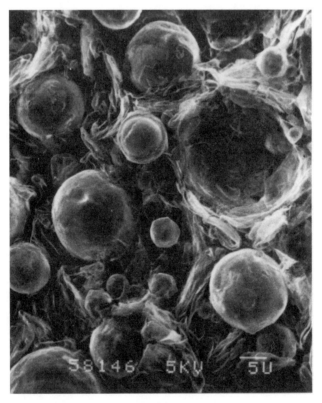

Figure 37.   Cross-section of the ternary cream of the composition designated by R in Fig. 28 as observed with a cryo-SEM. (Taken by K. Yoshida, Shiseido Research Center.)

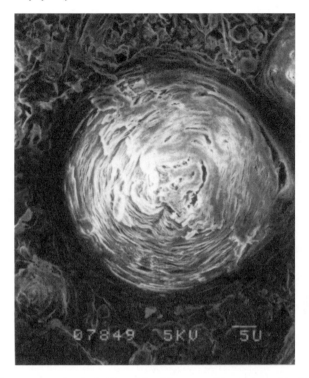

Figure 38.   Cross-section of a single M particle of the composition designated by R in Fig. 27 as observed with a cryo-SEM. (Taken by K. Yoshida, Shiseido Research Center.)

8.   Temperature and the Process Yielding the Liquid Crystalline Phase

In Section 7, two new phases consisting of hexagonal lamellar LC and spherulite were introduced. The lamellar LC phase, termed D_2 phase, forms in the region of Hexaoctadecanol (3:2):OE-15 = 3:1, and the M region, in which spherulite particles are closely packed, occurs in the region of Hexaoctadecanol (3:2):OE-15 = 10:1. This section describes the formation of these two phases and their location in the ternary systems.

8.1.   Temperature of the Formation of the Liquid Crystalline Phase

8.1.1.   Method of Phase Separation Experiments[127]

The present authors carried out phase separation experiments to investigate the nature of phases which formed at various temperatures using a ternary cream consisting of 25% Hexaoctadecanol (3:2), 10% OE-15, and 65% water, which corresponds to point Q in Fig. 27. Therefore, both phase D_2 and M particles coexist in the ternary cream.

In the developed procedure ~10 g of the ternary cream, which was once emulsified by the ordinary method, was placed into a glass test tube and the top was sealed by fusion. Next, the test tube was immersed into a water bath controlled at 70 °C and, after aging overnight, the external appearance of the phase separation was noted. The system was shaken vigorously and dipped into the water bath again, the temperature of which was then lowered in steps of 0.1 °C until it reached 40 °C. The content of the tube was inspected, and if any variation in appearance was observed, the reversibility was checked by again raising the temperature. Optical anisotropy was checked with crossed nicols.

The same experiment was repeated with dodecaethoxydodecylether (DE-12), a nonionic surfactant consisting of monodispersed alkyl and ethyleneoxide chains.

8.1.2. Results of the Phase Separation Experiment

Figure 39 shows the phase separation results as a function of temperature.[127] In the case of the OE-15 cream (upper part), separation into two phases of

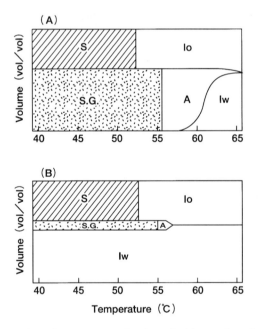

Figure 39.   Phase separations of ternary creams given in semi-arbitrary volume fractions as a function of temperature. (A) Hexaoctadecanol (3:2)/OE-15/water ternary cream; (B) Hexaoctadecanol (3:2)/DE-12/water ternary cream; S, white and waxy phase; IO, optically isotropic and transparent phase; S.G., optically anisotropic and translucent gel; A, optically anisotropic and transparent gel; I_W, Isotropic liquid.[127]

transparent liquids, I_O and I_W, took place at 70 °C, neither of which displayed optical anisotropy. The I_O phase consists of a molten higher alcohol solution and the I_W of an aqueous solution. Both phases contain surfactant OE-15. With time, the I_O phase decreased somewhat in volume without any change in the external appearance, possibly due to migration of the higher alcohol into the aqueous phase.

At 65 °C, a viscous phase designated by "A" appeared between I_O and I_W, which was also transparent, but showed optical anisotropy. The volume of this phase increased as the temperature was lowered, replacing the I_W phase, which ultimately completely disappeared at 55.7 °C.

At 55.3 °C, the A phase clouded up and was named the S.G. phase, since it was a suspension of a gelatinous matter. When centrifuged at ~107 m/s^2 it separated into two layers, one being optically anisotropic, translucent, and gelatinous (G phase), the same as D_2, and the other was an isotropic transparent aqueous phase, identified as I_W. With S is the described solid matter, which most likely consisted of M particles. Although this figure does not extend below 40 °C, the appearance of these systems remained unchanged on long-term storage at room temperature. The S.G. phase is thus a stable form to which the A phase has changed.

The DE-12 cream (lower part of Fig. 39) also separated into two transparent isotropic liquids at 70 °C. The transparent, optically anisotropic and gelatinous A phase appeared at 57 °C, and began to cloud at 55 °C. The I_O phase solidified at 52.5 °C.

It is evident from data given in Fig. 39 that the OE-15 cream differed from the DE-12 cream, mainly by the behavior of the I_W phase. Since the S.G. phase of the OE-15 cream separated on centrifugation, it could be assumed that this phase is a suspension of the G phase in I_W phase. The present authors believe that in practical application there are no essential differences between the OE-15 and DE-12 creams.

8.2. G Phase and S Phase

8.2.1. Microscopic Observation[1]

The micrograph of the S.G. phase of the OE-15 cream (Fig. 40) shows a pattern of randomly distributed platelets, suggesting that this system is a suspension of lamellar crystalline particles. The latter did not show optical anisotropy, apparently due to their thin nature. The micrograph of the G phase (Fig. 41), taken under polarizing light, indicates the presence of closely packed, elongated crystalline particles, which are agglomerates of the lamellar particles. A similar image was also observed with the S.G. phase, separated from the DE-12 cream, suggesting that the latter is almost the same with the G phase of OE-15 cream.

The G phase of both creams showed no deterioration on long-term storage at 25 °C. However, creams prepared using 1-hexadecanol or 1-octadecanol instead of Hexaoctadecanol (3:2) exhibited a pearly lustrous sheen after several days storage at the same temperature, due to the formation of rhombic particles, as was described

Figure 40. A microphotograph of the S.G. phase extracted from the hexaoctadecanol (3:2)/ OE-15/water ternary cream in Fig. 39(A).[127]

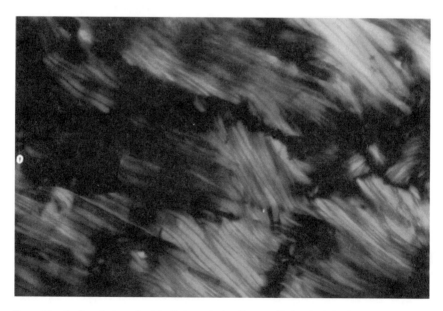

Figure 41. A microphotograph of the G phase obtained by centrifuging the S.G. phase, observed using crossed nicols.[127]

in Section 6. Therefore, it can be concluded that the deterioration of a cream is based on the breakdown of the G phase, i.e., on the crystallization of the β- or γ-alcohol in this phase.

The S phase of both OE-15 and DE-12 creams showed a similar appearance under a polarizing microscope as that displayed in Fig. 33. The creams containing this phase were also stable at room temperature, but became pearly lustrous when either 1-hexadecanol or 1-octadecanol was used instead of Hexaoctadecanol (3:2).

8.2.2. X-Ray Diffraction

The low-angle XRD of the G phases exhibited several peaks in the range from 3 to 10° and a single peak at 21.4°, similar to that of the D_2 phase shown in Fig. 28. The analysis of these diffraction peaks confirmed that the G phase had the same lamellar LC structure as the D_2 phase of the OE-15 ternary system, with long interplanar spacings of 11.8 and 9.5 nm for the G phases of OE-15 and DE-12 creams, respectively. The former value is longer than the interplanar spacing of the D_2 phase for the composition given by P in Fig. 28 (see also Table 27), which should be due to the inclusion of more water into the hydrophilic layer. The S phase gave only one peak at 21.4°, as did the M phase, leading to the conclusion that both were the same.

8.2.3. Differential Scanning Calorimetry

The differential scanning calorimetry (DSC) curves of the OE-15 cream (A), the G phase (B), and the S phase (C) are shown in Fig. 42. By comparison with (B), the peak at the highest temperature in (A) must be due to the solidification of phase G. On the other hand, by comparison with (C), the peak at the lowest temperature should be due to solidification of the IO phase. The other two weak peaks indicate solidification of different G phases made from different hydrophilic chain lengths as described below.

Creams made with surfactants which have polydispersed hydrophilic chain lengths, as OE-15, will contain G phases different in the long interplanar spacing, based on the difference of migration mobility of the surfactants from the I_O to the I_W phase in the process of cooling the temperature.

Figure 43 displays the DSC curves of the DE-12 cream (A) and the G phase (B). Unlike OE-15 cream, curve (A) has only two peaks, due to the monodispersed surfactant. By comparison with Fig. 39 (lower part), the peaks at lower and higher temperatures can be assigned to the solidification of the phases A and I_O, respectively.

8.2.4. Temperature at which the LC Phase is Formed

So far we have concluded that LC phases are formed under the condition at which higher alcohols appear in the α′-structure (hemihydrate). However, as shown

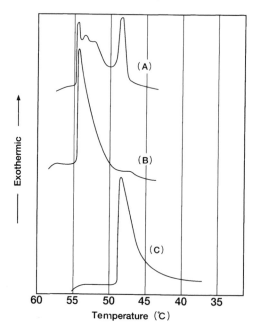

Figure 42.   Differential scanning calorimetry (DSC) cooling curves for (A) the OE-15 ternary cream, (B) the G phase extracted, and (C) the S phase.[127]

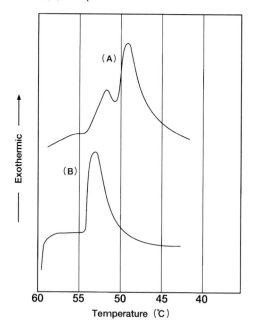

Figure 43.   DSC cooling curves for (A) the DE-12 ternary cream and (B) the extracted G phase.[127]

Table 28. Temperatures at which Phases A and I_0 of Ternary Creams Prepared with Different Surfactants Solidify[a(127)]

Surfactant formulated in the cream	Phase A solidies at (°C)	Phase I_0 solidifies at (°C)
OE-15	55.3	52.0
DE-12	55.0	52.0

[a]Hexaoctadecanol (3:2) hydrate melts at 54 °C.

in Table 28, the temperatures at which the phases I_0 and A solidify (resulting in the formation of phases S and G, respectively) are somewhat different from those at which the hemihydrate is formed (i.e., 54 °C). It should be noted that solidification of phase A occurs at a higher temperature than the melting point of the hemihydrate, and that the temperature differs due to the surfactant.

8.3. In situ Formation of G and M Phases

The series of micrographs in Fig. 44 illustrate the formation of the G and S phases of the DE-12 cream *in situ*, obtained under the following procedure. A small amount of the DE-12 ternary cream, prepared by the standard technique with 30 g Hexaoctadecanol (3:2), 7 g DE-12, and 70 g water, was used in the observation and was redispersed in a 0.1% aqueous solution of DE-12. Two or three drops of the dispersion were placed on a hollowed glass plate, covered with glass, and then mounted on the hot-stage of a microscope. After waiting for a few minutes at 70 °C until the globules coalesced to about 10 nm (to make the observation easier), the system was slowly cooled and inspected at intermittent periods of time.

At 70 °C, the system was a typical emulsion in which spherical globules were dispersed in a liquid (A). At ~60 °C, a third liquid phase enveloped the emulsion droplets (B). The new liquid phase increased in volume with the lowering of temperature (C), and began to solidify at 55 °C (D). At 52 °C, the emulsion droplets began to solidify (E).

Temperatures at which these variations occurred agreed well with those at which phase A appeared (57 °C), then changed into the S.G. phase (55 °C), and finally the I_0 phase changed into the S phase (52.5 °C) (Fig. 39). Consequently, emulsion droplets represent the I_0 phase, while the third phase, which appeared at the surface of the globules, is the A phase. Moreover, the latter changes to a suspension of the G phase (i.e., the S.G.) after it solidifies, and it can be assumed to correspond to the D_2 phase, based on the similarity of the thermal behavior and XRD results. Details on the structure of the A phase are unknown. On solidification of the I_0 phase, the S phase is formed and consists of densely packed M particles (spherulites) (Fig. 34), since the latter showed a skewed Maltese cross under a polarizing microscope.

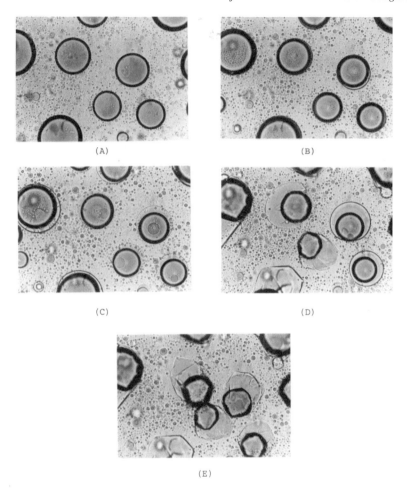

Figure 44. Micrographs showing the change in the appearance of the Hexaoctadecanol (3:2)/ DE-12/water ternary cream; (A) 70 °C, (B) 60 °C, (C) 58 °C, (D) 55 °C, and (E) 52 °C.[127]

Based on the above findings, it has become clear that D_2, which is a frozen lamellar LC phase, causes the aqueous phase to gel. Furthermore, M spherulites are formed in the Hexaoctadecanol (3:2)/DE-12/water ternary cream. During the cooling process a third phase, A, which shows optical anisotropy, appears at a given temperature at the molten Hexaoctadecanol (3:2) droplet/aqueous phase interface. At still lower temperature phase A solidifies to yield the D_2 phase and the latter separates from the Hexaoctadecanol (3:2) droplets and disperses into an aqueous phase. Thus, the D_2 phase increases the cream viscosity by forming a gel structure in the I_W phase.

An analogous microscopic observation was unsuccessful with the OE-15 cream, perhaps because of the polydispersity of hydrophilic chains of the surfactant. The relationship between the formation of the higher alcohol hemihydrate and LC phases with Hexaoctadecanol (3:2) is summarized in Fig. 45.[128] Similar systems could be established for other higher alcohols, such as 1-dodecanol, 1-tetradecanol,

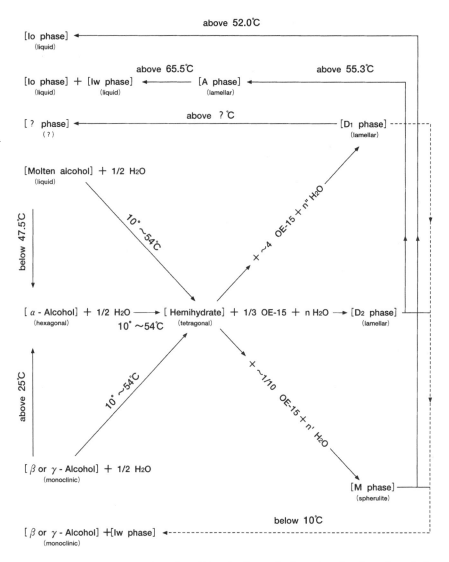

Figure 45. Schematic presentation of conditions leading to the formation of various phases in the ternary system consisting of Hexaoctadecanol (3:2)/OE-15/water.[128]

1-hexadecanol, 1-octadecanol, or their mixtures, although the dependence on the temperature may vary.

9. The Function of Liquid Crystalline Phases in O/W Creams

In previous sections, the characteristic nature of higher alcohols in the formation of LC phases has been explained from the viewpoint of the polymorphism of the alcohol. In summary, there is a very close relationship between the formation of hydrates of alcohols in the presence of water and the appearance of the LC phase. In ternary creams prepared with Hexaoctadecanol (3:2), a surfactant, and water, we found a lamellar LC (D_2) phase and a phase containing (M) spherulites. The molar ratio of Hexaoctadecanol (3:2) to surfactant for the former is 3 and for the latter is 10. The former causes the aqueous phase of the ternary cream to become viscous by dispersing in the aqueous phase and making it gelatinous. The spherulites are dispersed in the D_2, increasing the viscosity of the system by the increase in the internal (dispersed) phase.

The O/W creams or ointment bases in cosmetic or in pharmaceutical fields generally contain other oily components, such as liquid paraffin or vaseline, or various kinds of oils. It is of much interest to establish the properties of ternary creams when a fourth oily component is added. This section introduces various studies on the quaternary (O/W) creams with specific emphasis on cetostearyl alcohol, in order to elucidate the effect on the increase in viscosity of emulsions.

9.1. Studies on O/W Creams

There are many reports dealing with creams containing cetyl or cetostearyl alcohol in their formulations, particularly those by Talman et al. and Barry et al. Table 29 lists publications on emulsions containing higher alcohols including

Table 29.   Literature Survey on Emulsions or on Interfacial Films Containing Higher Alcohols

Higher alcohol	Surfactant	Oil	Ref.[a]
1-Octanol	Cetomacrogol 1000	LP[b]	129,130
	Cetrimide	LP	129,130
1-Decanol	Cetomacrogol 1000	LP	129,130
	Cetrimide	LP	129,130
1-Dodecanol	Ceromacrogol 1000	LP	129–132
	Cetrimide	LP	129–132
	K laurate	LP	131
	Na dodecylsulfate	LP	131–134
1-Tetradecanol	Cetomacrogol 1000	LP	129,130

Table 29. Continued

Higher alcohol	Surfactant	Oil	Ref.[a]
	Cetrimide	LP	129,130
1-Hexadecanol	Brij 35[c]	White oil	135
	Cetomacrogol 1000	LP	129,130
	Polyoxyethylene(2)stearate, Polyoxyethylene(40)stearate	LP, Vaseline	136
	Hexaoxyethyleneglycolmono-hexadecyl	Chlorobenzene	137
	Pentadecaethoxyoleylether	LP, Vaseline	138
	Cetrimide	LP	129,130
	Solumin FX 170 SD[d]	LP	132
	Na dodecylsulfate	Hexane	139
	Sorbester Q 12[e]	LP	131,132
	Texofor FX 170[f]	LP	132
	Texofor N 4[g]	LP	140
Oleyl alcohol	Cetomacrogol 1000	LP	131,132
	Cetrimide	LP	131,132
	K laurate	LP	131
	Na dodecylsulfate	Hexane	139
	Sorbester Q 12	LP	131,132
1-Octadecanol	Cetomacrogol 1000	LP	129,130
		PE[h], PE + CCl$_4$	141[*]
	Hexaethyleneglycol-monohexadecyl	Chlorobenzen	137
	Myrij 52[i]	PE + CCl$_4$	141[*]
	Pentadecaethoxyoleylether	LP, Vaseline	138
	Tween 60[j]	PE + CCl$_4$	141[*]
	Tween 80[k]	PE + CCl$_4$	141[*]
	Cetrimide	LP	129, 130, 140
		PE + CCl$_4$	141[*]
	Dodecyltrimethylammonium Br	PE + CCl$_4$	141[*]
	Tetradecyltrimethylammonium Br	PE + CCl$_4$	141[*]
	Hexadecyltrimethylammonium Br	PE + CCl4	141[*]
	Na dodecylsulfate	Light petroleum	132
		LP	143,144
		PE,PE + CCl$_4$	141[*]
	Na hexadecylsulfate	Light petroleum	142
		PE	141[*]
	Na octadecylsulfate	PE	141[*]
	Sipon WD[l]	PE	141[*]
	Texapon L-100[m]	PE,PE + CCl$_4$	141[*]
Octyldodecyl alcohol	Cetomacrogol 1000	LP	129,130
	Cetrimide	LP	129,130

Table 29.   Continued

Higher alcohol	Surfactant	Oil	Ref.[a]
Cetostearyl Alcohol	Cetomacrogol 1000	Arachis oil	131
		Castor oil	131
		IPM[n]	131
		LLP[o]	131
		LP	129,130,132,138,145, 146
	Pentadecaethoxyoleylether	LP + Vaseline	138
	Cetrimide	LP	129,130,132,140,143, 146,147,148
	Dodecyltrimethylammonium Bromid	LP	149
	Tetradecyltrimethylammonium Bromide	LP	149
	Hexadecyltrimethylammonium Bromide	LP	149
	Octadecyltrimethylammonium Bromide	LP	149
	Na laurylsulfate	LP	132,134,148
		LP + Vaseline	150
	Sorbester Q 12	LP	132
Hexaoctadecanol (9:1)	Pentadecaethoxyoleylether	LP + Vaseline	138
Hexaoctadecanol (3:1)	Na dodecylsulfate	LP	134
Hexaoctadecanol (7:3)	Pentadecaethoxyoleylether	LP + Vaseline	138
Hexaoctadecanol (6:4)	Pentadecaethoxyoleylether	LP	151,152
		LP + Vaseline	138
	Polyoxyethylenesorbitan-monostearate	LP	153
	Glycerolmonooleate	LP	153
Hexaoctadecanol (5:5)	Pentadecaethoxyoleylether	LP + Vaseline	138
	Na dodecylsulfate	LP	134
Hexaoctadecanol (3:7)	Pentadecaethoxyoleylether	LP + Vaseline	138
Hexaoctadecanol (1:3)	Na dodecylsulfate	LP	134
Hexaoctadecanol (1:9)	Pentadecaethoxyoleylether	LP + Vaseline	138

[a]Reference numbers carrying an asterisk indicate papers dealing with interfacial viscosity.
[b]LP;Liquid paraffin.
[c]Brij 35; ethyleneoxide condensation product with aluryl alcohol.
[d]Solumin F 170 SD; sodium salt of the corresponding sulfate material.
[e]Sorbester Q 12; polyoxyethylene-sorbitanmonolurate.
[f]Texofor FX 170; an ethoxylated alkylcresol.
[g]Texofor N 4; an ethoxylated aliphatic alcohol.
[h]PE; petroleum ether.
[i]Myrij 52; polyoxyethylene(40) stearate.
[j]Tween 60; polyoxyethylenesorbitanmonostearate.
[k]Tween 80, polyoxyethylene-sorbitanmonooleate.
[l]Sipon WD; a pure grade of Na dodecylsulfate.
[m]Texapon L 100; Na dodecylsulfate.
[n]Isopropylmyristate.
[o]LLP; light liquid paraffin.

cetostearyl alcohol. References carrying an asterisk specifically deal with inter-facial viscosity of such systems.

9.2. The Difference in the Viscosity-Increasing Effects due to the Nature of Higher Alcohols

Talman *et al.* studied the effect of various higher alcohols on the viscosity of O/W emulsions containing 50% liquid paraffin; a part of these results is given in Table 30.[131] Cetostearyl alcohol greatly increased the static yield values regardless of the nature of surfactants. Oleyl alcohol did not show such an effect, while lauryl alcohol increased the static yield value when used with Cetrimide (a cationic surfactant) or SDS. The melting and transition points of 1-dodecanol in the presence of excess water are 24.0 and 15.4 °C, respectively (Table 13). Therefore, it is

Table 30. Static Yield Values (SYV) of Emulsions Consisting of 50% of Liquid Paraffin, 0.5% of Various Surfactants, Different Concentrations of Various Higher Alcohols, and Water at 25 °C[131]

Higher alcohol (%)	Cetomacrogol 1000	Sorbester Q 12	Cetrimide	SDS	K laurate
Oleyl alcohol			SYV (Pa)		
1.0	0	0	0	0	0
2.0	0	0	0	0	0
4.0	0	0	0	0	0
6.0	0	0	0	0	0
8.0	0	0	0	0	0
10.0	0	0	0	0	0
Lauyl alcohol			SYV (Pa)		
1.0	0	0	15	6	0
2.0	0	0	35	36	0
4.0	0	0	48	45	0
6.0	0	0	50	57	0
8.0	0	0	69	68	*[a]
10.0	0	0	75	85	*[a]
Cetostearyl alcohol			SYV (Pa)		
0.25	0	0	0	0	0
0.75	0	0	0	0	0
1.5	28	13	25	25	15
2.5	91	113	100	48	75
4.0	234	141	241	134	264
7.0	390	264	666	653	**[b]

[a]Indicates pseudoplastic flow.
[b]Impossible to determine because of very high consistency.

possible for this alcohol to form a D_2 phase for 25 °C at which the viscosity was determined. Indeed, it was shown (Table 13) that the melting point of the hydrated 1-dodecanol is at ~25 °C.

9.3. The Difference in the Viscosity-Increasing Effect due to the Nature of Surfactants

The work by Talman *et al.* mentioned above suggests that the viscosity-increasing effect of higher alcohols does not depend on the nature of surfactants. Barry *et al.* also examined in detail the viscosity of quaternary creams consisting of liquid paraffin, cetostearyl alcohol, and water, in the presence of different surfactants such as Cetomacrogol 1000 (a nonionic surfactant), Cetrimide, dodecyl-trimethylammonium bromide, tetradecyltrimethylammonium bromide, hexadecyl-trimethylammonium bromide, octadecyltrimethylammonium bromide or sodium laurylsulfate, and reported no essential differences in the viscosity-increasing effect, analogously to the case of ternary creams, discussed earlier (Section 5.3.).

9.4. The Viscosity Change due to the Ratio of Cetostearyl Alcohol to Surfactant

Talman *et al.* found that at the constant concentration of cetostearyl alcohol, the viscosity or the static yield value reached a maximum for a given molar concentration of surfactants, regardless of their nature.[132] Figure 46 shows the maximum static yield value to be at about 0.017 mol/dm^3 surfactant for an alcohol concentration of 7%, corresponding to the molar ratio of alcohol to surfactant of ~16. At this ratio, the cream abounds in M particles. In the case of the ternary creams the corresponding molar ratio was ~10 (Section 5.3.2.). The difference may be due to the dissolution of cetostearyl alcohol in the oil phase, since creams studied by Talman *et al.* contained 50% liquid paraffin.

If this result is compared with the present authors' work on the phase diagram (Figs. 27 and 30), the viscosity or the static yield value is understood to reach the maximum at the molar ratio of cetostearyl alcohol/surfactant for which M particles are most abundantly formed. This ratio is about 10, if no oily phase is formulated. Table 31 shows the relationship between the molar ratios (MR) of cetostearyl alcohol/surfactant and the static yield values (SYV), calculated using the data reported by Talman *et al.*[132] The conditions yielding the maximum SYV values are boxed in, from which it is seen that the optimum cetostearyl alcohol/surfactant molar ratio lies between 7 and 20. The exceptions are found in the cases of 2.5% cetostearyl alcohol and 1% SDS, and of 7.0% cetostearyl alcohol and 0.25% Cetrimide, the MRs at these formulations being 2.85 and 40.3, respectively. However, it is impossible to say that the SYVs at these formulations are not necessarily the maximum values if compared with those at MR = 11.4 and 10.0.

Barry *et al.* examined in detail the rheological property of quaternary creams consisting of liquid paraffin, cetostearyl alcohol, surfactant, and water using various

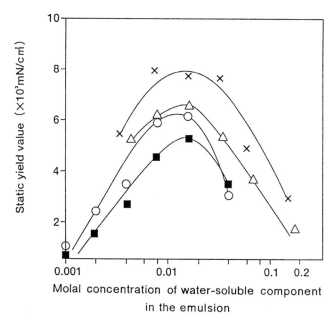

Figure 46.   The effect of surfactant concentration on static yield values of emulsions containing 7 wt% cetostearyl alcohol: Sorbester Q 12 (■), Cetomacrogol 1000 (○), Na dodecylsulfate (△), and Cetrimide (×).[132]

Table 31.   Relationship between Molar Ratio of Cetostearyl Alcohol to Surfactant and Static Yield Value (SYV) of Emulsions[132]

(1) Cetostearyl Alcohol 2.5 wt%

	Surfactant Concentration (%)					
Surfactant	0.125	0.25	0.5	1.0	2.0	5.0
Cetomacrogol 1000[a]						
Molar ratio[b]	92.3	46.2	**23.0[f]**	**11.5[f]**	5.74	2.30
SYV (Pa)	55.3	89.2	**98.0[g]**	**98.0[g]**	22.6	15.1
Sorbester Q 12[c]						
Molar ratio[b]	39.5	19.8	**9.88[f]**	4.94	2.47	0.99
SYV (Pa)	30.1	62.8	**79.2[g]**	40.2	42.7	36.4
Cetrimide[d]						
Molar ratio[b]	28.8	14.38	**7.21[f]**	3.59	1.80	0.72
SYV (Pa)	37.7	65.3	**100.5[g]**	70.4	45.2	10.1
Na dodecylsulfate[e]						
Molar ratio[b]	22.88	11.4	5.68	**2.85[f]**	1.42	0.57\
SYV (Pa)	49.0	94.2	93.6	**94.9[g]**	40.2	0.0

Shoji Fukushima and Michihiro Yamaguchi

Table 31. Continued

(2) Cetostearyl Alcohol 4.0 wt%

Surfactant	Surfactant concentration (%)					
	0.125	0.25	0.5	1.0	2.0	5.0
Cetomacrogol 1000						
Molar ratio	148	73.9	36.9	**18.4**[f]	9.19	3.69
SYV (Pa)	80.4	153	151	**271**[g]	149	27.6
Sorbester Q 12						
Molar ratio	63.2	31.6	15.8	**7.90**[f]	3.95	1.58
SYV (Pa)	50.2	77.9	105	**151**[g]	131	87.9
Cetrimide						
Molar ratio	46.1	23.0	**11.54**[f]	5.74	2.88	1.15

(3) Cetostearyl Alcohol 7.0 wt%

Surfactant	Surfactant concentration (%)					
	0.125	0.25	0.5	1.0	2.0	5.0
SYV (Pa)	156	221	**241**[g]	166	116	12.6
Na dodecylsulfate						
Molar ratio	36.4	18.2	**9.09**[f]	4.56	2.28	0.91
SYV (Pa)	121	171	**189**[g]	138	81.7	0.0
Cetomacrogol 1000						
Molar ratio	258	129	64.4	32.3	**16.1**[f]	6.45
SYV (Pa)	96.7	236	342	591	**616**[g]	302
Sorbester Q 12						
Molar ratio	110	55.4	27.7	13.9	**6.93**[f]	2.77
SYV (Pa)	60.3	159	264	452	**503**[g]	334
Cetrimide						
Molar ratio	80.7	**40.3**[f]	20.2	10.0	5.00	2.00
SYV (Pa)	534	**792**[g]	779	766	490	295
Na dodecylsulfate						
Molar ratio	63.8	31.9	**15.9**[f]	7.98	3.99	1.59
SYV (Pa)	515	616	**653**[g]	503	352	106

[a]Molecular weight (M.W.) = 1.166.
[b]Calculated by taking the M.W. of cetostearyl alcohol to be 253.
[c]M.W. = 500.
[d]M.W. = 364.
[e]M.W. = 288.
[f]The molar ratio at which the maximum SYV in the series was given.
[g]The maximum SYVs in the series.

surfactants. The weight ratio of cetostearyl alcohol to surfactant was kept constant at 9, regardless of the nature of the surfactant, based on the result that the viscosity of creams gave the maximum value at the molar ratio of 9 when SDS was used (Section 5.3.1.).

9.5. The Effect of Mixing 1-Hexadecanol with 1-Octadecanol

The present authors examined the viscosity of quaternary creams after one month of storage at 25 °C, using samples of compositions shown in Table 32. Creams prepared with either 1-hexadecanol or 1-octadecanol initially decreased in viscosity and reached a minimum within 3 days, and then remained constant. Creams formulated with commercial grade cetyl alcohol or the mixture of 1-hexadecanol and 1-octadecanol maintained high viscosities and high yield values during the entire storage period.[138] In all creams which decreased in consistency, crystals in the β- or γ-form of alcohol were found. Such a relationship between the viscosity lowering and crystallization of alcohols was the same as in the case of ternary creams described in Section 6.

Similar phenomena are also reported by Barry.[134] He prepared creams using mixtures of 1-hexadecanol and 1-octadecanol, or other higher alcohols as coemulsifiers to examine the viscosity variation with time of the composition listed in Table 33. The cetostearyl alcohol used contained 0.3% 1-decanol, 1.43% 1-dodecanol, 0.57% 1-tetradecanol, 21.0% 1-hexadecanol, and 76.7% 1-octadecanol.

Table 32.   Composition of Creams and their Rheological Properties after 1 Month Storage at 25 °C [138]

Cream	Composition								
	C_1	C_2	C_3	C_4	C_5	C_6	C_7	C_8	C_9
Material					(g)				
Cetyl alcohol[a]	15	—	—	—	—	—	—	—	—
1-Hexadecanol	—	15.0	13.5	10.5	9.0	7.5	4.5	1.5	—
1-Octadecanol	—	—	1.5	4.5	6.0	7.5	10.5	13.5	15.0
Vaseline	3	3	3	3	3	3	3	3	3
Liquid paraffin	12	12	12	12	12	12	12	12	12
OE-15[b]	5	5	5	5	5	5	5	5	5
Water	70	70	70	70	70	70	70	70	70
Rheological properties									
Apparent viscosity (Pa · s)	46	11	31	48	46	45	43	11	10
Yield value (10^{-5} Pa)	440	0	320	650	490	500	490	0	130

[a]For cosmetic use. 1-Hexadecanol:1-octadecanol was about 6:4.

[b]For cosmetic use. Pentadecaethoxyoleylether.

Shoji Fukushima and Michihiro Yamaguchi

Table 33. Composition of Creams Used in Barry's Experiments[134]

Cream	A_1	A_2	A_3	A_4	A_5	A_6	A_7	A_8
Material	(wt%)							
SDS	0.5	0.5	0.5	0.5	0.5	0.5	0.5	0.5
1-Decanol	—	—	—	—	—	—	—	0.01
Lauryl alcohol	—	—	—	—	—	—	—	0.06
Myristyl alcohol	—	—	—	—	—	—	—	0.03
Cetyl alcohol	4.5	3.38	2.25	1.13	—	—	0.97	0.95
Stearyl alcohol	—	1.13	2.25	3.38	4.5	—	3.53	3.45
Cetostearyl alcohol[a]	—	—	—	—	—	4.5	—	—
Liquid paraffin	30	30	30	30	30	30	30	30
Water	65	65	65	65	65	65	65	65

[a]Adopted in the British Pharmacopoeia.

Barry reported that the creams were firm and rheologically stable with ceto-stearyl alcohol, but were slightly less stable when (a) a mixture of cetyl and stearyl alcohol in the ratio present in cetostearyl alcohol or (b) a mixture of 1-decanol, lauryl, myristyl, cetyl, and stearyl alcohols in similar ratios to the cetostearyl alcohol was used in the formulation.[134] A more fluid, less stable cream was obtained with stearyl alcohol alone, but a stiff cream was formed, provided that the ratio of cetyl alcohol to total alcohol (cetyl alcohol plus stearyl alcohol) $\geqq 0.07$. He added that differences between creams were not related to the size distribution of emulsion droplets, but to the nature and content of the stearyl alcohol, which influenced the strength of the viscoelastic network formed.[154]

Davis and Smith examined the effect of the dispersed phase using different hydrocarbons and 1-hexadecanol on the stability of O/W emulsions made with SDS as an emulsifier.[139] The change in the mean droplet diameter and emulsion viscosity with time was correlated with solvent properties. An increase in oil polarity led to a decrease in stability, which increased with larger alkane chain lengths. In contrast, the interfacial film between the SDS solution and the same alkane became more expanded with the longer chain length of the oil. Small quantities of hexadecane, dodecane, or 1-hexadecanol increased the stability of unstable hexane, benzene, and cyclohexane emulsions. However, hexadecane had a far greater stabilizing effect on the bulk emulsion systems than 1-hexadecanol at the same molar concentration.

According to the present authors' opinion, these results may have been different if the experiments had been conducted at temperatures between 30 and 50 °C, at which 1-hexadecanol hemihydrate is formed. Similar comments refer to Carless and Hallworth's work on the oil/water interfacial viscosity.[141]

Okamoto and Oishi evaluated the stability of O/W creams containing 1-hexade-canol of 93% purity based on the separation of water, when these were centrifuged after stored at temperatures of 3, 25, 37 and 45 °C.[136] The creams which were kept at 3 and 25 °C separated water, which is predictable when data in Fig. 13 in Section 4 are taken into consideration.

9.6.   *Crystallization of Cetyl Alcohol in Cosmetic Creams*

Mapstone described in detail the crystallization of cetyl alcohol in cosmetic creams with basic composition shown in Table 34.[135]

When prepared by hand stirring, the initial viscosity of the creams was 20 Pa.s at 30 °C, which increased to 80 Pa.s at 30 °C after 6 days storage at 20 °C and to 86 Pa.s after an additional 3 days at 40 °C. After 3 weeks at 40 °C, the viscosity increased to about 400 Pa.s at 30 °C. Furthermore, creams stored at 5 °C first thickened and then remained apparently unchanged for at least 2 months, but broke down after 3.5 to 4 months with the development of sheen. When stored at 20 to 25 °C, the thickened creams remained stable and apparently unchanged after 8 months. However, almost every cream stored at 15 to 20 °C first thickened and then rapidly thinned to a viscosity of less than 5 Pa.s with the development of a pronounced sheen.

Creams that had been stored for awhile at 40 °C, or those seeded with cetyl alcohol crystals by the addition of a few drops of a previously crystallized batch, or prepared by stirring in a portion of the cream surface to dry, all thinned to a "milk" with a pronounced sheen within a few days at 15 to 20 °C. All such creams continued to develop the sheen, and finally broke down with the deposition of a scaly layer at the top.

Microscopic examination of the product at various stages showed an increasing abundance and size of flat crystals with age, particularly after thinning. These crystals were much larger than the drops of emulsified oil and appeared to be free of adhering oil droplets. To establish whether the crystals were made of cetyl alcohol, Mapstone conducted an examination by staining them with Rhodamine P and concluded that they consisted of cetyl alcohol. Unfortunately, the purity of the cetyl alcohol used was not given; it may be possible that the alcohol contained 5 to 10% of 1-octadecanol, since the cream was stable above 20 °C and unstable below this temperature.

Table 34.   Basic Composition of the Cream Mapstone Used[135]

Materials	wt%
Propylene glycol	3.1
Water	49.4
Brij 35[a]	1.7
Cetyl alcohol	1.7
White oil	2.1
Aluminum chlorohydroxide (50%)	42.0

[a]Ethyleneoxide condensation products with lauryl alcohol.

Mapstone also carried out experiments using other alcohols. Thus, when pure stearyl alcohol was substituted for cetyl alcohol in the formulation, the instability pattern of the creams was duplicated. Moreover, the use of a commercial 50:50 cetostearyl alcohol or a mixture of equal weights of the two pure alcohols gave a stable cream.

This experiment is of great interest and significance, since the cream contained much electrolyte and a low molecular polyol. The formation of liquid crystalline phases is likely unaffected by the presence of such materials.

10.   Internal Structure and Stability of O/W C Creams

10.1.   Internal Structure of O/W Creams

Yamaguchi *et al.* observed the internal structure of O/W creams using a cryo-scanning electron microscope (cryo-SEM), as described in Section 7.[155] The creams of the compositions shown in Table 35 were prepared by the ordinary procedure, i.e., by warming at 70 °C, stirring with a laboratory homogenizer, and cooling in the ice water mixture.

The cross-section of an M-phase particle, i.e., of a spherulite present in a ternary cream, was shown earlier (Fig. 38), and the entire system at a lower magnification was displayed in Figure 37. Figures 47 and 48 are the cross-sections of particles of samples X_2 and X_3, respectively. Both systems were prepared by substituting the same amount of liquid paraffin for a part of Hexaoctadecanol (3:2). By comparison with Fig. 38, it is seen that the centers of the particles are replaced with liquid paraffin.

The present authors have consistently used pentadecaethoxyoleylether (OE-15), which is a commercial nonionic surfactant, but it has not yet been established whether M particles or the D_2 phase are generated in a cream prepared with a different surfactant. However, the works by Barry *et al.*, Talman *et al.*, Suzuki *et al.*, and Yamaguchi and Noda suggest that the addition of liquid paraffin does not essentially affect the viscoelasticity of O/W creams, suggesting that both D_2 and M particles are formed in those systems.[156–161]

Table 35.   Composition of O/W Creams[155]

Cream Material	X_1^a	X_2	X_3
		(%)	
Liquid paraffin	—	5	10
OE-15[b]	5	5	5
Hexaoctadecanol (3:2)	30	25	20
Water	65	65	65

[a]Corresponds to the composition of point R in Fig. 27.

[b]Pentadecaethoxyolether (for cosmetic use).

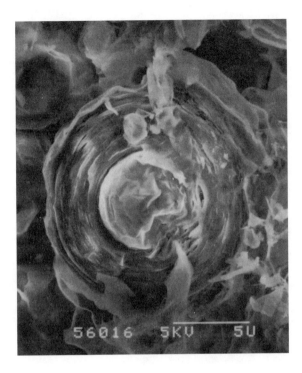

Figure 47. A cryo-SEM photograph of the cross-section of a particle having composition X_2 (Table 35).[155]

By considering various observations described in previous sections, the internal structure of an O/W quaternary cream formulated with cetostearyl alcohol can be illustrated schematically by Fig. 49.[155] Many D_2-phase particles consisting of crystalline, wafer-thin platelets, illustrated in hexagons to show that they are crystals (the actual size will be much smaller), are dispersed in the aqueous phase (I_w phase) to make it viscous. Emulsified oil droplets are enveloped by the shell of the M particles. The thickness of the latter will change with the quantity of the M particle contained, i.e., the amount of cetostearyl alcohol formulated, its ratio to surfactant, and the total surface area of oil droplets (i.e., average particle size of droplets) in the cream.

Moreover, the D_2 phase and M particles are formed near temperatures at which the higher alcohol used can assume the α'-structure (tetragonal structure), and at molar ratios of the higher alcohol to surfactant between 3 and 10. The viscosity of an ordinary cream reaches the maximum at a ratio of around 10 to 20, since a part of the formulated cetostearyl alcohol dissolves in the oil phase.

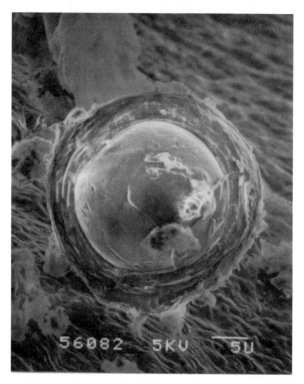

Figure 48.   A cryo-SEM photograph of the cross-section of a particle having composition X₃ (Table 35).[155]

The temperature range at which a *n*-higher alcohol may appear in the α′-form in the presence of excess water is 30–50 and 40–60 °C for pure 1-hexadecanol and 1-octadecanol, respectively, while for the mixture of both alcohols in the 3:2 wt% ratio it is 10–55 °C, which is quite close to the normal "room" temperatures. Thus, it is recommended to use a mixture of both alcohols in this ratio in the formulation of cosmetic creams or pharmaceutical emulsion-type ointments, rather than to use the pure respective alcohols.

It is of great interest that the "cetyl alcohol" obtained from a sperm whale and used to produce emulsion-type creams has such a ratio and specific property. However, with this knowledge, it is possible to produce stable O/W cream-type cosmetics or pharmaceuticals without sacrificing whales.

10.2. *A Novel Theory for the Stabilization of O/W Creams*

It is not the authors' intention to deal with the theories of the stability of "emulsions," since they can be found in many textbooks. However, in discussions

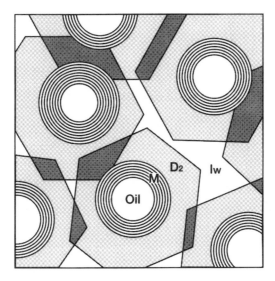

Figure 49.   Schematic illustration of the internal structure of an O/W cream prepared by formulation of cetostearyl alcohol, in which the D_2 phase and the M particles are being dispersed.[155]

on the stability of emulsions, the theories are based on the hypothesis that the latter consist basically of two phases, a dispersed and a continuous phase. Sometimes, the existence of the interfacial layer is taken into consideration.

In general, the instability of emulsions is explained by (1) creaming or sedimentation of the emulsion droplets, (2) flocculation of the emulsion droplets, (3) coalescence of the emulsion droplets, and (4) diffusion of smaller droplets and their coalescence with the larger ones (the Ostwald ripening).[162] Emulsions in this case have a liquid-in-liquid structure. Another explanation concerning the emulsion stability is based on the contribution of the LC phase.[163] However, the composition and structure of the LC phase are basically different from those described by the present authors.

Cosmetic creams we encounter in daily life are semi-solid. In order to prepare those which have an agreeable viscoelasticity, the use of "cetyl alcohol" is essential in addition to an oil, surfactant, and water. Indeed, they cannot be obtained without using it. Although it is customary to think that creams are a kind of emulsion, they are not emulsions as defined physicochemically, as described in the previous section. They contain a lamellar LC phase and spherulites in addition to the oil and aqueous phases. Especially, the lamellar LC phase (D2 phase) is dispersed in the aqueous phase to make the continuous (aqueous) phase viscous.

The movement of emulsified particles will be impaired if the continuous phase is viscous, resulting in less likely flocculation and a lower creaming velocity of oil droplets. The spherulites (M particles) surround the oil droplets, preventing their coalescence and the Ostwald ripening. Thus, stability of O/W creams stabilized with "cetyl alcohol" depends on that of the LC phases which formed in themselves.

Appendix 1: Additional References on Cetyl and Homologous Alcohols

Many other studies on cetyl alcohol have been published in addition to the references cited in the text. For the convenience of readers, these references are listed below with subtitles.

1. Production Method or Composition

(1) S. Alexrad, *J. Ind. Eng. Chem.* **9**, 1123–1125 (1917). A wool fat (lanolin) substitute and the preparation of cetyl alcohol.

(2) S. Alexrad and I. Hochstädter, *U.S. Patent 1,290,870.* Cetyl alcohol.

(3) T. P. Hilditch and J. A. Lovern, *J. Soc. Chem. Ind.* **48**, 365–368 (1929). The head and blubber oils of the sperm whale III. Quantitative determinations of the higher fatty alcohol present.

(4) M. Tsujimoto, *Report Tokyo Imp. Ind. Lab.* **15**, 1–80 (1920). Composition of the head oil from sperm whale (*Phyester macrocephalus L.*).

(5) M. A. Youtz, *J. Am. Chem. Soc.* **47**, 2252–2254 (1925). Rapid preparation of cetyl alcohol.

2. Structure or Physical Natures

(1) E. R. Andrew, *J. Chem. Phys.* **18**, 607–618 (1950). Molecular rotation in certain solid hydrocarbons.

(2) A. Eucken, *Z. Elektrochem.* **45**, 126–150 (1939). Rotation of molecules and ionic groups in crystals.

(3) G. E. Mapstone, *J. Soc. Cosmet. Chem.* **12**, 54–55 (1961). A simple test to differentiate between cetyl, cetostearyl and stearyl alcohols.

(4) J. D. Meyer and E. E. Reid, *J. Am. Chem. Soc.* **55**, 1574–1584 (1933). Isomorphism and alternation in the melting points of the normal alcohols, acetates, bromides, acids, and ethyl esters from C_{10} to C_{18}.

(5) H. Motz and J. J. Trillat, *Z. Kryst.* **91**, 248–254 (1935). Investigation of structures of extremely thin fatty films by means of electron diffraction.

(6) J. W. C. Phillips and S. A. Mumford, *J. Chem. Soc.,* 235–236 (1933). Aliphatic long-chain alcohols.

(7) J. W. C. Phillips and S. A. Mumford, *J. Chem. Soc.,* 1657–1665 (1934). Dimorphism of certain aliphatic compounds. V. Primary alcohols and their acetates.

(8) D. Precht, *Fette, Seifen, Anstrichmittel*, **78**, 145–192 (1976). Kristalstruktur-untersuchungen an Fettalkoholen und Fettsären mit Elektronen- und Rötgenbeugung I.

(9) A. Prietzschk, *Z. Phys. Chem.*, **40**, 482–501 (1941). Rötgenuntersuchungen an unterkühltem Äthylalkohol.

(10) Th. Schoon, *Z. Phys. Chem.* **39**, 385–400 (1940). Polymorphic forms of crystalline carbon compounds with long extended chains structure investigation by electron diffraction.

(11) C. P. Smyth, *Proc. Am. Phyl. Soc.* **76**, 485–489 (1939). Molecular rotation in solids.

(12) C. Stetiu, *Physica* **XXIV**, 69–73 (1984). The potential depth of primary alcohols and their mixtures.

(13) D. Vorlander and W. Selke, *Z. Physik. Chem.* **129**, 435–471 (1927). The uniaxial structure of soft solid crystalline masses and of crystalline liquids.

3. Dielectric Constant

(1) B. V. Hamon and R. J. Meakins, *Aust. J. Sci. Res.* **5**, 671–682 (1952). Dielectric absorption and D. C. conductivity in *n*-primary alcohols.

4. Solubility

(1) C. S. Hoffman and E. W. Anacker, *J. Chromatogr.* **30**, 390–396 (1967). Water solubilities of tetradecanol and hexadecanol.

(2) F. P. Krause and N. Lange, *J. Phys. Chem.* **69**, 3171–3173 (1965). Aqueous solubilities of *n*-dodecanol, *n*-hexadecanol, and *n*-octadecanol by a new method.

(3) I. D. Robb, *Aust. J. Chem.* **19**, 2281–2284 (1966). Determination of the aqueous solubilities of fatty acids and alcohols.

5. Films

(1) F. Emir, *Compt. Rend.* **189**, 239–240 (1929). Superficial solutions and molecular films.

(2) E. G. Goddard and J. H. Schulman, *J. Colloid Sci.* **8**, 329–340 (1953). Molecular interaction in monolayers. II. Steric effects in the nonpolar portion of the molecules.

(3) I. Langmuir, *Proc. Natl. Acad. Sci.* **3**, 251–257 (1917). The shape of group molecules forming the surfaces of molecules.

(4) E. N. Lawrence, R. Wall, and G. Adams, *J. Colloid Sci.* **13**, 441–458 (1958). Coalescence of liquid drops at oil–water interfaces.

(5) A. Marcelin, *Compt. Rend.* **189**, 236–238 (1929). Superficial films of water and molecular dimensions.

(6) B. Tamamushi, *Bull. Chem. Soc. Jpn.*, 161–165 (1934). Orientation of molecules on the water surfaces.

6. *Liquid Crystals or Interaction with Other Compounds*

(1) B. W. Barry and E. Shotton, *J. Pharm. Pharmacol.* **20**, 242–243 (1968). The penetration temperature of aqueous sodium dodecylsulphate into solid long chain alcohols.

(2) P. Ekwall, I. Danielsson, and L. Mandell, *J. Colloid Interface Sci.* **59**, 186–188 (1977). A study of the structure of the C phase in the sodium octanoate/decanol/water systems using electron microscopy.

(3) P. Ekwall, L. Mandell, and K. Fontell, *J. Colloid Interface Sci.* **29**, 639–646 (1969). The cetyltrimethylammonium bromide–hexanol–water system.

(4) P. Ekwall and K. Passinen, *Acta Chem. Scand.* **7**, 1098–1104 (1953). Studies on the interaction of paraffin chain alcohols and association colloids. I. The solubility of decanol-1 in sodium oleate and sodium myristyl sulfate solutions above the critical micelle concentration.

(5) P. Ekwall and C. F. Aminoff, *Acta Chem. Scand.* **10**, 237–243 (1956). Studies on the interaction of paraffin chain alcohols and association colloids. I. The sodium caprate concentration where interaction with long-chain alcohols begins and its dependence on the chain length of the alcohol.

(6) P. H. Elworthy and A. T. Florence, *J. Pharm. Pharmacol.* **21** suppl., 70S–78S (1969). Stabilization of oil-in-water emulsions by non-ionic detergents: The effect of polyoxyethylene chain length.

(7) M. Joly, *Nature* **158**, 26–27 (1946). The formation of molecular complexes in monolayers by penetration.

(8) Y. K. Kuchhal, R. N. Shukla, and A. B. Biswas, *Indian J. Chem.* **19A**, 1165–1170 (1980). Differential thermal analysis of mixtures of long chain primary alcohols and alkoxy ethanols.

(9) H. C. Kung and E. G. Goddard, *J. Phys. Chem.* **68**, 3465–3469 (1964). Molecular association in pairs of long-chain compounds. II. Alkyl alcohols and sulfates.

(10) A. Mlodziejowski, *Z. Phys.* **20**, 317–342 (1923). The formation of liquid crystals in the mixture of cholesterol and cetyl alcohol.

(11) K. Passinen and P. Ekwall, *Acta Chem. Scand.* **9**, 1438–1449 (1955). Studies on the interaction of paraffin chain alcohols and association colloids. IV. The effect of decanol-1 on the viscosities of some association colloids.

(12) J. Schubert and G. E. Boyd, *J. Chem. Phys.* **11**, 215 (1943). Energy relations in film penetration.

(13) J. H. Schulman and E. G. Cockbain, *Trans. Faraday Soc.* **36**, 651–661 (1940). Molecular interaction at oil-water interfaces. I. Molecular complex formation and the stability of oil-in-water emulsions.

(14) J. H. Schulman and E. K. Rideal, *Proc. Roy. Soc. (London)* **B122**, 29–45 (1937). Molecular interaction in monolayers.

(15) A. Trapeznikov, *Acta Physicochim. U.R.S.S.* **19**, 553–570 (1944). Temperature dependence of monolayer pressure as a method of investigating hydrates of higher aliphatic compounds. I.

(16) A. Trapeznikov, *Compt. Rend. Acad. Sci. U.R.S.S.* **47**, 417–420 (1945). Equilibrium between two- and three-dimensional hydrates of the higher alcohols and new phase transformations.

7. *Rheology of Emulsions*

(1) A. Axon, *J. Pharm. Pharmacol.* **8**, 762–773 (1956). Rheology of oil-in-water emulsions. I. Effect of concentration of constituents on emulsion consistency.

(2) A. Axon, *J. Pharm. Pharmacol.* **8**, 889–899 (1956). Rheology of oil-in-water emulsions. II. The microscopic appearance of emulsion in laminar flow.

(3) B. W. Barry, *J. Pharm. Pharmacol.* **20**, 483–484 (1968). Crystallization in polyhedral emulsion particles.

(4) B. W. Barry, *Manuf. Chem. Aerosol News* 1971, April, 27–33 (1971). Structure and rheology of some oil-in-water creams.

(5) B. W. Barry, *J. Soc. Cosmet. Chem.* **22**, 487–503 (1971). Evaluation of rheological ground states of semisolids.

(6) B. W. Barry and G. M. Eccleston, *J. Pharm. Pharmacol.* **25**, 394–400 (1973). Oscillatory testing of O/W emulsions containing mixed emulsifiers of the surfactant/long chain alcohol type: Influence of surfactant chain length.

(7) B. W. Barry and G. M. Eccleston, *J. Texture Stud.* **4**, 53–81 (1973). Influence of gel networks in controlling consistency of O/W emulsions stabilized by mixed emulsifiers.

(8) B. W. Barry, A. J. Grace, and P. Sherman, *J. Pharm. Pharmacol.* **21**, 396–397 (1969). Non-spherical emulsion particles.

(9) E. G. Cockbain and T. S. McRoberts, *J. Colloid Interface Sci.* **8**, 440–451 (1953). The stability of elementary emulsion drops and emulsions.

(10) T. de Vringer, J. G. H. Jooston, and H. E. Junginger, *Colloid Polym. Sci.* **264**, 691–700 (1986). A study of the gel structure in a nonionic O/W cream by differential scanning calorimetry.

(11) G. M. Eccleston, *J. Pharm. Pharmacol.* **31** (*S*), 5p (1979). The structure and stability of O/W emulsion stabilized by polyethyleneglycol 1000 monostearate/fatty alcohols.

(12) G. M. Eccleston, *Int. J. Cosmetic Sci.* **4**, 133–142 (1982). The influence of fatty alcohols on the structure and stability of creams prepared with polyethyleneglycol 1000 monostearate/fatty alcohols.

(13) M. J. Groves and D. C. Freshwater, *J. Pharm. Pharmacol.* **19**, 193–194 (1967). Polyhedral emulsion particles.

(14) G. W. Hallworth and J. E. Carless, *J. Pharm. Pharmacol.* **25** (*S*), 87P–95P (1973). Stabilization of oil-in-water emulsions by alkyl sulfates: The effect of long chain alcohol.

(15) F. A. J. Talman and E. M. Rowan, *J. Pharm. Pharmacol.* **20**, 810–811 (1968). Microscopical appearance of some oil-in-water emulsions.

(16) R. D. Vold and K. L. Mittal, *J. Colloid Interface Sci.* **38**, 451–459 (1972). The effect of lauryl alcohol on the stability of oil-in-water emulsion.

(17) I. von Eros und G. Kedvessy, *Pharm. Ind.* **47**, 777–781 (1985). Angewandte rheologische Forschung auf dem Gebiet der Salben glundlagen 1. Mit. Veränderung der Struktur Viskosität mit der Temperatur.

8. Physiology

(1) H. J. Channon and G. A. Collinson, *Biochem. J.* **22**, 391–401. Biological significance of the unsaponifiable matter of oils. IV. The absorption of higher alcohol.

(2) H. Goodman and A. Suess, *Urol. Cutaneous Rev.* **42**, 909–910 (1938). Cosmetic dermatology. Cetyl alcohol.

(3) T. Higuchi, *Japan Patent 129,206* (1939). Pharmaceutical preparation for treatment of skin.

(4) C. S. Hoffman and E. W. Anacker, *J. Chromatogt.* **30**, 390–396 (1967). Water solubilities of tetradecanol and hexadecanol.

(5) K. Imai, *Juzen (J. Juzenkai Japan)* **44**, 726–740 (1939). Effect of aliphatic alcohols on the growth of gonads of young male rat. I.

(6) K. Imai, *Juzen (J. Juzenkai Japan)* **44**, 941–946 (1939). Effect of aliphatic alcohols on the growth of gonads of young male rat. II.

(7) K. Imai, *Juzen (J. Juzenkai Japan)* **45**, 1927–1939 (1939). Effect of aliphatic alcohols on the growth of gonads of young male rat. III.

(8) K. Imai, *Juzen (J. Juzenkai Japan)* **45**, 568–584 (1939). Effect of aliphatic alcohols on the growth of gonads of young male rat. IV. V.

(9) S. Izumi, Y. Yamada, and S. Murata, *Juzen (J. Juzenkai Japan)* **43**, 309–318 (1939). Male hormones we separated. III. Sexual effects of aliphatic alcohols.

(10) S. Izumi, Y. Yamada, and S. Murata, *Juzen (J. Juzenkai Japan)* **45**, 352–356 (1940). Male hormones we separated. IV. Effects of aliphatic acids and alcohols on the testis of semi-castrated cocks.

(11) V. Natarajan and H. H. H. O. Schmid, *Lipids* **12**, 128–130 (1977). 1-Docosanol and other long chain primary alcohols in developing rat brain.

(12) V. Natarajan and H. H. H. O. Schmid, *Arch. Biochem. Biophys.* **187**, 215–222 (1978). Biosynthesis and utilization of long-chain alcohols in rat brain: Aspects of chain length specificity.

(13) K. Thomas and B. Flaschenrager, *Skand. Arch. Physiol.* **43**, 1–5 (1923). Is cetyl alcohol absorbed?

(14) S. Ugami and S. Suzuki, *J. Inst. Phys. Chem. Res. Jpn* **16**, 1464–1470 (1964). Physiological effects of a synthetic male hormone methyldihydrotestosterone. I. Increase in hormonal effects by addition of various aliphatic materials.

9. Application to Pharmaceuticals and Cosmetics

(1) A. Couleru, *Arch. Droguerie Pharm.* **7**, 32–36, 59–61 (1939). The aliphatic alcohols and their derivatives in perfumery.

(2) N. T. Farinacci and F. G. Firth, *Am. Perfumer* **48**, 41–45 (1946). X-ray spectroscopy in cosmetics.

(3) B. Filmer, *Fette u Seifen* **45**, 105–106 (1938). Walrat und Cetylalkohol in der Kosmetik.

(4) S. P. Jannaway, *Perfumery Essent. Oil Record* **27**, 154–156 (1936). The new cetyl alcohol cosmetics.

(5) H. Kaiho, Y. Takigawa, A. Ando, and Y. Kato, *J. Pharm. Soc. Jpn.* **99**, 1068–1072 (1979). Effects of long-chain alcohols on conversion of prednisolone crystal in oil-in-water type ointments.

(6) J. Kalish, *Drug Cosmet. Ind.* **37**, 739–740 (1935). Water-in-oil emulsions with cetyl alcohol.

(7) H. P. Kaufmann, *Fette u Seifen* **45**, 94–104 (1938). Whale products in pharmacy with particular attention to whale fats.

(8) H. Schwartz, *Seifensieder-Ztg.* **63**, 761–762 (1936). Cetyl alcohol.

(9) F. H. Sedwick, *Soap, Perfumery, & Cosmetics* **12**, 161–163 (1939). Cetyl and stearyl alcohols.

10. Miscellaneous

(1) E. G. Cockbain, *J. Colloid Sci.* **11**, 575–584 (1956). The adsorption of serum albumin and sodium dodecylsulfate at emulsion interfaces.

(2) A. Gascard, *Ann. Chim.* **15**, 332–389 (1921). The higher members of the saturated aliphatic series.

(3) F. C. Goodrich, *Proc. 2nd Int. Congress Surface Activity (London)* **1**, 85 (1957). Molecular interaction in mixed monolayers.

(4) A. N. Martin, in *Physical Chemical Principles in Pharmaceutical Science*, Lea and Febiger, Philadelphia, Pennsylvania, 1960, p. 629.

(5) A. J. Simko and R. G. Dressler, *Ind. Eng. Chem. Prod. Res. Develop.* **8** (4), 446–450 (1969). Investigation of C20 to C25 fatty alcohols and blends as water evaporation retardants.

(6) L. M. Spalton and R. W. White, in *Pharmaceutical Emulsions and Emulsifying Agents*, 4th ed., The Chemist and Druggist, London, 1964, p. 73.

(7) M. van den Tempel, *Proc. 2nd Int. Congress Surface Activity (London)* **1**, 439 (1957).

Appendix 2

Shortly after we published papers under the title of the effect of cetostearyl alcohol on stabilization of oil-in-water emulsion, a FAX letter was delivered from a European company. The content was as follows.

"We manufacture base of beeler and have a great problem with it: viscosity isn't high enough and it becomes liquid after a few weeks. You are working on cetyl alcohol crystallization and I thought you could suggest the right solution for our problems. Must we replace cetyl alcohol totally by cetostearyl alcohol or are there still other solutions? Has cooling velocity something to do with it, because we have greater viscosity with small batches? Please can you inform us? Many thanks."

A formula had been added at the end of the letters:

Cetyl alchol	15%
Beeswax	1
Sodiumlaurylsulfate	2
AQ PF AD	100

Of course, the authors recommended them to use cetostearyl alcohol instead of "pure" cetyl alcohols. After a few months another letter of appreciation was FAXed to us.

"Concerning our base of beeler: We replaced cetyl alcohol by cetostearyl alcohols. Stability is much better now and we have much hope that the problem is totally solved. Many thanks and best regards."

We feel very happy to hear that our works are useful globally in practical fields of emulsion industry. On the other hand, however, we are anxious that there may be a case where the theory is inapplicable. We would very much appreciate hearing of such cases.

ACKNOWLEDGMENTS. Our studies were first published in Japanese in the *Fragrance Journal*. We wish to thank the publisher for enabling us to explain our theory, and for consent to reproduce this work in English. We also acknowledge the assistance of Professor Egon Matijevic (Clarkson University) with the editing of the manuscript. We hope this text proves useful not only in industrial applications, but also advances our understanding of colloid chemistry.

References

1. *The Merck Index*, 9th edn., Merck & Co., Rahway, New Jersey, 1976, p. 254.
2. M. E. Chevreul, *Ann. Chim. Phys.* **2**, 155 (1817).
3. E. Andre and T. François, *Compt. Rend.* **183**, 663 (1926).

4. F. Krafft, *Ber. Dtsch. Chem. Ges.* **17**, 1627 (1884).
5. E. Ludwig, *Z. Phys. Chem.* **23**, 38 (1892).
6. R. von Zeynek, *Z. Phys. Chem.* **23**, 40 (1892).
7. F. Ameseder, *Z. Phys. Chem.* **52**, 121 (1907).
8. S. Izumi, K. Yamada, and S. Murata, *Juzen (J. Juzenkai Japan)* **43**, 309 (1938).
9. F. H. C. Stewart, *Aust. J. Appl. Sci.* **11**, 157 (1960).
10. R. G. Vines and R. J. Meakins, *Aust. J. Appl. Sci.* **10**, 190 (1959).
11. J. Kalish, *Drug Cosmet. Ind.* **37**, 595 (1935).
12. H. S. Redgrove, *Am. Perfumer* **38**(2), 32 (1939).
13. K. Tanaka, T. Seto, and T. Hayashida, *Bull. Inst. Chem. Res. Kyoto Univ.* **35**, 123 (1957).
14. K. Tanaka, T. Seto, A. Watanabe, and T. Hayashida, *Bull. Inst. Chem. Res. Kyoto Univ.* **37**, 281 (1959).
15. M. Tasumi, T. Shimanouchi, A. Watanabe, and R. Goto, *Spectrochim. Acta* **20**, 629 (1964).
16. S. Abrahamsson, G. Larsson, and E. von Sydow, *Acta Crystallogr.* **13**, 770 (1960).
17. T. Seto, *Mem. Coll. Sci. Univ. Kyoto, Ser.* **A30**, 89 (1962).
18. D. A. Wilson and E. Ott, *J. Chem. Phys.* **2**, 231 (1934).
19. J. C. Smith, *J. Chem. Soc.* **1**, 802 (1931).
20. J. D. Hoffman and C. P. Smyth, *J. Am. Chem. Soc.* **71**, 431 (1948).
21. D. G. Kolp and E. S. Lutton, *J. Am. Chem. Soc.* **73**, 5593 (1951).
22. D. Chapman, *Proc. Colloq. Spectroscopicum Int. VI, Amsterdam*, Pergamon Press, London, 1956, pp. 609–617.
23. R. G. Vines and R. J. Meakins, *Aust. J. Appl. Sci.* **10**, 190 (1959).
24. F. H. C. Stewart, *Aust. J. Appl. Sci.* **11**, 157 (1960).
25. E. J. Benton, in *Retardation of Evaporation by Monolayers,* V. LaMer (ed.), Academic Press, New York, 1962, pp. 235–244.
26. J. H. Brooks, in *Retardation of Evaporation by Monolayers,* V. LaMer (ed.), Academic Press, New York, 1962, pp. 255–258.
27. S. Fukushima and M. Yamaguchi, *J. Jpn. Oil Chem. Soc.* **29**, 933 (1980).
28. L. Pauling, *Phys. Rev.* **36**, 430 (1930).
29. J. D. Bernal, *Z. Kristallogr.* **83**, 153 (1932).
30. J. D. Bernal, *Nature* **129**, 870 (1932).
31. E. Frosch, *Ann. Phy.* **42**, 254 (1942).
32. E. Ott, *Z. Phys. Chem.* **193**, 218 (1944).
33. S. Fukushima, K. Yoshida, and M. Yamaguchi, *J. Pharm. Soc. Jpn.* **104**, 986 (1984).
34. T. Malkin, *J. Am. Chem. Soc.* **52**, 3739 (1930).
35. A. Watanabe, *Bull. Chem. Soc. Jpn.* **36**, 336 (1963).
36. J. D. Meyer and E. E. Reid, *J. Am. Chem. Soc.* **55**, 1574 (1933).
37. P. E. Vercade and J. Coops, Jr., *Res. Trav. Chim.* **46**, 903 (1927).
38. A. S. C. Lawrence, M. A. Al-Mamun, and M. P. McDonald, *Trans. Faraday Soc.* **63**, 2789 (1967).
39. A. Watanabe, *Bull. Chem. Soc. Jpn.* **34**, 1728 (1961).
40. F. Krafft, *Ber. Dtsch. Chem. Ges.* **15**, 1714 (1882).
41. L. Schon, *Pharm. J.* **168**, 360 (1952).
42. A. Trapeznikov, *Acta Physicochim. U.R.S.S.* **20**, 589 (1945).
43. W. O. Baker and C. P. Smyth, *J. Am. Chem. Soc.* **60**, 1229 (1938).
44. A. Gascard, *Compt. Rend.* **170**, 886, 1326 (1930).
45. H. M. Huffman, G. S. Parks, and M. Barmore, *J. Am. Chem. Soc.* **53**, 3876 (1931).
46. J. W. C. Phillips and S. A. Mumford, *J. Chem. Soc.,* 1732 (1931).
47. K. Higasi, and M. Kubo, *Sci. Papers Inst. Phys. Chem. Res., Tokyo* **36**, 286 (1939).
48. J. W. C. Phillips and S. A. Mumford, *J. Chem. Soc.,* 1657 (1934).
49. J. C. Smith, *J. Chem. Soc.,* 802 (1931).
50. B. W. Barry and E. Shotton, *J. Pharm. Pharmac.* **19**(S), 110S (1967).

51. Y. K. Kuchhal, R. N. Shukla, and A. B. Biswas, *Indian J. Chem.* **20A**, 837 (1981).
52. A. Gascard, *Ann. Chim.* **15**, 322 (1921).
53. W. Levene and van der Scheer, *J. Biol. Chem.* **20**, 521 (1915).
54. P. A. Levene and F. A. Taylor, *J. Biol. Chem.* **59**, 905 (1924).
55. F. Francis, F. J. E. Collins, and S. E. Piper, *Proc. Roy. Soc. London* **A158**, 691 (1937).
56. M. A. Al-Mamun, *J. Am. Oil Chem.* Soc. **51**, 234 (1974).
57. L. Schon, *Pharm. J.* **168**, 360 (1952).
58. R. G. Vines and R. J. Meakins, *Aust. J. Appl. Sci.* **10**, 190 (1959).
59. F. H. C. Stewart, *Aust. J. Appl. Sci.* **11**, 157 (1960).
60. T. P. Hilditch and J. A. Lovern, *J. Soc. Chem. Ind.* **48**, 365 (1929).
61. S. Fukushima, in *Physical Chemistry of Cetyl Alcohol,* Fragrance Journal Ltd., Tokyo, 1992, pp. 31–34.
62. E. J. Benton, in *Retardation of Evaporation by Monolayers,* V. La Mer (ed.), Academic Press, New York, 1962 , pp. 235–244.
63. M. A. Al-Mamun, *J. Am. Oil Chem. Soc.* **51**, 234 (1974).
64. M. Yamaguchi, M. Takahashi, F. Harusawa, and S. Fukushima, *J. Soc. Cosmet. Chem. Japan* **12**(2), 16 (1978).
65. K. Higasi and M. Kubo, *Sci. Papers Inst. Phys. Chem. Res. (Tokyo)* **36**, 286 (1939).
66. A. Trapeznikov, *Acta Physicochim.* U.R.S.S. **20**, 589 (1945).
67. F. H. C. Stewart, *Aust. J. Appl. Sci.* **11**, 157 (1960).
68. J. H. Brooks, in *Retardation of Evaporation by Monolayers,* V. La Mer (ed.), Academic Press, New York, 1962, pp. 255–258.
69. J. H. Brooks and A. E. Alexander, *J. Phys. Chem.* **66**, 1851 (1962).
70. W. Kauzmann and D. Eisenberg, in *The Structure and Properties of Water,* Oxford at the Clarendon Press, London, 1969.
71. A. S. C. Lawrence, A. Bingham, B. Capper, and K. Hume, *J. Phys. Chem.* **68**, 3470 (1964).
72. A. S. C. Lawrence, M. A. Al-Mamun, and M. P. McDonald, *Trans. Faraday Soc.* **63**, 2789 (1967).
73. S. Fukushima and M. Yamaguchi, *J. Jpn. Oil Chem. Soc.* **29**, 933 (1980).
74. S. Fukushima, in *Physical Chemistry of Cetyl Alcohol,* Fragrance Journal Ltd., Tokyo, 1992, pp. 35–49.
75. Y. K. Kuchhal, R. N. Shukla, and A. B. Biswas, *Indian J. Chem.* **20A**, 837 (1981).
76. K. Pachler and M. von Stackelberg, *Z. Krystal.* **119**, 15 (1963).
77. L. Mandell, in *Surface Chemistry,* P. Ekwall, K. Groth, and V. R. Runnstro (eds.), Proc. 2nd Scand. Symposium Surface Activity, Stockholm, 1964, Munksgaad, 1965, pp. 185–202.
78. P. Ekwall, L. Mandell, and K. Fontell, in *Liquid Crystals,* Proc. Int. Congress on Liquid Crystals, Gordon and Breach, London, 1965, pp. 325–381.
79. A. J. Hyde, D. M. Langbridge, and A. S. C. Lawrence, *Disc. Faraday Soc.* **18**, 239 (1954).
80. P. Ekwall, L. Mandell, and K. Fontell, *J. Colloid Interface Sci.* **29**, 542 (1969).
81. F. A. J. Talman, *J. Pharm. Pharmacol.* **22**, 338 (1970).
82. H. Kasai and M. Nakagaki, Nippon Kagaku Zassi **91**, 19 (1970).
83. J. S. Merland and B. A. Mulley, *J. Pharm. Pharmacol.* **24**, 729 (1972).
84. P. Ekwall, I. Danielsson, and L. Mandell, *Kolloid Z.* **169**, 113 (1960).
85. P. Ekwall, *J. Colloid Interface Sci.* **29**, 16 (1969).
86. H. C. Kung, and E. D. Goddard, *J. Phys. Chem.* **67**, 1965 (1963).
87. M. B. Epstein, A. Wilson, C. W. Jacob, L. E. Conroy, and Ross, *J. Phys. Chem.* **58**, 860 (1954).
88. F. A. J. Talman, P. J. Davies, and E. M. Rowan, *J. Pharm. Pharmacol.* **20**, 513 (1968).
89. A. M. Poskanzer and D. C. Goodrich, *J. Phys. Chem.* **79**, 2122 (1975).
90. F. A. J. Talman, P. J. Davies, and E. M. Rowan, *J. Pharm. Pharmacol.* **19**, 417 (1967).
91. M. B. Epstein, A. Wilson, J. Gershman, and J. Ross, *J. Phys. Chem.* **60**, 1051 (1956).
92. M. Yamaguchi and A. Noda, A., *Nippon Kagaku Kaishi* 1632 (1987).
93. B. W. Barry and E. Shotton, *J. Pharm. Pharmacol.* **19**(S), 121S (1967).

94. B. W. Barry, *J. Colloid Interface Sci.* **28**, 82 (1968).
95. M. Yamaguchi and A. Noda, A., *Nippon Kagaku Kaishi*, **26** (1989).
96. B. W. Barry and G. M. Saunders, *J. Colloid Interface Sci.* **34**, 300 (1970).
97. B. W. Barry and G. M. Saunders, *J. Pharm. Pharmacol.* **22**(S), 139S (1967).
98. B. W. Barry and G. M. Saunders, *J. Colloid Interface Sci.* **35**, 689 (1971).
99. B. W. Barry and G. M. Saunders, *J. Colloid Interface Sci.* **6**, 130 (1971).
100. B. W. Barry and G. M. Saunders, *J. Colloid Interface Sci.* **8**, 616 (1972).
101. B. W. Barry and G. M. Saunders, *J. Colloid Interface Sci.* **8**, 62 (1972).
102. S. Fukushima, M. Yamaguchi, and F. Harusawa, *J. Colloid Interface Sci.* **59**, 159 (1977).
103. S. Fukushima and M. Yamaguchi, *J. Jpn. Oil Chem. Soc.* **29**, 106 (1989).
104. M. Yamaguchi, K. Yoshida, M. Tanaka, and S. Fukushima, *J. Electron Micros.* **31**, 249 (1982).
105. S. Fukushima, K. Yoshida, and M. Yamaguchi, *J. Pharm. Soc. Jpn.* **104**, 986 (1984).
106. B. W. Barry and G. M. Eccleston, *J. Pharm. Pharmacol.* **25**, 244 (1973).
107. B. W. Barry and E. Shotton, *J. Pharm. Pharmacol.* **19**(S), 110S (1967).
108. S. Fukushima, in *Physical Chemistry of Cetyl Alcohol,* Fragrance Journal Ltd., Tokyo, 1992, p. 57.
109. S. Fukushima, M. Yamaguchi, and F. Harusawa, *J. Colloid Interface Sci.* **59**, 159 (1977).
110. B. W. Barry, *Rheol. Acta* **10**, 96 (1971).
111. B. W. Barry and G. M. Saunders, *J. Colloid Interface Sci.* **34**, 300 (1970).
112. B. M. Barry and G. M. Saunders, *J. Colloid Interface Sci.* **38**, 626 (1972).
113. B. W. Barry and E. Shotton, *J. Pharm. Pharmacol.* **19**(*S*), 110S (1967).
114. B. W. Barry and E. Shotton, *J. Pharm. Pharmacol.* **19**(*S*), 121S (1967).
115. G. E. Mapstone, *Aust. Chem. Process. Eng.* **25**(12), 18 (1972).
116. M. Yamaguchi, M. Takahashi, F. Harusawa, and S. Fukushima, *J. Soc. Cosmet. Chem. Jpn* **12**(2), 16 (1978).
117. S. Fukushima and M. Yamaguchi, *J. Pharm. Soc. Jpn* **101**, 1010 (1981).
118. K. Larsson, *Nature (London)* **191**, 383 (1967).
119. J. E. Bowcott and J. H. Schulman, *Z. Elektrochem.* **59**, 283 (1955).
120. J. H. Schulman and M. Stenhagen, *Proc. Roy. Soc. Ser. B* **126**, 356 (1938).
121. D. O. Shah and J. H. Schulman, *Lipid Res.* **8**, 215 (1967).
122. A. J. Simko and R. G. Dressler, *Ind. Eng. Chem., Prod. Res. Develop.* **8**(4), 446 (1969).
123. D. O. Shah, *J. Colloid Interface Sci.* **37**, 744 (1971).
124. M. Yamaguchi and A. Noda, *Nippon Kagaku Kaishi* **26** (1989).
125. B. W. Barry and E. Shotton, *J. Pharm. Pharmacol.* **19**(9), 110s (1967).
126. M. Yamaguchi, K. Yoshida, M. Tanaka, and S. Fukushima, *J. Electron Microsc.* **31**, 249 (1982).
127. S. Fukushima and M. Yamaguchi, *J. Jpn Oil Chem. Soc.* **29**, 106 (1980).
128. S. Fukushima, in *Physical Chemistry of Cetyl Alcohol*, Fragrance Journal Ltd., Tokyo, 1992, pp. 95–96.
129. F. J. A. Talman and E. M. Rowan, *J. Pharm. Pharmacol.* **22**, 338 (1970).
130. F. J. A. Talman and E. M. Rowan, *J. Pharm. Pharmacol.* **22**, 417 (1970).
131. F. J. A. Talman, P. J. Davies, and E. M. Rowan, *J. Pharm. Pharmacol.* **19**, 417 (1967).
132. F. J. A. Talman, P. J. Davies, and E. M. Rowan, *J. Pharm. Pharmacol.* **20**, 513 (1968).
133. B. W. Barry, *J. Colloid Interface Sci.* **28**, 82 (1968).
134. B. W. Barry, *J. Colloid Interface Sci.* **32**, 551 (1970).
135. E. G. Mapstone, *Aust. Chem. Process. Eng.* **25**(12), 18 (1977).
136. K. Okamoto and H. Oishi, *Yakuzaigaku* **37**, 52 (1977).
137. P. H. Elworthy, A. T. Florence, and J. A. Rogers, *J. Colloid Interface Sci.* **35**, 34 (1971).
138. S. Fukushima, M. Takahashi, and M. Yamaguchi, *J. Colloid Interface Sci.* **57**, 201 (1976).
139. S. S. Davis and A. Smith, in *Theory and Practice of Emulsion Technology*, A. L. Smith (ed.), Academic Press, London, 1976, pp. 325–346.
140. G. M. Eccleston, *J. Colloid Interface Sci.* **57**, 66 (1976).

141. J. E. Carless and G. W. Hallworth, *J. Colloid Interface Sci.* **26**, 75 (1968).
142. G. W. Hallworth and J. E. Carless, in *Theory and Practice of Emulsion Technology*, A. L. Smith (ed.), Academic Press, London, 1976, pp. 305–324.
143. B. W. Barry and G. M. Saunders, *J. Colloid Interface Sci.* **34**, 300 (1970).
144. B. W. Barry and G. M. Saunders, J. *Pharm. Pharmacol.* **22**(*S*), 139S (1970).
145. S. Fukushima and M. Yamaguchi, *J. Jpn. Oil Chem. Soc.* **29**, 106 (1980).
146. B. W. Barry and G. M. Eccleston, *J. Pharm. Pharmacol.* **5**, 244 (1973).
147. B. W. Barry and G. M. Eccleston, *J. Pharm. Pharmacol.* **25**, 394 (1973).
148. B. W. Barry and G. M. Saunders, *J. Colloid Interface Sci.* **41**, 331 (1972).
149. B. W. Barry and G. M. Saunders, *J. Colloid Interface Sci.* **35**, 689 (1971).
150. C. Fuhrer, H. Junginger, and S. Friberg, *J. Soc. Cosmet. Chem.* **29**, 303 (1978).
151. M. Yamaguchi, K. Yoshida, M. Tanaka, and S. Fukushima, *J. Electron Microsc.* **31**, 249 (1982).
152. S. Fukushima, K. Yoshida, and M. Yamaguchi, *J. Jpn. Pharm. Soc.* **104**, 986 (1984).
153. T. Suzuki, H. Tsutsumi, and A. Ishida, *Nippon Kagaku Kaishi* 337 (1983).
154. B. W. Barry and E. Shotton, *J. Pharm. Pharmacol.* **20**, 242 (1968).
155. S. Fukushima, K. Yoshida, M. Yamaguchi, *J. Pharm. Soc. Jpn.* **104**, 986 (1984).
156. B. W. Barry, *J. Colloid Interface Sci.* **28**, 82 (1968).
157. B. W. Barry and G. M. Saunders, *J. Colloid Interface Sci.* **41**, 626 (1972).
158. B. W. Barry and G. M. Saunders, *J. Colloid Interface Sci.* **41**, 331 (1972).
159. T. Suzuki, H. Tsutsumi, and A. Ishida, *Nippon Kagaku Kaishi* 337, (1983).
160. M. Yamaguchi and A. Noda, *Nippon Kagaku Kaishi* 26 (1989).
161. F. J. A. Talman, P. J. Davies, and E. M. Rowan, *J. Pharm. Pharmacol.* **20**, 513 (1968).
162. B. W. Barry and E. Shotton, *J. Pharm. Pharmacol.* **19**, 110S (1967).
163. S. Friberg, *J. Colloid Interface Sci.* **37**, 291 (1971).

2

Ionization Processes and Proton Binding in Polyprotic Systems: Small Molecules, Proteins, Interfaces, and Polyelectrolytes

Michal Borkovec, Bo Jönsson, and Ger J. M. Koper

1. Introduction

Binding of ions to various materials, such as small molecules, proteins, polymers, colloid particles, and membranes, represents a central theme in basic and applied chemistry. Particularly, the case of proton binding to these substances (i.e., their acid–base behavior) has been a focus of research in many branches of chemistry since the turn of the century. One important topic in physical, analytical, and inorganic chemistry is the measurement, compilation, and prediction of acid–base properties of simple molecules or solvated metal ions.[1–8] These topics remain of much relevance for the development of new analytical techniques and tailoring of buffering or complexing agents.[7,9] Accurately known ionization constants also represent a rather stringent testing ground of our *ab initio* simulation capabilities of simple molecules in water.[10] Acid–base properties of proteins have been also investigated from early on.[11–14] This field has now matured into an active area of modern biochemistry with implications to the current view of protein folding, enzyme action, and photosynthesis.[15–17] Similar studies of weak polyelectrolytes were initiated in polymer science almost simultaneously.[18–21] These systems represent an ongoing challenge to our understanding of acid–base equilibria.[22–25] The substantial interest in polyelectrolytes is due to their use as complexing, flocculating, or stabilizing agents, and their importance in various applications in catalysis, material engineering, biochemistry, and water purification.[26–28] More recently, several studies have focussed on the acid–base properties of various mineral particles and polymeric materials of synthetic or natural origin, most

Michal Borkovec • Department of Chemistry, Clarkson University, Potsdam, NY. *Bo Jönsson* • Theoretical Chemistry, Chemical Center, University of Lund, Lund, Sweden. *Ger J. M. Koper* • Leiden Institute of Chemistry, Leiden University, Gorlaeus Laboratories, Leiden, The Netherlands.

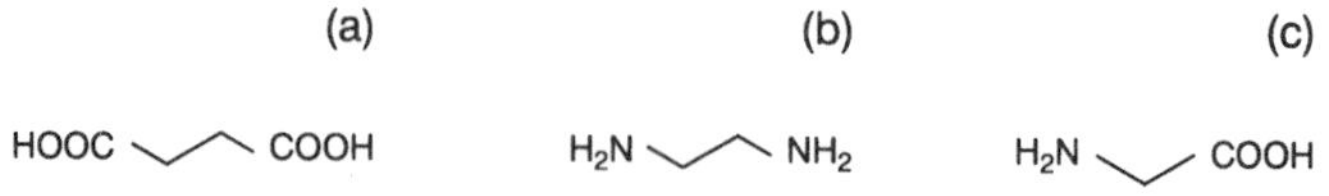

Figure 1.   Organic diproptic acids and bases. (a) Succinic acid, (b) ethylenediamine, and (c) glycine.

notably within the material science and environmental chemistry communities. In these disciplines, one of the central themes is the protonation behavior of the water–solid interface as it controls the particle charge and thereby the stability of colloidal suspensions.[29,30] Applications of such systems are diverse. In material engineering, mineral particles are used in paints, paper, cosmetics, or as ceramic precursors, while functionalized latex particles are employed in chemical industry as coatings, synthetic rubbers, adhesives, and in biotechnology for immuno-assays and cell-labeling.[31] In environmental sciences, such studies are mostly motivated by the ubiquitous presence of oxide particles and humic substances in aquatic and subsurface environments and their important role in controlling pollutant pathways.[29] One should also mention interesting developments focussing on acid–base properties of surfactant membranes as realized in Langmuir–Blodget films, micelles, microemulsions, or vesicles.[32–37]

Figures 1–7 summarize the broad variety of systems, which are considered in this context. One common feature is that a single proton binds to an individual, localized site. These ionizable sites usually are oxygen, nitrogen, and sulfur atoms. The number of ionizable sites, which are attached to a single molecule or particle,

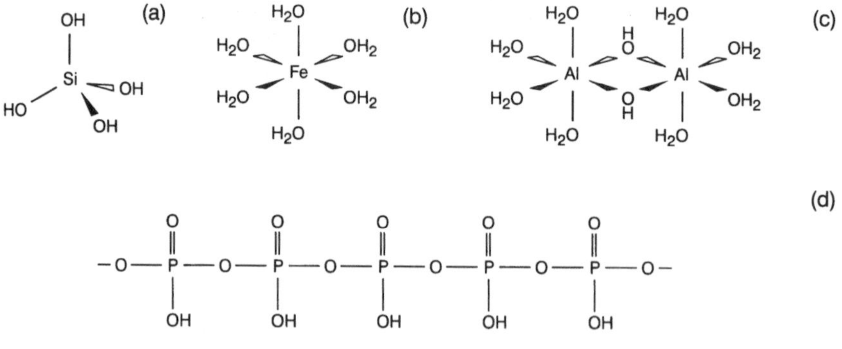

Figure 2.   Various inorganic systems exhibiting acid–base behavior. Protons bind exclusively to oxygen atoms. In principle, each oxygen atom can be protonated in two steps, in the first step as the oxo group ($-O^-$) and in the second step as the hydroxo group ($-OH$). There are two kinds of oxygen atoms, in singly and doubly coordinated positions. (a) Silicic acid, (b) iron(III) complex, (c) binuclear aluminum complex, and (d) linear poly(phosphoric acid).

can be very different. Small molecules carry just a few sites, maybe 1–10, proteins typically 20–100 sites, while for colloidal systems such as polyelectrolytes, latex, or oxide particles the number of sites may exceed 10^3 by several orders of magnitude. For systems with a small number of sites, acid–base properties are sensitive to this number. On the other hand, as this number of sites becomes large, the acid–base properties become independent of the system size, and one has reached the *large system limit*.

Let us recall the classical Brønsted definition of acids and bases.[1,4,29,38] An *acid* is a donor of protons, while a *base* is a proton acceptor. If a molecule has a few binding sites, such as a diprotic acid, base, or a metal complex (see Figs. 1 and 2), the intermediate forms may act as a proton acceptor or a proton donor and are called *ampholytes*. While these concepts are extremely useful for small molecules, discussing macromolecules with many ionizable groups, essentially all protonation forms are ampholytic in terms of the above definition. Therefore, a macromolecule with acidic functional groups is referred to as a *polyacid*, such as poly(acrylic acid) (PAA, see Fig. 3a), while basic functional groups make up a *polybase*, such as linear poly(ethylene imine) (LPEI, see Fig. 4a). A macromolecule with acidic and basic groups is referred to as a *polyampholyte*, such as a protein (see Fig. 5). Note that a polyacid is uncharged in its fully protonated state, a polybase is uncharged in its fully deprotonated state, while a polyampholyte is uncharged at some intermediate protonation state.

We can also classify these systems as *homogeneous* or *heterogeneous*. A *homogeneous* system contains only one type of site in one type of environment,

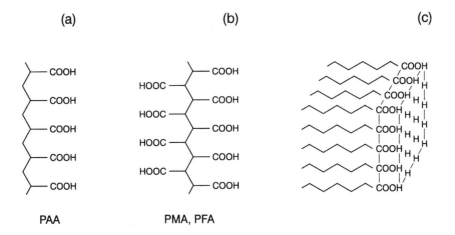

(a)   **(b)**   **(c)**

PAA   PMA, PFA

Figure 3. Various organic polyprotic acids. (a) Poly(acrylic acid) (PAA), (b) poly(maleic acid) (PMA) or poly(fumaric acid) (PFA), and (c) ordered fatty acid monolayer. The latter structure is similar to a surface of a carboxylated latex particle, which has a random arrangement, however.

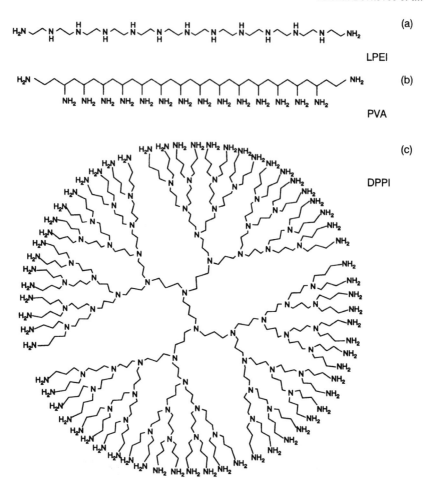

Figure 4. Various organic polyprotic bases. (a) Linear poly(ethylene imine) (LPEI), (b) linear poly(vinyl amine) (PVA), (c) dendritic poly(propylene imine) (DPEI). Reprinted from Ref. 312 with permission from Steinkopff Verlag.

such as the secondary amine groups in LPEI (see Fig. 4a). A *heterogeneous* system has different sites in different environments, as for example primary and tertiary amine groups in dendritic poly(propylene imine) (DPEI) as depicted in Fig. 4c.[28,39] From this point of view, even a perfect oxide surface, such as for example the rutile (TiO_2) 110 crystal face, must be considered as *heterogeneous*. As evident from Fig. 6, there are namely three types of ionizable sites: singly, doubly, and triply coordinated oxygen atoms.[40,41] In the case of heterogeneous systems, one may distinguish between *regular* arrangement as in the case of the dendrimer or 110

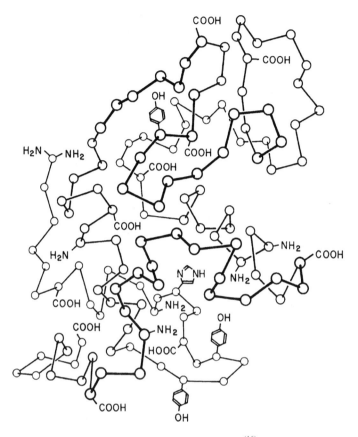

Figure 5. Ionizable groups of hen egg white lyzosyme (HEWL).[16] The individual amino acids are presented by circles on the protein backbone. The ionizable residues are imidazole in His 15, carboxylic acid groups at chain terminal and in Glu 7, Glu 35, Asp 18, Asp 48, Asp 52, Asp 66, Asp 87, Asp 101, and Asp 119; phenolic groups in Tyr 20, Tyr 23, and Tyr 53; amino groups at chain terminal and in Lys 1, Lys 13, Lys 33, Lys 96, Lys 97, and Lys 116.

rutile face (see Figs. 4c and 6) and *irregular* (or random) arrangement as in the case of a humic acid (see Fig. 7b), where each carboxylic or phenolic group has its own particular environment.[42] (For surfaces, the latter case is also referred to as *molecular* heterogeneity, which is contrasted to *patch-wise* heterogeneity. We shall not employ this definition here, as "heterogeneity" shall be used in the sense defined above.) In the case of *irregular* arrangement of the sites, we usually also face sample polydispersity, but not in all cases. For example, a properly purified protein solution is *monodisperse* as it consists of one type of molecule, while a humic acid is *polydisperse* as it consists of a complex mixture of similar, but individually different molecules.

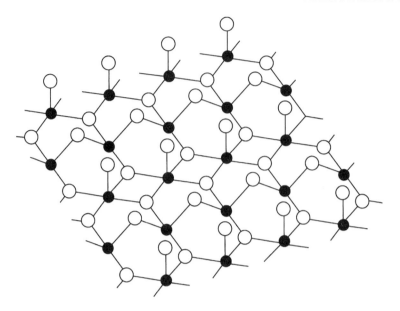

Figure 6.   Rutile 110 surface (TiO_2, • titania and ○ oxygen atoms). Oxygens represent binding sites for protons, and they occur in singly, doubly, and triply coordinated positions.

In the present review, we shall elucidate the current understanding of acid–base properties of these various systems from a unified point of view. In spite of the substantial research effort to understand acid–base equilibria for all different systems, there seems to be no general framework to describe such processes. Quite on the contrary, for each type of system a specialized approach has been developed and little effort was made to understand its relations to alternative treatments. (i) Description of small molecules is commonly based on successive protonation equilibria, whereby *macroscopic* as well as *microscopic* ionization constants are considered.[4,7,29,38] (ii) For humic substances and biological materials, many authors favor the concept of a statistical distribution of sites.[43–47] (iii) The water–oxide interface is approached based on electrochemical concepts, such as surface potential, surface capacitance, and potential determining ions.[40,48–51] (iv) Protonation behavior of linear polyelectrolytes has been described on the basis of a site binding model, where interactions between nearest neighbors are taken into account.[18,19,21]

These approaches will be discussed in more detail in the following. However, the relation between them is far from clear from the available literature. If all these seemingly different models are valid, they must be related and should follow as special cases from a general theory of acid–base equilibria. Probably only with the

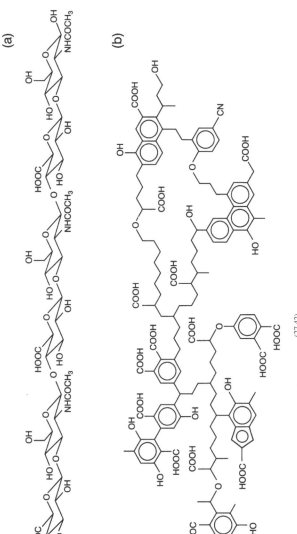

(a)

(b)

Figure 7.   Structure of natural polyelectrolytes.[27,42] (a) Hyaluronic acid and (b) hypothetical structure of a humic acid.

exception of the protein literature, this general framework is not "well-known" and these concepts are usually not applied outside of these fields.

The general framework for the treatment of ionization equilibria for complex systems was laid down in 1952 by Kirkwood.[52] To the authors' knowledge, this report is the first where discrete state variables were introduced for the enumeration of all possible protonation states of a polyprotic molecule. Shortly thereafter, but without reference to Kirkwood, a large number of papers were published which have exploited this idea.[18–21,53,54] Steiner[53] was probably first to point out the fundamental analogy between a one-dimensional chain of magnetic dipoles discussed initially by Ising[55] and the protonation of a linear polyelectrolyte. However, his paper remained unnoted and the application of the Ising model to protonation of linear polyelectrolytes is usually referred to the later works of Harris and Rice[18] and Marcus,[19] and to the subsequent work due to Lifson[20] and Katchalsky.[21] The appreciation of the equivalence to the Ising model was crucial, since such models did represent a central theme in statistical mechanics.[56–59] A recent discussion of the Ising model from the statistical mechanical point was given by Robertson.[60] However, already in the fifties, many established techniques for the solution of the Ising model were available, and these techniques could be readily applied to the problem of acid–base properties of polyelectrolytes.[18,19,21]

Within the same time period, Hill[54] was investigating the ionization of proteins, but much more influential was the later paper by Tanford and Kirkwood.[61] This famous paper discusses the essentials of the modern view of these processes in proteins, and has deeply influenced this branch of biochemistry for decades to come.[15–17,62–68] Quite surprisingly, however, the relation to the Ising model was appreciated in the protein literature much more reluctantly. Similarly, in the case of the ionization of the water–oxide interface, a statistical mechanical treatment was only touched upon[50] and discussed only recently.[69,70]

In the central part of this review article, we shall discuss the various approaches for the description of acid–base reactions. After a short description of the pertinent experimental techniques in Section 2, we shall introduce the main theoretical tools in Section 3, which provide the basis for understanding the ionization behavior in these systems. The rest of the article will focus on various systems of increasing complexity. Small molecules with a small number of ionizable sites will be discussed in Section 4. Systems with a large number of ionizable sites such as proteins, polyelectrolytes, and ionizable interfaces will be discussed in Sections 5, 6, and 7, respectively. Proteins, polyelectrolytes, and interfaces have a number of common features, which will be considered in some detail below. Nevertheless, all sections can be read independently.

There is an important conceptual difference between the dilute solution of small molecules and macromolecules (or colloidal particles). For a solution of small molecules one can perform the extrapolation to zero ionic strength without difficulties. For a solution of highly charged macromolecules, it is only in the presence of a high salt concentration that the macromolecules can be considered as independent (*high-salt regime*). For low salt concentrations, the counterions, which balance the charge of the macromolecule, become important, and the interactions between the macromolecules can be no longer ignored even for a highly diluted sample (*low-salt regime*).[71,72] The so-called *suspension effect*, which refers to difficulties of pH measurement in particle suspensions, is just another manifestation of the same phenomenon.[73] Here we shall mainly focus on the high-salt regime, but also some implications for the low-salt regime will be discussed.

The review focuses on aqueous solutions exclusively and does not consider the interesting aspects that emerge from the use of nonaqueous solvents.[38] We shall also not cover monoprotic acids and bases in much detail, but refer the reader to classical monographs.[4,6,30,74,75]

2. Experimental Techniques

Two broad classes of techniques for measuring binding of protons are available. The first class includes macroscopic measurements of the proton partitioning, and these methods are sensitive to the average degree of protonation of the substrate. The second class involves spectroscopic methods. In favorable cases, these techniques can be used to obtain microscopic information about the protonation reactions of interest, and yield the average degree of protonation of individual sites.

2.1. Macroscopic Techniques

Proton adsorption isotherms can be determined in a straightforward manner with potentiometric titrations. Various other techniques, which are sensitive to the proton partitioning, include, for example, chromatographic methods, electrical conductivity, or measurements of surface tension.

2.1.1. Definition and Measurement of pH

The pH value of a solution is defined as

$$pH = -\log_{10} a_{H^+} \tag{1}$$

where a_{H^+} is the activity of the proton in M with pure water taken as the reference state.[76–78] This notation was introduced by Sørensen 1909 for the "pondus Hydrogenii" (hydrogen exponent).[79] Many authors have attempted to replace this actually somewhat unfortunate notation, but without success.[80]

While any commercial pH meter gives measurements accurate to the first decimal place, more accurate pH measurements require some care. The difficulties are related to the fact that electrochemical methods can only yield mean activity coefficients, where the activities of single ions are, strictly speaking, inaccessible. Proper construction of the electrochemical cells and knowledge of the behavior of activity coefficients at high dilution permit more accurate measurements of pH. Such methods give the primary pH standards, which are then used to calibrate secondary standards or to check the pH measuring devices in use. Here we shall only summarize the most relevant aspects; details are given in specialized monographs.[76,77]

An accurate method for the pH determination was developed by Bates.[78] His technique relies on an electrochemical cell without a liquid junction, where two electrodes are immersed into a buffer solution (e.g., phosphate buffer), which contains KCl as the excess monovalent electrolyte. The cell consists of a standard hydrogen electrode, where pure hydrogen gas is bubbled over a platinum black electrode, and an Ag/AgCl reference electrode, in which silver is coated with silver chloride. The potential of the hydrogen electrode with respect to the reference electrode is given by

$$U = -U_{AgCl} + \Delta U \log_{10} a_{H^+} a_{Cl^-} \qquad (2)$$

where $\Delta U = kT \ln 10/e \simeq 59$ mV at 25 °C (kT being the thermal energy and e the elementary charge). The potential of such a cell is measured by a high-impedance voltmeter and investigated as a function of the salt concentration. The activity of the hydrogen ions in the pure buffer solution is obtained by extrapolating the electrode potential to zero salt concentration. Thereby, one uses the known dependence of the activity coefficients from the Debye–Hückel expression [see Section 3, Eq. (33)].

While this technique can be accurate to the third decimal, it is not very practical for routine use. The most convenient setup consists of a standard glass electrode and a reference electrode with a salt bridge (liquid junction). The liquid junction is often combined with the reference electrode into a single unit. For routine measurements, both electrodes are integrated together with the liquid junction into the *combination glass electrode*. However, for the determination of titration curves the two-electrode setup is more convenient.

The glass electrode consists of a glass tube whose bottom is closed with a hemispherical glass membrane cap of 50–500 μm thickness. The electrode is filled with a buffer solution containing an indifferent electrolyte (mostly KCl). A silver wire coated with silver chloride provides an internal reference electrode. The glass membrane is semipermeable for protons, and the electrode potential is related to the pH difference between the inner (reference) solution and the outer (sample) solution. While the glass electrode is very simple to use, it is not free of artifacts. The most important problem is the presence of the alkali error, which may become important at pH > 10. Since the glass acts mainly as a cation exchanger, the electrode also responds to the major cations present in the system, and the apparent pH is lower than the actual one. The problem is most severe with Na^+, but can be reduced to a certain extent by working with K^+. For highly acidic solutions, the electrode response is also nonideal, and the apparent pH turns out to be higher than the actual one.[76]

As the outer reference electrode, one typically uses the standard silver/silver chloride electrode as described above. Another very stable reference electrode system is the mercury/calomel electrode. The reference electrode can be directly immersed into the solution, but in order to reduce contamination of the sample solution, one introduces a liquid junction to separate the sample solution from the electrolyte in the reference electrode. From the various liquid junction setups three are of interest. For routine measurements a *porous ceramic plug*, which is fused into the shaft of the reference electrode, is the most convenient. In order to avoid contamination of the plug and of the reference electrode, it is necessary to maintain a slight overpressure inside the reference electrode to ensure a small, but steady flow of electrolyte from the reference into the sample solution (a typical value is > $0.5 \ \mu Lh^{-1}mm^{-2}$ per plug area). However, one must carefully check whether the introduced traces of the reference electrolyte do not alter the sample solution in an undesired fashion. This problem can be reduced by using a *double junction*, where both junctions can be integrated within the reference electrode. The *free-flowing junction* avoids the contamination problems completely. In this junction the electrolyte from the reference electrode and from the sample solution are continuously pumped into y-shaped or concentric capillaries at flow rates of $0.5 - 2 \ \mu L/h$, and the outflow is discarded. While the solutions mix under flow, a liquid junction with a sufficiently high electrical conductivity builds up. This junction causes essentially no contamination of either the sample solution or the reference electrolyte. The only disadvantage is that the solutions are continuously consumed.

In all these setups, the electrical potential of the glass electrode relative to the reference electrode is approximated by

$$U = U_0 + U_d + \Delta U' \log_{10} a_{H^+} \tag{3}$$

where U_0 and $\Delta U'$ are constants, and U_d is the diffusion potential. Note that $\Delta U'$ is usually somewhat below the theoretical value of 59 mV. This expression does not include the possibility of the alkali error, which can be incorporated in an approximate fashion by considering the exchange between protons and major salt cations in the glass membrane.

The diffusion potential of a single liquid junction can be estimated by approximating the steady-state concentration profiles in the plug by straight lines. The result reads[76]

$$U_d = \frac{kT}{e} \frac{\Sigma_i \lambda_i |Z_i| (c_i' - c_i)/Z_i}{\Sigma_i \lambda_i |Z_i| (c_i' - c_i)} \ln \frac{\Sigma_i |Z_i| \lambda_i c_i'}{\Sigma_i |Z_i| \lambda_i c_i} \tag{4}$$

where Z_i is the ion valency, λ_i the equivalent ionic conductivities, and the sums run over all species in solution i. The bulk ionic species concentrations are denoted by c_i and c_i' for the sample solution and for the electrolyte in the reference electrode, respectively. In the case of two liquid junctions, the overall diffusion potential can be estimated from the sum of diffusion potentials obtained from Eq. (4) applied to each junction separately.

In many practical situations, this potential can be taken to be approximately constant for a given ionic strength. In that case, the calibration of the glass electrode involves the determination of two independent constants, namely of the electrode slope $\Delta U'$ and of the intercept $U_0 + U_d$. The simplest way to determine these quantities is by means of standard buffer solutions. For accurate measurements, however, one must take the diffusion potentials properly into account. This correction can be carried out either by using a junction, which eliminates this contribution to a good degree of approximation, or by estimating these contributions by means of Eq. (4). Another calibration procedure uses a blank titration of the pure supporting electrolyte and will be discussed below.

2.1.2. Potentiometric Titration Techniques

Potentiometric titrations are carried out most conveniently with a separate glass and reference electrode with an integrated liquid junction. In the simplest setup one uses a single automatic burette and a high-impedance voltmeter (pH meter). To an initially acidified solution one successively adds small aliquots of a strong base and, after the solutions have been mixed, the potential of the glass electrode is monitored until a constant value is reached. Depending on the application, the equilibrium is established when the potential drift remains below $10-100\,\mu V/min$. In order to keep the ionic strength constant during a titration run, the added base should have the same ionic strength as the sample solution. Such an apparatus is controlled most conveniently with a microcomputer.

When one uses a commercial computer-controlled titration instrument, it is essential to ensure that the appropriate drift criterion is properly satisfied. Commercial titration instruments are usually designed to detect titration endpoints, and often they do not allow the selection of an appropriate drift criterion. Particularly, for the titration of polyelectrolytes and colloidal particles, a very helpful extension of such an automatic titration setup is a second automatic burette for a reverse titration, whereby the reversibility of the titration run can be checked. Two further automatic burettes filled with distilled water and concentrated salt solution permit the adjustment of the ionic strength after each addition of base or acid. When controlled by a computer, such a four-burettes setup runs an entire series of forward and backward titrations at different ionic strengths in a single experiment.[81] Such titrations were also performed up to 250 °C with a different instrument.[82]

A convenient electrode calibration can be carried out by a blank titration of the pure supporting electrolyte acidified with an accurately known amount of a strong acid. The titration curve of the pure electrolyte follows from the charge balance. If the supporting electrolyte is NaCl, the ions in the solution are H^+, OH^-, Na^+, and Cl^-. Since the concentrations of Na^+ and Cl^- from the supporting electrolyte are equal, the excess concentrations of these ions are fully determined by the amounts of added NaOH and HCl. These concentrations are balanced by the concentrations of H^+ and OH^- and one obtains

$$c_a - c_b = c_{H^+} - c_{OH^-} \qquad (5)$$

where c_{H^+} and c_{OH^-} denote free ion concentrations while c_a and c_b are the total concentrations of strong acid and base added. We use the ionic product of water

$$K_w = a_{H^+} a_{OH^-} \qquad (6)$$

recognize that $a_i = \gamma_i c_i$, where γ_i is the activity coefficient, and assume a constant diffusion potential and equal activity coefficients for cations and anions. All remaining parameters can be determined by fitting the solution of Eqs. (3), (5), and (6) to the experimental blank titration curve. The only known input is the initial concentration of the titrated strong acid. Fitting these relations to the blank titration curves, one cannot only determine accurately the base concentration, but also find all electrode parameters, namely the ionic product of water, and the corresponding (mean) activity coefficient. The latter quantities serve as a check of the entire procedure. The calibration is typically accurate within 0.05 pH units. The accuracy can be pushed down to < 0.01, but the experimental effort becomes substantially larger.

Titrations of sample solutions are performed in the same way as the blank titrations. The amount of bound or released protons is obtained from the charge balance, which now reads

$$c_a - c_b = c_{H^+} - c_{OH^-} + q \qquad (7)$$

where q is the number of bound protons per unit volume. The free concentrations c_{H^+} and c_{OH^-} can be calculated or taken from the experimental record of the blank titration [cf. Eq. (5)]. This experiment yields the charge q, which corresponds to the amount of adsorbed protons, as a function of pH (see Fig. 8a).

The conversion of the experimentally accessible quantity q to the degree of protonation θ may not be straightforward. In principle, q will approach constant plateau values, q_{max} and q_{min}, at extreme pH values (see Fig. 8b). If these two values are known, the degree of protonation is given by

$$\theta = \frac{q - q_{min}}{q_{max} - q_{min}}. \qquad (8)$$

Note that the denominator $q_{max} - q_{min}$ represents the total concentration of ionizable sites. This quantity can be therefore checked against an independent measurement (e.g., the total nitrogen concentration in the case of a polyamine). In many cases, one of the plateau values q_{max} and q_{min} (or both) lie outside the experimental window and the degree of protonation must be obtained by an alternative method. As illustrated in Fig. 8b, one of these plateau values may be still accessible, and the

(a)

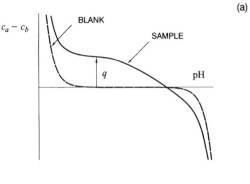

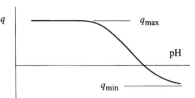

(b)

Figure 8. Schematic representation of titration curves. (a) Experimental record of potentiometric titrations of a sample and of a blank. (b) The difference gives the excess charge which can be converted to the degree of protonation from both plateau values q_{min} and q_{max}.

other can be obtained from the total concentration of ionizable sites. In such a situation, it may also be helpful to prepare the sample in a defined ionization state, for example by dialyzing it against pure water. If the sample contains acidic or basic groups only (e.g., polycarboxylate, polyamine), then the measured titration curve will start at the point of zero charge (PZC) of the material. If the sample contains both kinds of groups (e.g., polyampholytes, most metal oxides), it is no longer possible to find the fully protonated or deprotonated state. However, one can use the PZC as a reference point, which shows up as a common crossing point of titration curves measured at different ionic strengths. In literature, titration curves are reported in various different ways, such as the degree of deprotonation $1 - \theta$, the number of bound protons $N\theta$ (where N is the total number of dissociable sites), the surface charge density, or the total molecular charge $N\theta - Z$ (where Z is the charge in the fully deprotonated form).

As an example consider the titration curve shown in Fig. 9 from van Duijven-bode *et al.*[83] The substrate is dendritic poly(propylene imine) (DPPI), and its structure is shown in Fig. 4c. The titration curves shift toward lower pH with decreasing ionic strength, which is a typical trend for a weak polybase. The absolute degree of protonation was determined by titrating a pure solution of the dendrimers (i.e., in the fully deprotonated state), while the position of the plateau at low pH was determined by total nitrogen analysis of the sample. This procedure brings the plateau at low pH very close to the theoretical value of unity, providing a good check of the analysis. Two further consistency tests are shown in the figure as open

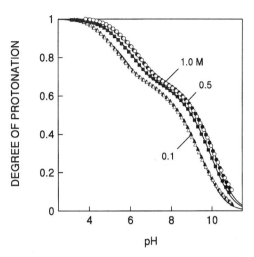

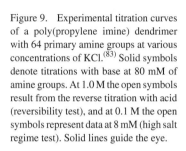

Figure 9. Experimental titration curves of a poly(propylene imine) dendrimer with 64 primary amine groups at various concentrations of KCl.[83] Solid symbols denote titrations with base at 80 mM of amine groups. At 1.0 M the open symbols result from the reverse titration with acid (reversibility test), and at 0.1 M the open symbols represent data at 8 mM (high salt regime test). Solid lines guide the eye.

symbols. In the data set at 1.0 M the open symbols represent the backward titration, and one observes that the titration curve is fully reversible. In the data set at 0.1 M we compare the titration curves performed at 80 mM and 8 mM dendrimer concentration. The coincidence of these curves shows that the titration was carried out in the excess of salt.

Such a titration setup can routinely achieve an accuracy in the molar balance of < 50 μM. Therefore, one can titrate samples where the total concentration of the ionizable groups is in the mM range. However, for this accuracy it is essential that all solutions are free of ionic impurities, the major problem being carbon dioxide, which must be eliminated at various stages. The deionized water used to prepare the solutions must be boiled and cooled under CO_2-free nitrogen. Particularly sensitive are the strong bases, which typically contain several percent of carbonate upon delivery from the manufacturer. Various procedures are available to prepare CO_2-free NaOH or KOH.[84] Once the solutions are purified, they must be kept in a CO_2-free atmosphere. Another problematic inorganic impurity is silica, which usually originates from dissolution of glass, particularly in solutions of high pH. For this reason, NaOH or KOH should be kept in plastic containers, and one must not use glass titration vessels for precision measurements at high pH.

Another problem, which may arise during the work with concentrated polyelectrolytes or solid suspension at low salt levels, is the presence of the so-called *suspension effect*.[71,73] This *low salt regime* will be discussed in more detail in Section 3. At low salt levels, the counterions of the macromolecules will also contribute to the ionic strength, and due to interactions between macromolecules the titration curve now also depends on the macromolecular concentration. Furthermore, there is a finite electrostatic potential difference (Donnan potential) between the macromolecular solution and a pure salt solution. This potential drop does affect the electrode calibration and must be properly included in the data analysis.

Problems of this kind can be simply avoided by working at sufficiently high concentration of the background electrolyte (i.e., *high salt regime*). As rule of thumb, the concentration of the added salt should be larger than the total concentration of ionizable groups at least by a factor of two. To check whether one is indeed within the high salt regime one should repeat the titration at a lower concentration of the macroions. When the same titration curve is recovered, one can be assured that one is operating within the high salt regime. While interesting, the regime of low salt is much harder to study experimentally.

2.1.3. Other Macroscopic Techniques

Information about proton adsorption characteristics cannot only be obtained from batch potentimetric titration, but other experimental methods can be used as well. These techniques include various flow-through methods,[85,86] contact angle or capillary techniques,[87–92] as well as conductivity[93–97] and electrophoretic measurements.[98,99] Some of these techniques have the advantage of higher sensitivity, which opens the possibility to measure proton binding characteristics to substrates with small site densities or even for planar macroscopic interfaces. The disadvantage is that these methods are not generally applicable and the data analysis can be more difficult.

As an example of a flow-through method, let us mention a chromatographic technique, which was used to measure the titration curve of a silica sand.[85] The surface area of this material is so low that the amount of adsorbed and released protons cannot be detected with a conventional batch titration. However, when the material was packed into a chromatographic column, retarded pH breakthrough curves could be readily measured. The charging curve was extracted from such a chromatographic experiment using standard chromatographic analysis. The result is shown in Fig. 10 as a solid line, where the points represent batch titration data of amorphous silica.[100] The good agreement between these two data sets indicates that the silica sand has the same charging behavior as amorphous silica.

Another interesting means to investigate the charging properties of interfaces are contact angle and capillary rise techniques.[87–91] The free energy of the solid–solution interface depends on the surface charge (i.e., the degree of protonation), and the surface tension will also vary accordingly. Therefore, contact angle or the capillary size measurements as a function of pH may yield information about the charge of the interface. For example, Chatelier *et al.*[90] could infer quite detailed information about the charging characteristics of a planar polymeric surface with this method. A novel approach to study air–water interfaces combines neutron reflection with surface tension measurements.[92] Unfortunately, the effects are not easily measurable, and impurities must be carefully excluded. While these tech-

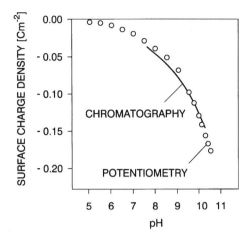

Figure 10. The surface charge density for a low surface area silica sand determined by a chromatographic method at an ionic strength of 0.01 M (solid line).[85] The data are compared with potentiometric titration results for amorphous silica (points).[100]

niques are far from being established, they provide a possibility to approach ionization reactions on planar surfaces, which are difficult to study otherwise.

Another class of methods for the study of ionization equilibria are based on the analysis of the electrical conductivity of the sample. Titration end points are readily detected through the appearance of minima in conductometric titration curves. This method is therefore well suited for the determination of the total number of dissociable groups, such as, for example, functionalized latex particles.[93–96] However, the detailed analysis of the shape of conductometric titration curves is rather involved,[95–97] and such techniques are not too easily applicable to measure proton adsorption isotherms.

Electrophoresis represents another related technique, which can be used to study the charge of colloidal particles down to the nanometer-size range. An external electric field is applied across a colloidal suspension, and the resulting particle migration velocity is measured. Various commercial instruments differ in the way the velocity is being measured, including videomicroscopy, laser Doppler velocimetry, and photon correlation spectroscopy.[98,99] The electrophoretic mobility is obtained as the ratio of the velocity to the applied electric field, and is independent of the field strength for sufficiently low electric fields (typically < kV/cm). Mathematical models are used to extract information about the electrostatic surface potential from the measured mobilities. A popular approach is due to O'Brien and White,[101] who have solved the problem of a uniformly charged sphere migrating in an electrolyte solution in a uniform external field in the Poisson–Boltzmann approximation. The problematic part in the analysis is the location of the so-called *plane of shear*, at which the shear vanishes. Since this plane is displaced outward, the potential at this plane (denoted as ζ-potential) is typically smaller than the actual surface potential. Examples will be given in Section 7. Some authors have suggested that contributions due to surface conductivity are important and should be taken into account.[99,102] Related techniques involve streaming potentials,[98,99] the dielectric response of colloidal suspensions,[102] or, as proposed recently, electroacoustics.[103]

2.2. Spectroscopic Methods

More detailed information than from macroscopic measurements can be obtained by spectroscopic techniques. In favorable cases, one can measure the average degree of protonation for individual ionizable groups, and learn about the molecular mechanisms of ionization reactions. Nuclear magnetic resonance (NMR) is by far the most powerful technique for such studies. Ultraviolet/visible (UV/VIS) and infrared (IR) spectroscopy are somewhat more limited, but can be very useful in certain situations. The present review will not discuss the background and experimental details of all these techniques exhaustively; here we refer to the appropriate monographs for NMR[104–106] and spectroscopic techniques.[76,107] We shall only focus on the important aspects in the use of these methods to elucidate ionization reactions in aqueous media.

2.2.1. Nuclear Magnetic Resonance (NMR) Techniques

NMR is the most powerful spectroscopic technique for the determination of the ionization behavior of complex molecules. This technique investigates transitions between individual energy levels of an atomic nucleus with a nonzero spin. These levels are normally close to degenerate, but can be split through an externally applied magnetic field B. For the common spin 1/2 nuclei, one obtains two states, whose energy splitting ΔE is proportional to the magnetic field. If such a nucleus is placed into an oscillating electromagnetic field of frequency ν, energy will be absorbed when the following resonance condition is satisfied:

$$\Delta E = 2\mu B = h\nu \tag{9}$$

where h is the Planck constant and $\mu = geh/(8\pi m)$ is the nuclear magnetic moment with g being the dimensionless nuclear factor, and m the proton mass. A few properties of the most important spin 1/2 nuclei are summarized in Table 1. The NMR experiment is typically carried out at a fixed radiation frequency ν, and the applied magnetic field B is tuned through the resonance. The absorbance spectrum is plotted as a function of the magnetic field.

The resonance condition occurs at a slightly different magnetic field for a free atomic nucleus and for the same nucleus in a molecular environment. The reason is that the nuclei are weakly shielded from the external magnetic field due to induction currents within the electron cloud of the molecule. Such chemical shifts

Table 1.   Some Properties of Common Nuclei Spin 1/2 used in NMR Studies[105]

Nucleus	g	Natural abundance (%)	Sensitivity[a]
^{1}H	5.585	99.98	63
^{13}C	1.405	1.11	1
^{15}N	−0.567	0.36	0.066
^{19}F	5.257	100	52
^{29}Si	−1.111	4.7	0.49
^{31}P	2.263	100	4.2

[a]Sensitivity is given as the relative response to an equal number of ^{13}C nuclei.

are rather small, but extremely sensitive to the chemical and electronic environment near the atom whose nucleus is being probed. For this reason the NMR spectrum is usually reported as a function of the chemical shift $\delta = (B - B_0)/B_0$, which is the shift of the magnetic field relative to the magnetic field B_0 of a reference compound, and is usually reported in ppm. In modern spectrometers, typical resonance frequencies are 100–800 MHz, and typical chemical shifts lead to frequency differences of <300 Hz. The chemical shift is commonly used to distinguish nonequivalent atoms in a molecule. Each nonequivalent nucleus leads to an absorption line, which may be split due to spin–spin coupling. The total area under the line is proportional to the number of nuclei in the molecule. While the assignment of NMR spectral lines is usually straightforward for smaller molecules, it can represent a formidable task for complex molecules, such as proteins.

With Fourier transform (FT) techniques the sensitivity of modern NMR spectrometers could be increased dramatically. The sample is excited with a short intense electromagnetic pulse, and the decay of the magnetization of the sample is recorded. The FT of this signal yields the spectrum. The advantage of these techniques is that the process can be repeated many times, and the sensitivity can be increased by adjusting the acquisition time. There are many variants of the pulsed FT/NMR techniques, which employ adjustments of the frequency, intensity, duration of the pulse, and the application of various pulse sequences.[104,105]

The chemical shift can also be used to probe the protonation state of an individual atom. As an illustration, let us first discuss the simplest case of a ^{15}N-atom within an amine group. This nitrogen can bind a proton, and depending on its protonation state the chemical shift of the nucleus will be different. For a chemical ionization equilibrium, the protonated and deprotonated species will be present simultaneously, and one expects two different resonance peaks, whose areas would reflect the relative population of each species. However, due to the smallness of the nuclear level splitting, and because of the finite lifetime of the individual protonation states, this simple picture is not quite correct. For sufficiently slow proton exchange, one would indeed observe two separated peaks. For fast exchange, however, these peaks collapse into a single line, which is centered in between the two original peaks. The transition between slow and fast exchange is well under-stood[104] and is illustrated schematically for two equally populated states in Fig. 11. The relevant dimensionless parameter is $\Delta\nu\tau$, where $\Delta\nu$ is the peak splitting and τ is the lifetime of the protonated state. When $\Delta\nu\tau \approx 1$, the technique can be used to measure exchange rates of protons. However, with a typical NMR splitting $\Delta\nu$ around 100–300 Hz and lifetimes $\tau < 2$ ms, we typically have $\Delta\nu\tau < 0.5$, and the resulting spectrum usually collapses into a single line. The chemical shift of this line is given by the weighted average of the relative populations according to[104]

$$\delta = \delta_{HB}\theta + \delta_B(1 - \theta) \qquad (10)$$

where θ is the degree of protonation, while δ_{HB} and δ_B are the chemical shifts of the protonated form, HB, or the deprotonated form, B. (Figure 11 shows the case of $\theta = 1/2$, and the resulting line lies in the center.) Thus, the relative degree of protonation of an amine group can be measured through the chemical shift of a ^{15}N-nucleus by using the equation

$$\theta = \frac{\delta - \delta_B}{\delta_{HB} - \delta_B} \qquad (11)$$

The key point is that this type of analysis cannot only be applied to a monoprotic acid or base but, more importantly, to a complex molecule. The chemical shift is a

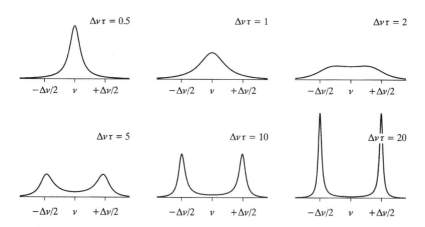

Figure 11.   NMR line shapes for a nucleus influenced by an ionizable group, for which the protonated and deprotonated states have equal probability. If the protonated state is long-lived, there are two equal resonance lines, which are centered at ν and split by $\Delta\nu$. When the lifetime τ becomes shorter, the two lines collapse into a single one centered at ν. The spectra are shown for different values of the dimensionless parameter $\Delta\nu\tau$.

very local probe, and a given nitrogen nucleus will almost exclusively respond to the protonation state of this particular atom and remains largely unaffected by the protonation state of the other sites in the molecule. This observation is the basic tenet behind most NMR titration studies in complex molecules. As will be discussed below, this method can be extended to other nuclei, and was used extensively to study the ionization of individual ionizable sites in proteins and in other molecules.[15,68,108–121]

As an illustration consider a polyamine with three types of equivalent nitrogens,[108] and consequently give rise to three peaks in the ^{15}N NMR spectrum (denoted as α, β, and γ, see Fig. 12). They can be easily assigned based on the individual peak areas. As the solution pH is changed, all three types of nitrogen are

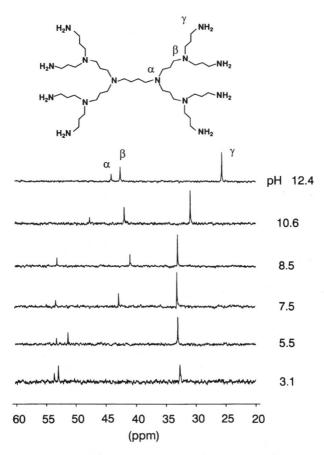

Figure 12. ^{15}N NMR spectra of the shown poly(propylene imine) dendrimer for different pH values.[108]

being protonated and the resonance peaks shift from the right to the left. The resulting chemical shifts are plotted as a function of pH in Fig. 13. In this particular case, the plateau values are clearly read off at low pH, but a certain ambiguity in the assignment of these values remains at high pH. Once the latter are decided upon, one can deduce the degree of protonation as a function of pH for each type of nitrogen atom from Eq. (11). The accurate determination of the plateau values δ_{HB} and δ_B often represents a problem in NMR titrations, which is similar to the determination of plateau values in potentiometric titrations. But the chemical shift mostly responds to a single ionizable site, and the transitions occur often over a narrow pH range, which facilitates their unambiguous determination.

The effects of the protonation of an atom on the chemical shift of a nucleus are, of course, the largest when the nucleus belongs to the same atom which is being protonated. In practice, one faces this situation only for ^{15}N within an amine group, and this technique therefore represents a very selective means to probe protonation reactions in polyamines.[108,114] However, a nucleus of a different atom, which is in close proximity to the protonating atom, will respond similarly, albeit the effect of the protonation on the chemical shift will be smaller. For this reason, one can probe the protonation of oxygens in carboxylic groups by ^{13}C NMR,[109] in phosphates by ^{31}P NMR,[110,111] and through ^{1}H NMR of an adjacent $-CH_2-$ or $-NH-$ group.[112,113,121] Similarly, the protonation of amines can be also probed by ^{13}C and

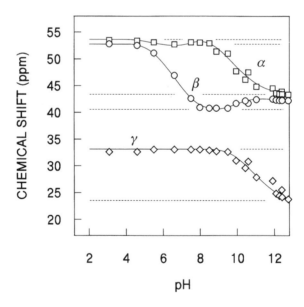

Figure 13.   Chemical shifts from ^{15}N NMR spectra as a function of pH for poly(propylene imine) dendrimer.[108] The data are extracted from the spectra shown in Fig. 12. Solid lines guide the eye. Dotted lines represent the plateau values for δ_{HB} and δ_B.

^{1}H NMR of adjacent atoms.[68,114,115] Fortunately, this effect is rather short-ranged, and a nucleus typically responds to the protonation of just a single ionizable site. In some cases, a given nucleus may also respond to the protonation state of two (or more) neighboring ionizable sites, which may occur in molecules with closely spaced ionizable sites. This aspect represents a fundamental problem in NMR titration studies in complex molecules, as one has to ensure that changes in a certain chemical shift originate solely from one particular titrating group. If several ionizable sites affect the chemical shift of a nucleus under consideration, data analysis is still possible, but becomes much more complicated.[112]

NMR was used extensively to monitor protonation of ionizable amino acid residues within proteins.[15,68,116–121] When the sequential assignments of the protein NMR spectrum have been made, the atomic level resolution of the NMR spectrum provides the molecular characteristics to a detail unsurpassed by any other technique. In spite of the recent advances in NMR techniques, a complete determination of all titrating groups in a protein remains a formidable task. Nevertheless, it has now become possible to follow the titration behavior of all titrating residues in proteins of up to 10 kDa or even higher molecular weights.

The line assignment was greatly facilitated with the invention of two-dimensional NMR methods.[105,122,123] Traditionally, the NMR spectrum is represented as a function of a single frequency variable (or the chemical shift). But even with the strongest magnets available, one often cannot avoid the overlap of spectral lines in a complex biological molecule. This problem can be often resolved with two-dimensional NMR spectroscopy. The spectrum is represented as a function of two independent frequencies, both of which correspond to a time variable in the time domain. The spectra are recorded through a pulse sequence, where the first variable corresponds, for example, to the evolution time between applied pulses, and the second variable to the time during the detection period. Homonuclear NMR methods, such as correlated spectroscopy (COSY), analyze couplings between nuclei of the same kind, and permit the detection of spin–spin coupling between particular protons. Nuclear Overhauser enhancement spectroscopy (NOESY) utilizes three pulses and offers further possibilities in the determination of nuclear proximity in macromolecules. Total correlation spectroscopy (TOCSY) represents another important homonuclear two-dimensional method to assign all protons in a protein.[122] With heteronuclear methods one can study couplings between nuclei of different kinds. The most important techniques include heteronuclear single quantum coherence (HSQC) and heteronuclear multiple quantum coherence (HMQC). These methods have been extensively used for the analysis of titrating groups in proteins. As an example, we show the ^{13}C-^{1}H HSQC spectra for outermost $-CH_2-$ groups of lysine side chains in Fig. 14. The measurements were carried out in 0.5 mM solutions of the protein calbindin (D_{9k} in the Ca^{2+} form) with a sequence specific assignment of the peaks.[68] The protonation of the amine groups was monitored through changes of the chemical shift in the outermost $\varepsilon-CH_2-$ group by ^{13}C NMR and by ^{1}H NMR. The pulse sequence can also be designed to contain

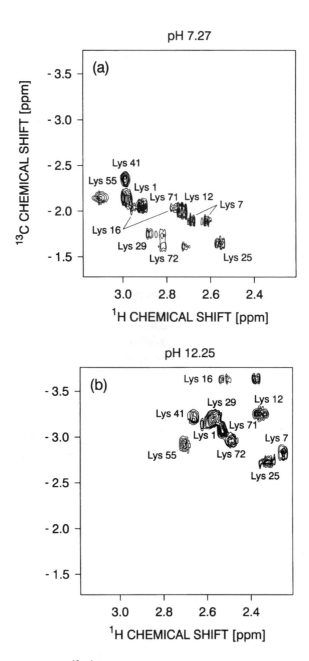

Figure 14.   Two-dimensional ^{13}C-^{1}H HSQC NMR spectrum of the lysine residues in 0.5 mM solution of calbinidin D_{9k} in the Ca^{2+} form.[68] (a) pH 7.27 and (b) pH 12.25.

pulses for both HSQC and TOCSY. For example, two-dimensional ^{15}N-^{1}H HSQC-TOCSY experiments show resonances for the protons in amino acid side chains in addition to the correlation between ^{15}N and ^{1}H resonances for N–H groups both in backbone of the protein as well as in side chains.[123]

Such two-dimensional methods greatly facilitate the determination of site-specific titration curves from NMR. The most versatile method involves the analysis of the chemical shift of several nonzero spin nuclei in close neighborhood to the ionizable site. Such nuclei are expected to respond to the same site, and should thus yield the same titration curve. An example is shown in Fig. 15, where we display site-specific titration curves of two amino side chains of lysine 16 and 55 in the protein calbindin.[68] These shifts were extracted from spectra such as shown in Fig. 14. One indeed observes that the site-specific titration curves obtained from different nuclei are rather similar, thus confirming that one indeed measures the ionization of the particular site of interest. Table 2 shows that this analysis yields a consistent set of pK values for all lysine residues within the protein. A similar consistency test was carried out with NOESY on a mutant of human insulin. Sørensen and Led[120] could determine the pK values of seven carboxylic groups, whereby they used up to four different nuclei to monitor their protonation.

The difficulties one might encounter in this kind of analysis are exemplified in Fig. 16. Full symbols indicate the usual one-step behavior of lysine 29 in the Ca^{2+} form, as discussed above. To the apo form no Ca^{2+} is bound, and ^{1}H NMR of the same residue shows two steps in the chemical shift as a function of pH. Blind application of Eq. (11) would lead to the incorrect conclusion, that this residue does protonate in two steps. However, ^{13}C NMR shows the usual one-step behavior, and the two steps in the ^{1}H NMR data must be rather attributed to the protonation of a

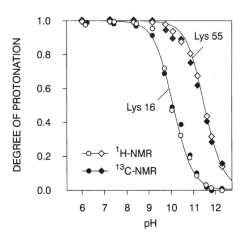

Figure 15. Site-specific titration curves of two amino side chains, lysine 16 and 55, in the protein calbindin (D_{9k} in the Ca^{2+} form). The protonation was monitored through changes of the chemical shift in the outermost ε–CH$_2$– group by ^{13}C NMR and by ^{1}H NMR. The solid line is the titration curve for a monoprotic base.[68]

Table 2.   Determination of Apparent pK Values by ^{1}H and ^{13}C NMR Titration^a

Residue	^{1}H NMR	^{13}C NMR	Mean
Lys1	10.61 ± 0.04	10.58 ± 0.07	10.60 ± 0.05
Lys7	11.39 ± 0.03	11.31 ± 0.03	11.35 ± 0.03
Lys12	10.39 ± 0.08	11.06 ± 0.07	11.00 ± 0.07
Lys16	10.06 ± 0.05	10.12 ± 0.06	10.09 ± 0.06
Lys25	11.90 ± 0.05	11.72 ± 0.06	11.81 ± 0.06
Lys29	11.05 ± 0.03	10.88 ± 0.06	10.99 ± 0.04
Lys41	10.95 ± 0.06	10.83 ± 0.05	10.89 ± 0.06
Lys55	11.43 ± 0.04	11.33 ± 0.06	11.39 ± 0.05
Lys71	10.72 ± 0.04	10.73 ± 0.05	10.72 ± 0.05
Lys72	11.01 ± 0.05	10.91 ± 0.07	10.97 ± 0.06

^{a}The measurements were done in 0.5 mM solutions of the protein calbindin (D_{9k} in the Ca^{2+} form).[68]

nearby carboxylic residue.[68] On the other hand, such two-dimensional NMR methods can also firmly establish a two-step protonation curve of an ionizable residue within a complex protein environment. Figure 17 shows the chemical shifts reflecting the protonation of Asp 10 in ribonuclease HI isolated from *Escherichia Coli*.[121] One observes the ^{13}C chemical shift of the carboxyl resonance (cf. Fig. 17a), and the ^{1}H chemical shifts of both methylene protons attached to the β-C of the side chain (cf. Fig. 17b). The common trends in all three data sets prove that the Asp 10 residue protonates in two steps.

Modifications of amino acid residues can represent an alternative when trying to interpret complex spectra from biomolecules. For example, lysine residues can be modified by reductive methylation with ^{13}C-enriched formaldehyde.[124] Such a modification of the protein facilitates the spectra considerably, but it has to be ensured that the structural perturbation caused by the additional methyl group is minimal. Zhang *et al.*[124] have attempted to quantify the effect of methylation by comparing mono- and dimethylated samples. For all the lysine groups in calbindin they found an average pK shift between the two forms of about 0.8, the dimethylated lysines being more acidic.

Another type of complication may arise when the titration is accompanied by a conformational change. A slow exchange between two conformations, having different ionization constants, may result in two separate resonance lines (see Fig. 11). If the environments of the group in the two conformational states are similar, a broadening of the resonance is observed. If only one of the sites titrates, slow conformational exchange results in separate but pH independent lines. In such a case, ionization constants may be obtained by measuring the area under the peaks

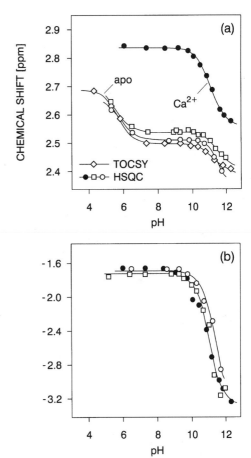

Figure 16. Chemical shift as a function of pH of lysine 29 in the protein calbindin D_{9k}. Solid data points correspond to the Ca^{2+} form, while open symbols indicate the apo form.[68] The protonation was monitored through changes of the chemical shift in the outermost ε–CH_2– group by (a) ^1H NMR and (b) ^{13}C NMR. ($\diamond$) 1.65 mM apo form (TOCSY), ($\square$) 1.5 mM apo form (HSQC), ($\circ$) 0.5 mM apo form (HSQC), and ($\bullet$) 0.5 mM Ca^{2+} form (HSQC).

as a function of pH.[125] Conformational changes during the titration of ionizable residues in a protein have been accounted for by a complete quantitative spectral analysis of the pH dependence of the ^{13}C NMR spectrum of 26 leucine residues in the human growth hormone with ^{13}C-labeled backbone carbonyls.[126]

For smaller molecules, one can test the reliability of site-specific titration curves obtained by NMR through potentiometric titration data. An example of this type of analysis can be found in the work of Mernissi *et al.*[111] The average degree of protonation was extracted for three different individual phosphate groups using ^{31}P NMR. Averaging the site protonation curves, the average degree of protonation of the molecule was obtained, and could then be directly compared with the classical potentiometric titration experiment [see Section 3, Eq. (83)]. The result is shown

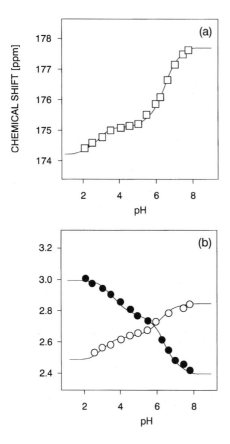

Figure 17. Chemical shifts of Asp 10 in Ribonuclease HI from *Escherichia Coli* in 0.1 M NaCl measured by HSQC NMR spectroscopy.[121] (a) ^{13}C chemical shift of the carboxyl resonance, and (b) ^{1}H chemical shifts of both methylele protons attached to the β-C of the side chain.

in Fig. 18, and the good agreement between both sets of data confirms the correctness of the site titration curves determined by NMR.

The suppression of artifacts, such as the huge water resonance, which obscures interesting signals, is a severe problem in multidimensional NMR of biomolecules. The spectra can be recorded in D_2O in order to minimize the influence of the solvent peak, but then slowly exchanging NH-resonance peaks disappear. The water resonance can also be suppressed by saturation in the pulse sequence prior to data acquisition. Alternatively, a relatively new method in heteronuclear correlation experiments is available, where the suppression of undesired signals, such as the water resonance, can be achieved by the application of pulsed field gradients. This technique improves tremendously the quality of spectra.[116,127]

In spite of all strengths of the NMR, it is important to mention one serious limitation, namely low sensitivity. For this reason, long accumulation times are necessary and one is also forced to work at high concentrations of macromolecules.

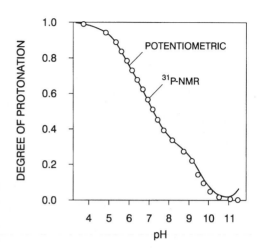

Figure 18. Comparison of the average degree of protonation for a D-*myo*-inositol 1,2,6-tri(phosphate) in 0.1 M Et$_4$NBr at 25 °C obtained by ^{31}P NMR and by potentiometric titrations.[111]

For example, each spectrum shown in Fig. 12 had to be accumulated for 24 h and the solution did contain about 50% of the amines. The low sensitivity is not only due to the inherent NMR sensitivity of an individual nucleus, but moreover some of the important spin 1/2 nuclei have rather low natural abundance (see Table 1). The latter problem can be circumvented by enriching the samples with the appropriate spin 1/2 isotopes, but the costs of such a procedure might be prohibitive.

2.2.2. Optical and Other Spectroscopic Methods

Another classical class of methods to probe ionization reactions are the IR, UV/VIS, or fluorescence spectroscopies. Various protonation states of a given molecule will have a different electronic structure, which could be discerned by changes of the spectral characteristics of the molecule. By monitoring the latter as a function of pH, information about protonation reactions can be extracted.[76,107]

The investigation of a monoprotic acid or base represents a straightforward application. At extreme pH values the spectrum corresponds to either the fully protonated form or the fully deprotonated form. When both forms are present simultaneously, the overall spectrum is a weighted sum of the spectra of the individual species. At a given wavelength the absorbance is given by[76]

$$A = A_{HB}\theta + A_B(1 - \theta) \tag{12}$$

where θ is the degree of protonation, and A_{HB} and A_B are the absorbances of the protonated form, HB, or the deprotonated form, B. When the absorbance for both species happens to be the same at one particular wavelength, the spectra cross in a single point. This point is sometimes referred to as the *isosbestic* point for absorption spectroscopy (meaning equal extinction), and the *isostilbic* point for fluorescence (meaning equal brightness). A typical example of such a spectral evolution is shown in Fig. 19, which displays the UV/VIS spectra of a surface-active anionic probe incorporated in a micelle.[37] As the solution pH is decreased, the spectrum evolves from one of the basic species B to one of the acidic species BH. This example actually shows two common crossing points.

Measuring the absorbance as a function of pH can yield the relative concentration of an ionizable size in a molecule from

$$\theta = \frac{A - A_B}{A_{HB} - A_B}.$$ (13)

In the case of identical groups, this relation can be applied for monitoring a protonation reaction in a straightforward fashion. The method is most sensitive at those wavelengths where the variation in the absorbance are the largest (i.e., near the spectral peaks). Again it is essential to determine the absorbance of A_{HB} and A_B accurately; the problems are quite analogous as discussed previously for potentiometric titrations or NMR. An important check of the procedure is to perform the analysis at different wavelengths. Identical functional forms of the degree of protonation as a function of pH must be obtained. An example of such an analysis is shown in Fig. 20, where the degree of protonation is plotted as a function of pH for an indicator, which is incorporated into a monolayer at the air–water interface.[128] These data were obtained by fluorescence spectroscopy at two different wavelengths. The collapse of the data onto a common curve indicated the correct-

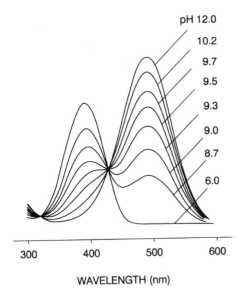

Figure 19. Evolution of the UV/VIS absorption spectrum for different pH values of a 25 µM aqueous solution of 1-hexadecyl-4[(oxocyclohexadienylidene)ethylidene]-1, 4-dihydropyridine, which is a zwitterionic surface active indicator, in a 10 mM solution of *n*-dodecyloctaoxyethylene glycol monoether ($C_{12}E_8$). Adapted from Ref. 37.

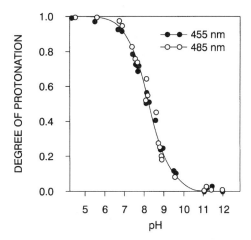

Figure 20. Degree of protonation measured by fluorescence spectroscopy at different emission wavelengths of 4-heptadecyl-7-hydroxycoumarin (HHC) in a methylarachidate monolayer in contact with an aqueous 10 mM sodium phosphate solution.[128] The solid line guides the eye only.

ness of the analysis; only at low pH slight deviations hint at minor problems. If the data do not overlap, the analysis requires a reconsideration as some unidentified species come into play. Such a species would also manifest itself in the absence of a common crossing point.

For a monoprotic acid or base, the same information is, of course, accessible by a classical potentiometric titration experiment. In many situations, however, the spectroscopic method can offer substantial advantages, such as in the measurement of the protonation state of an ionizable monolayer, in investigations of ionization constants of fluorescent indicators in monolayers, micelles, or microemulsion droplets, as well as in the determination of the ionization constants of molecules in the excited state by means of fluorescence.[128–131]

Attenuated total reflection (ATR) techniques now provide excellent sensitivity in the UV/VIS and IR regime even for monolayers.[33,34,88] A somewhat different method for the study of monolayers is second harmonic generation, which directly responds to the surface potential.[132] These reactions would be mostly inaccessible with classical potentiometric titration technique due to the minute differences in the mass balance. IR spectroscopy has always been popular in order to obtain semiquantitative information on the protonation of oxygen atoms.[18,33,34,88,133,134] ATR FT/IR has now become the method of choice for IR work in aqueous media, as it provides sufficiently short path lengths that make it possible to subtract the aqueous background absorption.[135,136]

Direct spectroscopic methods are much more difficult to apply to complex molecules. In marked contrast to NMR, where the chemical shift of a nucleus represents a very local probe, the absorption, vibrational, or fluorescence spectra are determined by the overall electronic structure and are therefore influenced by the molecular properties in a much more global manner. While the observed changes can be sometimes associated with a particular molecular environment, this assignment is often difficult to do.

These difficulties can be illustrated by ATR FT/IR spectra of oxalic acid[137] (see Fig. 21a). While some of the peaks can be assigned to molecular properties, the overall pH dependence is rather complex. In order to extract all three spectra of different oxalate species one has to combine the appropriate mass action and mass balance relationships with deconvolution techniques. The spectra calculated in this way are shown in Fig. 21b. While some characteristic features are obvious, such as the sharp peak at 1320 cm^{-1} originating from the fully deprotonated oxalate ion,

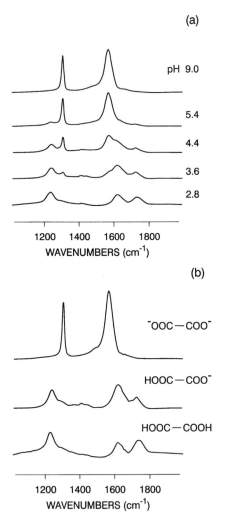

Figure 21. IR spectra of an aqueous 0.1 M solution of oxalic acid. (a) ATR FT/IR spectra at different pH values.[137] (b) The spectra of the three different species obtained from a deconvolution analysis, which also yields 1.52 and 3.84 for the macroscopic pK values.

there are several subtle differences in the spectra of the singly and doubly protonated species. These spectra cannot be easily guessed based on some simplifying assumptions, and therefore it is difficult to obtain additional molecular information, including site-specific titration curves. Similar deconvolution techniques can be used in UV/VIS or fluorescence spectroscopy, but suffer from the same limitations.

Despite these difficulties, molecular information about ionization equilibria can be extracted in certain situations. One relevant application is the selective measurement of the protonation of ionizable aromatic groups in a complex molecular environment by means of the UV/VIS spectroscopy. This method was used in the classical work of Edsall *et al.*[138] to determine the microscopic ionization constants for tyrosine. (A detailed discussion of the microscopic constant will be given in Section 4.) The side chain of the amino acid tyrosine carries a phenolic ring, which can deprotonate. This deprotonation reaction is accompanied by a delocalization in the conjugated electronic π-system in the ring and results in a change of the UV absorption spectrum. However, this π-system is localized to the ring, and is decoupled from the rest of the molecule. Consequently, the UV/VIS spectrum responds selectively to the protonation on the phenolic ring and is independent of the protonation state of the remaining ionizable groups within the molecule. The same type of analysis was used in the classical work of Nozaki and Tanford[139] to probe the protonation of tyrosine side chains within a protein (see Section 5).

In conclusion, IR, UV/VIS, and fluorescence methods represent very valuable and sensitive methods for quantitative measurements of protonation reactions in systems with a single type of ionizable site, such as monolayers, air–water interfaces, and micelles. However, the ability of these techniques to measure site-specific titration curves in complex molecules is limited to rather special situations, and cannot compete with the versatility and resolution of NMR.

3. Modeling of Ionizable Systems

Ionization properties of macromolecules and other colloidal systems such as water–solid interfaces, proteins, and polyelectrolytes have repeatedly been addressed on various levels of approximation. While much progress has been made, we are still far from being able to predict the ionization properties of macromolecules reliably.

3.1. General Considerations

We shall mainly deal with the *high-salt regime*. This regime is attained by adding sufficient amounts of salt to a macromolecular solution. The salt ions screen the long-range Coulomb forces, and macromolecule–macromolecule interactions can be neglected up to rather high concentration of macromolecules. In such solutions, one can focus on the ionization of a *single* macromolecule, which represents a substantial simplification. At low salt concentrations, the counterions neutralizing the charge of the macroions also contribute to the screening, but the diffuse layers originating from different macroions do overlap. In this *low-salt regime*, the counterion concentration as well as macromolecule–macromolecule interactions can no longer be neglected even at very low concentrations of macromolecules.

For most applications, water is the solvent of interest. Aqueous electrolyte solutions are difficult to treat in full atomistic detail. If water molecules are considered explicitly, one usually models the water molecule by a rigid and nondissociating entity.[140,141] Most often, however, the solvent is treated as a dielectric continuum and the molecular nature of the solvent is neglected entirely. The frequently treated *primitive model* only considers salt ions as individual entities, whereby Coulomb interaction and hard-core repulsion is being assumed.[72,142–144] Rigorous treatment of the primitive model can be achieved by means of computer simulations. The well-known Poisson–Boltzmann (PB) and Debye–Hückel (DH) approximations represent a mean-field treatment of the primitive model.

Besides a proper description of the electrolyte solution (solvent), the problem calls for an appropriate treatment of the ionizing system of interest (solute). The latter might be a small molecule, but here we are mainly interested in macromolecules, or colloidal particles. Only recently have we started to approach such problems in their full atomistic detail by means of computer simulations.[10,140,145–147] Usually, however, one resorts to simplified treatments, where one considers a collection of point charges placed on rigid structures, which mimic the charged residues on the macromolecule.[16,61,64,68,148] The final step in the abstraction from the molecular reality is achieved by considering idealized planar and spherical geometries with uniform charge density distributions.[50,149,150] Sometimes, such simplified models capture the ionization process in a surprisingly accurate fashion.

Figure 22 illustrates the commonly used levels of approximation for a protein molecule. A caricature of the all-atom description, where every individual atom of the protein, electrolyte ions, and solvent are treated explicitly, is shown in Fig. 22a. Figure 22b represents the situation where the solvent is replaced by a dielectric continuum, but the salt ions and the protein molecule are still treated explicitly. Figure 22c shows the next level of approximation, where the protein is approximated by a sphere, but the individual ionizable residues are taken into account explicitly—this is the basic picture behind the Kirkwood–Tanford model for proteins.[61] Figure 22d displays the simplest model of a uniformly charged sphere in a continuum electrolyte. The fully atomistic model can only be solved using computer simulations, while the second and third models can also be solved within the mean-field approximation using either the PB or DH equation. The last model is solved trivially within the DH approximation. Clearly, the various models and approximations for the protein and solvent can be chosen independently. For

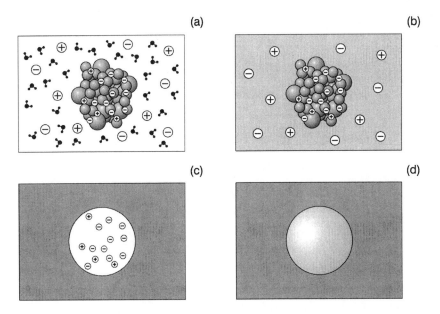

Figure 22. Various levels of description of a protein molecule in an electrolyte solution. (a) All-atom model with balls and sticks, (b) primitive model with an atomistic model of the protein, (c) primitive model description of the electrolyte and discrete point charges for an idealized geometry of the protein, and (d) a smeared-out charge model of the protein.

example, it is customary to consider a molecularly detailed model for the protein molecule in an electrolyte treated within the PB approximation.[16,68,151]

All the mentioned approaches and techniques are applicable to all kinds of charged macromolecules with fixed charged residues (strong polyelectrolytes). Here we shall mainly focus on macromolecules, whose ionization state depends on solution pH through chemical equilibrium reactions (weak polyelectrolytes). For small molecules, the ionization process can be adequately treated in terms of successive chemical equilibria.[4] For macromolecules or any other systems with a larger number of ionizable sites, the ionization process represents a many-body problem, which can be only approached rigorously with statistical mechanics and computer simulation techniques.[19,54,65,152] A mean-field treatment may represent a useful approximation to the solution of the many-body problem.[70,148,150] A typical example is the treatment of ionizable interfaces; the effect of site–site interactions is incorporated into the mean-field electrostatic term.[40,50,69] On the other hand, the mean-field description may break down, as is often observed for certain types of weak polyelectrolytes.[21,25,153] In the following, we shall summarize the important concepts underlying the modeling of such systems.

3.2. Computer Simulation Techniques

Computer simulation has recently emerged as a powerful tool for the modeling of electrolyte solutions. Let us briefly introduce some of the concepts here. For a detailed discussion of computer simulation methods the reader is referred to the literature.[154,155]

The most intuitive way to model such systems is the *molecular dynamics* method. One specifies the interaction potentials between individual atoms and integrates Newton equations of motion of all atoms in a box. If the system is sufficiently large, one expects that the observables are independent of system size and correspond to the thermodynamic limit. For example, the temperature is related to the kinetic energy of the system. Observables of interests can be evaluated from the appropriate time averages of the corresponding quantities.

An alternative way to calculate such averages is by means of *Monte Carlo* simulation. The configurations of the system are generated by changing the translational and internal coordinates by random increments. Such a move is accepted or rejected according to the change in the total energy of the system. If the total energy decreases the move is always accepted, while if the energy increases by ΔE the move is accepted with a probability $e^{-\beta \Delta E}$ where $\beta = 1/(kT)$, with T being the absolute temperature and k the Boltzmann constant. After an initial transient, this sequence of configurations represents realizations of the thermal ensemble of temperature T. Thermal averages can be computed in a straightforward fashion from this sequence. Different strategies for implementing such simulations exist, but the original Metropolis recipe[156] sketched above turns out to be surprisingly efficient.

The Monte Carlo simulation can in principle be performed in any ensemble, that is in systems where various external constraints are being fixed. In the commonly used *canonical ensemble* the number of particles, temperature, and volume are held constant. In the *grand canonical ensemble*, the number of particles is allowed to fluctuate while the chemical potential is kept fixed. To a certain extent, different ensembles can be also treated with the molecular dynamics technique.

Many simulations use different forms of periodic boundary conditions in order to mimic an infinite system. These conditions also imply that the intermolecular potential has to be cut off at some point. The long-range character of the Coulomb interaction requires a special consideration of boundary effects in the calculations. The number of particles in any simulation is limited and, in particular, in charged systems one cannot assume the potential to have decayed to zero at the boundaries. Various techniques are available to minimize the boundary effects, such as the Ewald summation, reaction field method, or different mean field corrections.[154] Such problems are much less severe in simulations with a screened Coulomb potential, which decays rapidly with distance.

Molecular dynamics and Monte Carlo methods yield basically equivalent results. Nevertheless, these methods are complementary. Since the molecular dynamics method relies on the use of Newton equations to simulate the time evolution of the system, the results reflect the actual time-dependent processes. The Monte Carlo method, on the other hand, does not yield actual time-dependent information, but does not require a system whose time-evolution equations are known, and it can be therefore applied to any thermal ensemble. As we shall discuss later, the Monte Carlo method can be also used to simulate protonation and deprotonation reactions of individual binding sites.

3.3. Simple Electrolyte Solutions

An electrolyte solution is typically a three-component system, which consists of the cations, anions, and solvent (water) molecules. While electrolyte solutions are sometimes treated by considering water molecules explicitly, in most cases the solvent is replaced by a structureless, dielectric continuum. Within such a *primitive model*, the salt ions are treated as charged hard spheres.[142-144] The interaction potential between two ions at a distance r is given by the Coulomb potential

$$W(r) = \frac{Q_i Q_j}{4\pi\varepsilon_0 D_w} \frac{1}{r} \tag{14}$$

where Q_i is the charge of ion i, ε_0 is the dielectric permittivity of vacuum, and D_w is the relative permittivity of water. Hard-core repulsion is introduced by assuming that the potential is infinitely large for distances smaller than the sum of the hard-sphere radii. Equation (14) represents a good approximation for the interaction of two ions in water, also at rather high electrolyte concentrations.

The force calculated from the ion–ion potential given by Eq. (14) is not a simple mechanical force but a thermodynamic force. This force can be calculated as the thermal average of the force felt by two ions held at a distance r in water. The potential of this force, which is referred to as the *potential of mean force*, is a free energy since it involves the averaging over solvent degrees of freedom, and depends on temperature and density.

From a molecular point of view, the ions will interact according to the Coulomb potential in vacuum. However, the electric field induced by the ion charges will polarize the individual water molecules and the resulting polarization field will weaken the interaction substantially. This polarization effect is incorporated in Eq. (14) through the relative permittivity of water D_w, and represents, in fact, an excellent approximation. At shorter distances, the molecular structure of the solvent needs to be taken into account and so-called *structural forces* come into play. The resulting interaction potential may even become oscillatory, and is strongly repulsive at short distances.[141,149,157] Within the primitive model this repulsion is mimicked by the hard-core interaction.

3.3.1. *Poisson–Boltzmann (PB) and Debye–Hückel (DH) Approximations*

The simplest treatment of the primitive model of strong electrolytes involves the PB or DH approximation.[74,149,158] The model considers the average concentration profiles of counterions and coions around a selected ion, which is placed at the coordinate origin. The starting point is the Poisson equation, which relates the total charge density $\rho(\mathbf{r})$ at the location $\mathbf{r}$ in space, and the electrostatic potential $\psi(\mathbf{r})$ as

$$\nabla^2 \psi = -\frac{\rho}{\varepsilon_0 D_w} \tag{15}$$

where ∇ is the gradient operator. The electrostatic potential is related to the electric field by $\mathbf{E} = -\nabla\psi$. The DH theory of electrolyte solutions considers the spherically symmetric potential distribution around a selected ion. Other geometries will be considered later.

The charge density ρ has two contributions, $\rho = \rho^{(\text{fix})} + \rho^{(\text{diff})}$, the contribution due to fixed charges $\rho^{(\text{fix})}$ and the diffuse contribution $\rho^{(\text{diff})}$ of the charges originating from the remaining salt ions. Within the DH theory of electrolyte solutions, $\rho^{(\text{fix})}$ gives only a contribution at the coordinate origin, while away from the center only $\rho^{(\text{diff})}$ has to be considered.

For monovalent ions, as will be dealt with here exclusively, we may write the charge distribution around the central ion as

$$\rho(\mathbf{r}) = e[c_+(\mathbf{r}) - c_-(\mathbf{r})] \tag{16}$$

where e is the elementary charge and $c_\pm(\mathbf{r})$ are the concentrations of cations and anions, respectively. The PB approximation can be introduced in a number of different ways.[159] For example, we may assume that the chemical potential, which is constant throughout the sample, can be written as

$$kT \ln c_\pm(\mathbf{r}) \pm e\psi(\mathbf{r}) = \text{const.} \tag{17}$$

This leads to

$$c_\pm(\mathbf{r}) = c_s \exp[\mp\beta e\psi(\mathbf{r})]. \tag{18}$$

Here, c_s is the concentration of salt ions where the electrostatic potential vanishes. (Recall that $\beta = 1/kT$.) Inserting Eq. (18) into Eqs. (15) and (16) leads to the famous PB equation[149]

$$\nabla^2 \psi(\mathbf{r}) = \frac{ec_s}{\varepsilon_0 D_w}\{\exp[\beta e\psi(\mathbf{r})] - \exp[-\beta e\psi(\mathbf{r})]\}. \tag{19}$$

Note that the electrostatic potential $\psi(\mathbf{r})$ entering this equation is an averaged or mean-field potential.

Equation (19) is nonlinear and therefore difficult to solve analytically. Debye and Hückel have shown that for a sufficiently low electrostatic potential this equation can be simplified by invoking the power series $\exp x = 1 + x + \cdots$ in Eq. (19). One obtains the so-called DH equation (or linearized PB equation)

$$\nabla^2 \psi(\mathbf{r}) = \kappa^2 \psi(\mathbf{r}) \tag{20}$$

where κ is the inverse Debye screening length

$$\kappa^2 = \frac{2e^2 \beta c_s}{\varepsilon_0 D_w}. \tag{21}$$

Since the DH equation is linear in the potential, it can be solved more easily than the nonlinear PB equation. The Debye screening length κ^{-1} is the basic length scale. A few representative values for κ^{-1} in water at 25 °C are given in Table 3.

For sufficiently low potentials the DH equation is fully equivalent to the PB equation. The derivation suggests that the potentials must be smaller than $kT/e \approx$ 25 mV. However, comparison of various numerical results indicates that the DH approximation is accurate up to electrostatic potentials, which are comparable to this threshold value. Clearly, substantial deviations may occur for higher potentials. Another criterion for the validity of the DH equation is that the DH screening length is large with respect to the so-called Bjerrum length [$\kappa^{-1} \gg L_B$, where $L_B = \beta e^2/(4\pi\varepsilon_0 D_w) \approx 0.7$ nm]. This condition states that L_B should be small compared to the average ion–ion distances.[56]

Given appropriate boundary conditions, the solution of either the PB or the DH equation determines the electrostatic potential. The corresponding concentration profiles then follow from Eq. (18). Recall that in the DH approximation, the exponentials entering this equation are being linearized.

To obtain the free energy of charging, various procedures have been discussed. It is essential to realize that the free energy of a system described by the PB equation contains energetic and entropic contributions. The energy part contains the Coulombic attractions and repulsions, while the excess entropic part is related to the

Table 3. *Representative Values of the Debye Screening Length κ^{-1} for Different Concentrations of Aqueous Monovalent Electrolytes at 25 °C*

Salt concentration c_s (M)	Screening length κ^{-1} (nm)
10^{-4}	30
10^{-3}	10
10^{-2}	3
10^{-1}	1
1	0.3

nonuniform distribution of the salt ions. For a linear system, the electrostatic energy can be calculated from the electrostatic potential as[160]

$$F = \frac{1}{2} \int \rho(\mathbf{r}) \psi(\mathbf{r}) \, d^3\mathbf{r} \tag{22}$$

where the integration extends over the entire volume of the system. This equation can also be used to evaluate the charging free energy within the DH approximation, but ρ entering Eq. (22) must be interpreted as the charge density of fixed charges.[149,161] The charges originating from the salt ions enter only implicitly through the electrostatic potential.

Due to the nonlinear nature of the PB equation, Eq. (22) is no longer applicable. Among various procedures to evaluate the free energy, the simplest is the continuous charging approach. Thereby, one evaluates the integral[149,161]

$$F = \int\limits_0^1 \int \lambda\rho(\mathbf{r}) \psi(\mathbf{r}, \lambda) d^3\mathbf{r} d\lambda \tag{23}$$

where $\psi(\mathbf{r}, \lambda)$ is the solution of the PB equation for a free charge density $\lambda\rho(\mathbf{r})$ $(0 < \lambda \le 1)$. This prescription is advantageous, if an analytical solution of the PB is available. Integration of the corresponding energy and entropy contributions over the entire space represents one of various other possibilities to calculate the electrostatic free energy.[161–164]

If one considers a uniformly charged object (such as a plane or sphere), it follows from Eq. (23) that the electrostatic free energy can be obtained by simple charging integral, namely

$$F = \int\limits_0^Q \psi_0(Q') dQ' \tag{24}$$

where $\psi_0(Q')$ is the surface potential of the object with total charge Q'. Conversely, the surface potential is given by

$$\psi_0 = \frac{\partial F}{\partial Q} . \tag{25}$$

This relation is important in the mean-field treatment of ionization reactions, and its use will be discussed later.

3.3.2. An Illustrative Example

Let us consider a simple example, which forms the basis of the DH theory of electrolyte solutions.[74,149,158] However, we shall present the argument in a slightly more general form, in order to clarify the concept of the self-energy of a point charge.[160]

A charged sphere is immersed into an aqueous electrolyte solution with a relative permittivity D_w and a Debye screening length κ^{-1}. Due to spherical symmetry, the electrostatic potential only depends on the radial distance r. To solve Eq. (20) we observe that $\nabla^2\psi = r d^2(r\psi)/dr^2$ and obtain

$$\psi(r) = \psi_d a \frac{e^{-\kappa r}}{r} \quad \text{for} \quad a \leq r \tag{26}$$

where $\psi_d = \psi(a)$ is the as yet unknown potential at the sphere surface.

Consider the uniformly charged sphere with a surface charge density σ and no charges inside. In this case, the potential inside the sphere satisfies $\nabla^2\psi = 0$ and equals the surface potential ψ_d throughout. This potential is determined by the requirement that the change in the dielectric displacement across the surface be identical to the surface charge density. This condition leads to

$$\psi(r) = \frac{Q}{4\pi\varepsilon_0 D_w} \frac{e^{\kappa a}}{1 + \kappa a} \frac{e^{-\kappa r}}{r} \quad \text{for} \quad a \leq r \tag{27}$$

where $Q = 4\pi a^2 \sigma$ is the total charge of the sphere. A few quantities of interest are shown in Fig. 23. The charge–potential relationship is obtained by comparing Eqs. (26) and (27). The resulting *total capacitance* of the system is

$$C = Q/\psi_d = 4\pi\varepsilon_0 D_w a(1 + \kappa a). \tag{28}$$

The linearity of the charge–potential relationship reflects the linearity of the DH equation. One often introduces a *specific capacitance* per unit area

$$\tilde{C} = \sigma/\psi_d = \varepsilon_0 D_w(a^{-1} + \kappa) \tag{29}$$

where σ is the surface charge density. The free energy can be evaluated from Eq. (22) and reads

$$F = \frac{Q^2}{8\pi\varepsilon_0 D_w a} - \frac{Q^2}{8\pi\varepsilon_0 D_w} \frac{\kappa}{1 + a\kappa}. \tag{30}$$

The occurrence of two terms on the right hand side can be interpreted as follows. The first term is the free energy of a uniformly charged sphere of radius a in water of relative permittivity D_w without the electrolyte, and represents the Born energy contribution.[8,61,74,149] The second term represents the free energy of this point charge in an electrolyte solution ($\kappa > 0$) relative to water ($\kappa = 0$). The latter term is the DH free energy.[158] The free energy can also be written as

$$F = \frac{Q^2}{2C} \tag{31}$$

where the capacitance is given by Eq. (28). Note that Eq. (31) is generally applicable for any linear system.

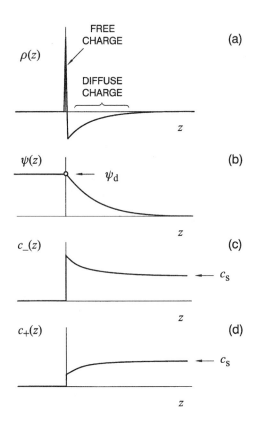

Figure 23. Schematic representation of the Debye–Hückel (DH) solution for a sphere. (a) Charge density, (b) electrostatic potential, (c) concentration profile of counterions, and (d) coions.

The electrical free energy is related to the excess chemical potential $\mu_i^{(ex)}$ of an ion i with respect to the pure solvent.[158] The activity of an ion is

$$a_i = \gamma_i c_i \tag{32}$$

where c_i is the concentration and γ_i the activity coefficient. The excess chemical potential is obtained by taking the difference between Eq. (30) and the same equation for $\kappa = 0$, with the result

$$\mu_i^{(ex)} = kT \ln \gamma_i = -\frac{Z_i^2 e^2}{8\pi\varepsilon_0 D_w} \frac{\kappa}{1 + \kappa a} \tag{33}$$

where Z_i is the charge of the ion i in units of the elementary charge e.

Let us now consider a dielectric spherical cavity of relative permittivity D_d with a point charge of magnitude Q at its center. The potential outside remains identical to Eq. (27), but inside the sphere the electrostatic potential will have the form

$$\psi(\mathbf{r}) = \frac{Q}{4\pi\varepsilon_0 D_d} \frac{1}{r} + \psi' \quad \text{for} \quad 0 < r \le a \tag{34}$$

where ψ' is a constant background potential, to be determined. The first term in Eq. (34) represents the Coulomb contribution from the point charge. Applying the boundary condition of continuity of the electric potential and of the normal component of the dielectric displacement, the background potential can be evaluated. In contrast to the smeared charge problem, the dielectric properties inside the sphere become significant, since the relative permittivity D_d now enters.

The charge–potential relationship remains unchanged and is given by Eq. (28). However, if one applies Eq. (22) to evaluate the electrostatic free energy, the result turns out to be infinite. This divergence originates from the presence of a point charge, and it is absent when the charge is uniformly smeared out. In order to remove this divergence one can consider the difference in free energies between the point charge in the dielectric cavity and an isolated point charge in an infinite dielectric. This procedure eliminates the divergent Coulomb term and isolates the background potential ψ' entering Eq. (34). The result is the so-called *self-energy* of a point charge

$$W^{(s)} = \frac{Q\psi'}{2} = W_0^{(s)} + \Delta W^{(s)}. \tag{35}$$

For later convenience, we have introduced a constant term ($\kappa = 0$)

$$W_0^{(s)} = \frac{Q^2}{8\pi\varepsilon_0 a} \left(\frac{1}{D_w} - \frac{1}{D_d} \right) \tag{36}$$

and the ionic strength dependent term

$$\Delta W^{(s)} = -\frac{Q^2}{8\pi\varepsilon_0 D_w} \frac{\kappa}{1 + a\kappa} \tag{37}$$

which vanishes for zero ionic strength. The constant term $W_0^{(s)}$ can be viewed as the difference of two Born energy terms, while $\Delta W^{(s)}$ gives the standard DH ionic strength dependence.

Solution of the DH equation outside the sphere [cf. Eq. (27)] gives a way of estimating the interaction energy of two ions in an electrolyte solution. Consider two ions i and j with charges Q_i and Q_j at a distance r. If this distance is large compared to the Debye length ($r \gg \kappa^{-1}$) the result can be obtained by means of the *superposition approximation*. If one assumes that the electrostatic potential profile generated by ion i is not perturbed by the presence of the other ion j, one can simply use Eq. (27) to calculate the interaction potential through the charging procedure. In this situation, the interaction potential is the screened Coulomb potential[149]

$$W(r) = \frac{Q_i Q_j}{4\pi\varepsilon_0 D_w} \left(\frac{e^{\kappa a}}{1 + \kappa a} \right)^2 \frac{e^{-\kappa r}}{r}. \tag{38}$$

This expression is valid for sufficiently large distances. For smaller distances, other approximation schemes or numerical solutions of the DH equation must be used.

The interaction potential given by Eq. (38) is also valid for large distances on the PB level. In this case, one only has to reinterpret the charges Q_i as effective charges, which must be evaluated from the solution of the PB equation for an isolated sphere. Such effective charges are typically smaller than the actual charges of the particles.[71,165–168]

3.3.3. Beyond the Poisson–Boltzmann (PB) Approximation

The PB treatment represents a mean-field approximation of the *primitive model* of electrolytes.[72,142–144] The latter model treats ions as charged hard spheres interacting according to the Coulomb potential [cf. Eq. (14)]. The solvent is "primitively" modeled as a structureless, dielectric continuum. This model can be studied by molecular dynamics and Monte Carlo simulations. Due to the long-range Coulomb interactions, such simulations tend to be rather difficult and time-consuming.

Comparison of computer simulation results with DH or PB theory shows that such approximations are surprisingly accurate for monovalent electrolytes and sufficiently low concentrations (say <0.01 M). For higher concentrations, and particularly for divalent electrolytes, the deviations may become substantial.[72,169]

The breakdown of the PB approximation can be ascribed to ionic correlation effects omitted in the mean-field theory but naturally included in, for example, Monte Carlo simulations. Mean-field theories represent a standard approach in the description of interacting systems, and it is well established that in sufficiently strongly coupled systems this approach becomes invalid due to the presence of ion–ion correlations. Since the PB equation assumes that the charge distribution is given by the mean potential, we expect the errors to increase with increasing electrostatic coupling. Strong coupling can be brought about in different ways, for example, by lowering the temperature or the relative permittivity of the solvent, or, more abruptly, by exchanging monovalent with divalent ions.

Due to the mentioned difficulties with computer simulations in charged systems, several analytical theories were proposed in the attempt to overcome the limitations of the PB theory. Two important classes of such theories must be mentioned, namely *integral equations* and *density functional theories*. It must be realized that both approaches are computationally somewhat expensive, albeit, to a much lesser extent than a molecular dynamics or a Monte Carlo simulation.

Integral equations are based on the Ornstein–Zernike equation, which relates the pair correlation function to the so-called direct correlation function.[158] The latter function is then approximated via a *closure relation*. A particularly successful closure for charged systems is the *hypernetted chain approximation*. It has been applied to simple electrolyte solutions by Rasaiah and Friedman[143,170] and compared to results obtained via Monte Carlo simulations.[171] The agreement is usually very good, in particular for 1:1 and 1:2 aqueous electrolytes, and holds to rather high concentrations (say <1 M). However, for 2:2 electrolytes this approach seems to deteriorate.[144]

A criticism raised against integral equation type approximations is the lack of a physical interpretation. Density functional theories were recently proposed as interesting alternatives, where the origin of different approximations is more clearly spelled out.[167,172,173] The DH hole theory due to Nordholm[174] approximates correlation effects by considering the fact that the ionic radial distribution function cannot become negative. Despite the overt simplicity, the free energies obtained with the hole theory are in surprisingly good agreement with more accurate treatments.

The excess chemical potentials of ions, which is equivalent to the activity coefficient, was evaluated on various levels of approximation.[175–180] As a recent example, consider the calculation of the solvation free energy by Monte Carlo simulations for ions on the primitive model level.[175] The results for the excess chemical potential as a function of ionic strength is shown in Fig. 24. Simple DH theory is given for comparison [cf. Eq. (33)]. One observes that DH theory gives an adequate representation for the divalent ion, while it performs much more poorly for the monovalent ion. This behavior can be understood as follows. The absolute error of the DH equation is about the same, since the contributions of ion–ion

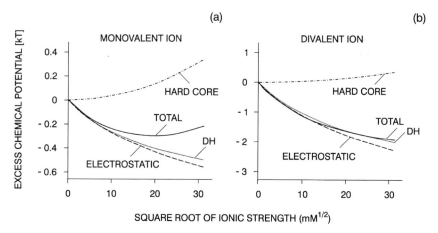

Figure 24. Excess chemical potential of an ion in a monovalent electrolyte solution calculated in the primitive model.[175] Solid lines are Monte Carlo results decomposed into hard-core (dotted) and electrostatic (dash-dotted) contributions. Results of simple DH theory is given for comparison [Eq. (33)]: (a) monovalent and (b) divalent ion.

correlations and hard-sphere interactions are comparable. The very large overall electrostatic free energy for divalent ions leads to a much smaller relative error, and the DH theory performs better in the divalent situation.

3.4. Charged Molecules and Macromolecules in Water

The treatment of ionization processes requires the free energy of a charged macromolecule or colloidal particle (solute) in an electrolyte solution (solvent). Such free energies have contributions originating from solute–solvent interactions as well as internal contributions of the solute, which can be evaluated on various levels of approximation. Analytical solutions are available for the simplest geometries within DH and PB approximations. For many situations, one must resort to numerical solutions.

3.4.1. Debye–Hückel (DH) and Poisson–Boltzmann (PB) Treatment

The DH and PB approximations can be used for various geometries. In most cases, the solute is modeled by a rigid dielectric cavity with a particular charge distribution. The geometry and charge distribution within the cavity is chosen to model the system of interest. The level of molecular detail, which is incorporated into the description, may vary greatly.

Consider the protein molecule shown in Fig. 22. A spherical dielectric cavity with a smeared charge density on its surface represents the simplest model, which is analogous to the model discussed as the illustrative example above, but the radius of the cavity is interpreted as the radius of the protein molecule. A more detailed description can be achieved by still considering a spherical cavity, but now describing the charge distribution as a collection of point charges, whose positions correspond to the actual locations of the charged residues within the protein. In terms of the DH approximation, this description is known as the Tanford–Kirkwood approximation in the protein literature.[61] The most detailed description within the PB approximation can be achieved by assuming a dielectric cavity, whose shape models the molecular shape to a reasonable degree of accuracy.[151] The actual charge distribution is mimicked by a collection of point charges within this cavity; the charges are often chosen to be fractional, in order to represent the potential distributions properly. Sometimes, a *Stern layer* of finite thickness around the cavity is introduced in order to decrease the effective interactions between the surface and the ions at short separations. Note, however, that this layer does not properly introduce the finite size effects of ions.[149] Various numerical approaches have recently been developed to solve the DH and PB equation for such realistic molecular geometries and charge distributions. These techniques have been applied to proteins extensively and will be discussed in Section 5.

Within the DH approximation, the free energy of an arbitrary object with a uniform charge density on the surface is given by Eq. (31). The actual value of the

capacitance depends on the geometry of the system. The free energies of nonuniformly charged objects have been analyzed within the DH approximation, and
analytical results for planar geometries are available.[181–183]

For a collection of point charges, the free energy can be also evaluated
analytically for certain geometries within the DH approximation. The electrostatic
contribution to the free energy always has the generic form[61]

$$F = \frac{1}{2} \sum_{i \neq j} \xi_i \xi_j W(\mathbf{r}_i, \mathbf{r}_j) + \sum_i W_i^{(s)} \xi_i^2 \qquad (39)$$

where ξ_i is the magnitude of the point charge i (in units of elementary charge). The
coefficient $W(\mathbf{r}_i, \mathbf{r}_j)$ represent the interaction energy of two elementary charges i
and j located at $\mathbf{r}_i$ and $\mathbf{r}_j$ and $W_i^{(s)}$ is the self-energy of charge i. For a single site,
only a single self-term in Eq. (39) remains, and we recover Eq. (31). Explicit expres-
sions for $W(\mathbf{r}_i, \mathbf{r}_j)$ and $W_i^{(s)}$ for a sphere, cylinder, and plane are available[181–183] and
summarized in Appendix A. For more complicated geometries, these quantities can
be evaluated numerically.

On the PB level, the free energy becomes a nonlinear function of the charge
density and is dependent on the actual geometry. Even for the simplest situation of
uniform charge density, an analytical solution exists for the planar geometry only.
The standard textbook solution of the PB equation leads to the potential profile[149,184]

$$\psi(z) = \frac{4}{e\beta} \text{arctanh}[\tanh(e\beta\psi_d/4)e^{-\kappa z}] \qquad (40)$$

where z is the distance from the wall. The corresponding charge–potential relationship is given by the Gouy–Chapman equation

$$\sigma = \frac{2\varepsilon_0 D_w \kappa}{e\beta} \sinh(e\beta\psi_d/2) \qquad (41)$$

where σ is the surface charge density and ψ_d the potential at the surface. This
relation can be used to evaluate the free energy from Eq. (24). Integrating the inverse
of Eq. (41) gives an explicit expression for the electrostatic free energy of a charged
plate in an electrolyte solution.[149] An analytical perturbative solution of the PB
equation for nonuniformly charged planes was proposed recently.[185] For a perfect
sphere and cylinder the PB equation must be solved numerically. However, such
calculations can be performed very quickly. Even complicated geometries can
nowadays be treated numerically on this level of approximation.[16,66,186]

3.4.2. High-Salt versus Low-Salt Regime

The interesting regime of low salt concentrations can be also addressed within
the PB approximation. Consider an aqueous suspension of charged macroions,

which might be proteins, colloidal particles, or polyelectrolyte molecules. Each macroion has a total charge Z_{tot} (in units of the elementary charge, $|Z_{tot}| \gg 1$), which is neutralized by counterions. The concentration of these counterions is $c^* = |Z_{tot}|n$, where n is the number concentration of the charged particles. Usually one considers the situation with excess of added salt, which obeys the condition $c^* \ll c_s$, where c_s is the concentration of added salt. By diluting a dispersion of charged particles in this *high-salt regime*, one soon reaches conditions where the macroions no longer interact, and there is no overlap between the diffuse layers of the different macroions. A different situation may occur in the *low-salt regime*, where $c^* \gg c_s$. In this situation the double layers of the macroions do overlap and particle–particle interactions cannot be neglected, not even at low concentrations of macroions. This regime must be treated in a different fashion.

The simplest realization of this problem is the planar geometry, as approximately realized in a lamellar liquid crystal. Two charged plates are considered to interact across an electrolyte solution. The cylindrical and spherical geometry has an interesting interpretation within the so-called *cell model*, illustrated in Fig. 25. Charged colloidal particles are distributed within imaginary cells. The electric field at the cell boundary will be close to zero, and the cells may be considered to be independent to a good degree of accuracy. One would also expect that the precise shape of such a cell does not play a major role, and thus one replaces each cell with an equivalent sphere with the macroion at its center. The radius of the sphere is determined by the concentration of the macroions. A similar interpretation is possible for a polyelectrolyte solution, where the polyelectrolyte chains are stretched out and ordered in a particular direction. Such a case can be pictured as a cut through the solution perpendicular to the orientation axis. In this situation, the cell is assumed to be cylindrical.

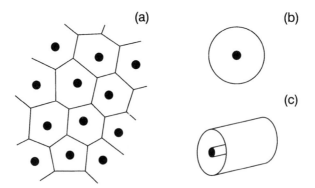

Figure 25.   Illustration of the cell model. (a) The colloidal suspension is subdivided into imaginary cells. These cells are replaced by an idealized geometry. (b) Spheres for spheroidal macroions and (c) cylinders for polyelectrolytes.

The problem of two charged plates interacting across an electrolyte solution has attracted substantial attention, and represents the cornerstone for testing the range of validity of the PB approximation.[149,150] Let us therefore discuss this situation in more detail. Consider two parallel, positively charged surfaces with a charge density σ separated by a distance $2L$. The midplane is taken as the origin ($z = 0$). In this geometry, the PB Eq. (19) can be written as[71,187]

$$\frac{d^2\psi}{dz^2} = \frac{e}{\varepsilon_0 D_w} [c_+^{(0)} \exp(-\beta e\psi) - c_-^{(0)} \exp(+\beta e\psi)] \tag{42}$$

where $c_\pm^{(0)}$ represents the cation and anion concentrations at the midplane ($z = 0$). Note that we have chosen the midplane as the reference point for the electrostatic potential [i.e., $\psi(0) = 0$].

Equation (42) is precisely equivalent to Eq. (19). The equivalence can be established by choosing the reference point for the electrostatic potential in an appropriate fashion, namely by taking $\psi(0) \neq 0$. Within the cell model, the equivalence is readily seen in the high-salt regime, since we have $\psi(0) \simeq 0$ and the ion concentration at the midplane corresponds to the concentration of added salt (i.e., $c_\pm^{(0)} \simeq c_s$). In the low-salt regime, $c_+^{(0)} \neq c_-^{(0)}$ and c_s can be interpreted as the salt concentration in a reservoir, which is in osmotic equilibrium with the planar charged system.

For the case of no added salt, Eq. (42) can be solved analytically. Let us assume the surface to be positively charged. Anions represent the neutralizing counterions and therefore one may set $c_+^{(0)} = 0$. In this particular case, the resulting concentration profile $c(z) = c^{(0)} \exp(\beta e\psi)$ can be evaluated analytically and reads[157,159]

$$c(z) = \frac{2\varepsilon_0 D_w}{\beta e^2 L^2} \frac{s^2}{\cos(sz/L)} \tag{43}$$

where s is a dimensionless quantity which parametrizes the solution. Given this parameter, the surface charge density σ can be evaluated from the equation

$$s \tan s = \frac{\beta \sigma e L}{2\varepsilon_0 D_w} . \tag{44}$$

Conversely, for given surface charge, the parameter s can be obtained by simple iteration. Figure 26 shows the corresponding counterion profile and its comparison with Monte Carlo simulations of the primitive model. One observes that the PB approximation is excellent in this particular case.[157,159]

Other system properties, such as the free energy or the force between the plates, can be obtained within the PB approximation as well. One can show that the force (or the swelling pressure) is given by the simple expression[149,157]

$$\Pi = kTc_-(0). \tag{45}$$

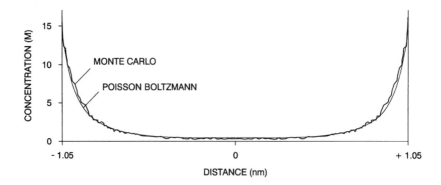

Figure 26.   Concentration profile of counterions between charged plates with a surface charge density of 0.22 C m^{-2} and a distance of 2.1 nm.[157,159] The solid line is obtained from Monte Carlo simulation, and the dotted line from PB theory Eq. (43).

In the no-salt case, the swelling pressure is proportional to the ion concentration at the midplane, which clearly demonstrates that the double layer repulsion is only an entropic force. Note that this relation is only valid within the PB approximation.

Figure 26 indicates that PB theory slightly overestimates the midplane concentration, but the difference is hardly visible on the scale of the graph. Another representation is given in Fig. 27, where the swelling pressure obtained from Monte Carlo simulations is directly compared to the prediction of the PB theory. For monovalent counterions, Fig. 27a shows that the PB approximation is accurate up to charge densities of about 50 mC m^{-2}, and substantial deviations occur for higher charge densities. For divalent counterions, as evident in Fig. 27b, the PB approximation fails except for very low charge density.

3.4.3.   Toward Detailed Molecular Models

PB theory is a mean-field approximation, and Fig. 27 indicates its typical failures. In this theory the force has a purely entropic origin, and is therefore always repulsive. However, there is also an attractive force due to ion–ion correlations. This force is in complete analogy with the quantum mechanical London forces[72]; as a matter of fact they have the same origin. In the monovalent case, the correlation term is relatively small and the interaction between the surfaces remains repulsive. For divalent counterions, the repulsive force component is much smaller and the attractive correlation force starts to dominate. The general rule of thumb is that one should be careful using the PB equation when divalent counterions are present, when the salt concentration is very high (>1 M), or when the surface charge densities are large.

The treatment of this two-plate problem with hypernetted chain closure was pioneered by Kjellander and Marcelja,[188] who solved the corresponding integral

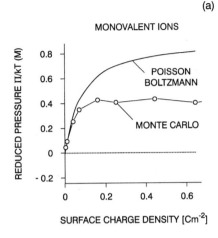

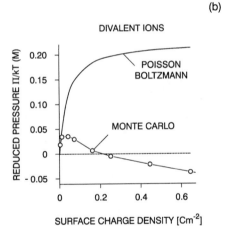

Figure 27. Swelling pressure between charged plates as a function of the surface charge density for a wall–wall separation of 2 nm.[72] Solid line is PB theory [cf. Eqs. (45) and (43)], and points are Monte Carlo simulations. (a) Monovalent counterions and (b) divalent counterions.

equations. While the analysis is rather complicated, the results are generally in very good agreement with Monte Carlo simulation results.[189] A recent extension including *bridge diagrams* has further improved the accuracy of this integral equation approach.[190] More complicated molecular structures can be treated within an integral equation theory on the basis of the reference interaction site model (RISM).[191,192] Integral equation theories have also been applied to study macroparticles in electrolyte solution where water molecules are considered explicitly.[141] The density functional theory represents another interesting alternative to integral equation theories.[167,172]

A fully molecular description of highly charged macromolecules in solution still presents insurmountable problems. The most important one is the size and charge asymmetry. In a computer simulation, the number of macromolecules ought to be on the order of 100; every macromolecule requires a few hundred counterions to be neutralized, in addition to a few thousand water molecules. This setting adds up to $10^5 - 10^6$ particles, which is too much even for the supercomputers of today. A second limiting factor is the flexibility of the macromolecule. For example, a polyacrylic acid chain with 1000 monomers turns out to be very difficult to equilibrate in a simulation. The final problem is that accurate force fields for the macromolecule under consideration are necessary. Substantial progress has been made in developing such force fields for proteins, but much remains to be done.

The simplest approach is to treat the macromolecule as a rigid entity. Such models were used extensively for the description of proteins, but also for some smaller molecules. The detailed geometric structure is usually taken from X-ray crystallography, and to each atom center the appropriate atomic radii are assigned. The positions of the protons are usually generated from known proton–atom distances. Within the resulting rigid body, point charges can be placed in order to mimic charged residues or molecular dipoles. The most popular use of such models is in conjunction with a DH or PB description of the electrolyte.[16,17,61]

If a molecule is not treated as a rigid body, one must specify the internal force field. In other words, one must know the potential energy of the molecule as a function of the internal coordinates. In the most detailed description, all atoms interact with a pairwise or more complicated interaction potential, but treating the chemical bonds as rigid. Such an approximation considers the rotational degrees of freedom of the chemical bonds, and pairwise interaction potentials between chemical groups. Rotational energies are represented by a sinusoidally varying potential energy. An example is shown in Fig. 28, which displays the potential energy of succinic acid as a function of the dihedral angle. The site–site interaction potentials entering the description are usually represented as a superposition of a short-range repulsive potentials and long-range Coulomb-type interaction with properly chosen partial charges of the molecules. Such force fields are largely empirical constructs, but if properly calibrated they are able to represent various molecular properties, such as the various conformations, dipole moments, and infrared spectra. A widely used force field was developed by Karplus and coworkers.[193,194]

As a simplified treatment of the conformational degrees of freedom one can consider a discrete model of the rotational isomerism. Recall the succinic acid as shown in Fig. 28. The potential energy of the internal rotations has three well-defined minima. The lower one corresponds to the *trans* conformer, while the two higher ones represent the two equivalent *gauche* conformers. The force-field picture does not necessarily distinguish between these conformational states; they are implicitly contained in the description. A molecular dynamics simulation will, in principle, sample both of them according to their relative probabilities. Due to the finite simulation times, however, it may be difficult to sample all conformers in practice. Within the Monte Carlo technique, various methods have been developed to circumvent such difficulties. One example is the so-called umbrella sampling,

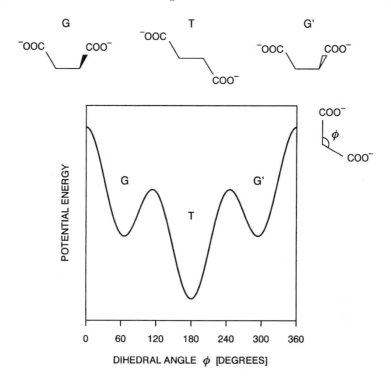

Figure 28.   Schematic representation of the torsional potential energy and the different conformational degrees of freedom for succinic acid.

where one introduces an artificial potential, which partly flattens the potential energy landscape.[155]

Another way to view the conformational degrees of freedom is to consider three different conformational species explicitly. In this case, one has the *trans* conformer (T) and the two *gauche* conformers (G and G'). The relative probabilities of these conformers can be deduced from a chemical equilibrium relation

$$T \rightleftharpoons G \rightleftharpoons G'$$ (46)

where the equilibrium constant for the first reaction $K < 1$, while for the second reaction $K' = 1$ due to symmetry (see Fig. 28).

The description in terms of different conformational states is only meaningful if the energy barriers between the different conformers are high compared to the thermal energy kT. In this case, the dihedral angles attain values, which fluctuate around the corresponding energy minima. In such a situation, individual conformational states are well defined, and the chemical equilibrium constants can be calculated unambiguously. In the reverse case, where the energy barriers are comparable to kT (or even smaller), the various conformers are not well-defined, and such a description does not make sense.

For well-defined conformational states, the internal degrees of freedom can be also modeled as a sequence of discrete states. For polymers, this approach was put forward by Flory.[195] The various conformers can be distinguished by the appropriate discrete rotational states of each bond, and from the corresponding free energy the relative conformer probabilities can be calculated. The approach becomes equivalent to the treatment of a mixture of the various conformers. This treatment of internal rotations does not incorporate the possibility of the overlap of individual segments. In polymer science, this aspect is known as the excluded volume problem.

The essential quantity for the consideration of ionization equilibria is the free energy of a solute. The calculation of this quantity on the DH and PB level is reasonably straightforward. However, its calculation from a computer simulation is more involved. The basic problem is that the free energy cannot simply be written as an average of a thermodynamic quantity, and thus is not directly accessible in a computer simulation.

All methods for the calculation of free energies rely on the simulation of a large number of states along an appropriate thermodynamic path. One possibility is to use the Widom particle insertion method and calculate the free energy of a solute in the canonical ensemble.[196] The Widom method is a general perturbation technique and it can be used to approach various problems within the primitive model.[197] In the traditional Widom method one inserts a fictitious test particle, and calculates the corresponding energy difference. The free energy is then obtained from the expectation value of the Boltzmann factor of this free-energy difference. Charge neutrality has to be obeyed, otherwise one faces convergence problems. A simple charge scaling procedure can correct much of this error.[175,197] This method was used to evaluate the solvation energy of various geometries of the primitive model level. For example, it was shown that a macroion of 1.4 nm in diameter and charge $-4e$ can be also well modeled by DH theory.[175] For a highly charged object, the charging contribution is very large and DH theory is often successful in estimating this contribution. The same effect is evident in Fig. 24.

3.5. Treatment of Ionization Equilibria

The focus of this review is on ionization equilibria involving more than one ionizable site. Nevertheless, for completeness we shall mention some essential aspects of the ionization of molecules involving a single ionizable site.

3.5.1. Single Ionizable Site

The dissociation equilibrium of a monoprotic acid or base is commonly expressed as the dissociation reaction

$$HA \rightleftharpoons A + H \qquad (47)$$

with the mass action law

$$K_{\mathrm{diss}} = \frac{c_A a_H}{c_{HA}} \qquad (48)$$

where K_{diss} is the (mixed) dissociation constant in the presence of salt, a_H is the activity of protons, and the concentration of the species HA and A are denoted by c_{HA} and c_A. The solution pH is given by $pH = -\log_{10} a_H$. One can also consider the association reaction

$$A + H \rightleftharpoons HA \qquad (49)$$

with the mass action law

$$K_{\mathrm{ass}} = \frac{c_{HA}}{c_A a_H} \qquad (50)$$

where K_{ass} is the conditional association constant. The commonly discussed pK value is defined as

$$pK = -\log_{10} K_{\mathrm{diss}} = +\log_{10} K_{\mathrm{ass}}. \qquad (51)$$

In the following, we shall always refer to the association constants and drop the index "ass." The association constant

$$K = \frac{c_{HA}}{c_A a_H} = \frac{Q'_{HA}}{Q_H Q'_A} \qquad (52)$$

can be expressed in terms of canonical partition functions Q'_{HA}, Q'_A, and Q_H of the corresponding species.[54,158] The partition function Q'_i is evaluated in the presence of salt, while Q_H is evaluated in pure water. The partition functions are related to the free energies by

$$F_i = -k_B T \ln Q'_i \qquad (53)$$

for each species i. The same relation applies to the proton, where Q_H has to be used. The calculation of an equilibrium constant is equivalent to the calculation of the free energies of the species HA, A, and H in an electrolyte solution.

These free energies can be easily evaluated in an approximate fashion. A frequently used approach is to split the free energy

$$F_i = F_i^{(\mathrm{ch})} + F_i^{(\mathrm{el})} \qquad (54)$$

into a chemical and electrostatic contribution. The chemical contribution is assumed to originate from short-range interactions, and to be independent of the dielectric environment of the sample. The electrostatic contribution incorporates all long-range electrostatic contributions and can be discussed within a continuum model.

This picture can provide insight concerning pK values of the same group in different media. The DH activity corrections, which lead to a salt dependence of the ionization constants [cf. Eq. (33)], represent a classical example. But more general results can be obtained from similar arguments as well.

Consider the same dissociating group in two different media (I) and (II), and denote the corresponding ionization constants by $pK^{(I)}$ and $pK^{(II)}$. Within the DH regime, the electrostatic free energy has the form

$$F_i^{(el)} = Z_i^2 W^{(s)} \tag{55}$$

where Z_i is the charge of species i in units of elementary charge and $W^{(s)}$ is the electrostatic self-energy of an elementary charge. While this electrostatic contribution will be different in each medium, we assume that the chemical contribution $F_i^{(ch)}$ remains the same. Denoting the charge of the basic species A as Z, the charge of the acidic species HA becomes $Z + 1$. Using Eqs. (52) and (54) we find

$$pK^{(II)} = pK^{(I)} - (1 + 2Z) \frac{\beta(W^{(s,II)} - W^{(s,I)})}{\ln 10} \tag{56}$$

where $W^{(s,I)}$ and $W^{(s,II)}$ are the electrostatic self-energies of an elementary charge in medium (I) and (II). Introducing the constant and ionic strength dependent terms [cf. Eq. (35)] we can also write the above equation as

$$pK = pK^{(0)} - (1 + 2Z) \frac{\beta \Delta W^{(s)}}{\ln 10} \cdot \tag{57}$$

The constant $pK^{(0)}$ corresponds to the ionization constant extrapolated to zero ionic strength, while the $\Delta W_i^{(s)}$ term describes the ionic strength dependence. This equation applies to both media I and II. When going from medium (I) to medium (II), the shift in the zero ionic strength constant can be written as

$$pK^{(0,II)} = pK^{(0,I)} - (1 + 2Z) \frac{\beta(W_0^{(s,II)} - W_0^{(s,I)})}{\ln 10} \tag{58}$$

where $W_0^{(s,I)}$ and $W_0^{(s,II)}$ refer to the self-energies at zero ionic strength in the corresponding media (i.e., $\kappa = 0$). Applications of Eqs. (57) and (58) will be discussed later. More general expressions of this kind were recently derived from cavity function theory.[198]

For the calculation of the free energies, one usually does not treat the problem on an all-atom level but rather investigates the individual species separately. If all interaction potentials are properly specified on an all-atom level, the formation of different species is taken into account implicitly. Different species do not enter the description, but they rather arise as different configurations of the same system.

However, the energy barriers separating the different energy wells (corresponding to individual species) are high, and therefore a full computer simulation of a chemical equilibrium is computationally too expensive. For this reason, one focusses on the simulation of different species, which are defined, by specifying fixed bond lengths, or by constraining the coordinates of the individual atoms.[140]

The calculation of the free energies of individual species then follows the same principles as discussed above. An additional problem is that the equilibrium constant is related to the free-energy difference of the acid and base species. Typically, these contributions are very large and almost cancel each other. In order to obtain just modestly accurate estimates of equilibrium constants, one requires extremely accurate values for the free energies. This requirement makes these calculations rather difficult.[145–147]

3.5.2. Localized versus Delocalized Binding

Proton binding appears to be localized. Each proton is localized within a particular energy well and bound to a single well-defined site. Once this site is occupied, no further protons can be bound. (Exceptions to this rule are rare, but will be mentioned in Section 4.) For a protein, for example, one can clearly specify whether a given ionizable amino acid residue is protonated or not. In other words, the effective interaction potential of the deprotonated solute with a proton will have several well-defined minima near each binding site. Upon protonation of a given site, the corresponding minimum will disappear.

This concept of *localized binding* only makes sense if the corresponding potential energy minima near each site are sufficiently deep compared to the thermal energy kT, such that the protons feel well-defined binding sites on the surface. If the depth of any of these minima is comparable to or smaller than kT, the concept of localized binding is no longer appropriate. In this situation, the overall binding process may still be well-defined, since the adsorbing species may experience a deep attractive potential well in one direction. However, there may be no well-defined potential energy minima present in the lateral direction. In such a case, one refers to *delocalized binding*. A typical example of delocalized binding is the adsorption of surfactant molecules on an air–water interface. The number of adsorbed surfactants on the interface is a well-defined quantity, but the surfactant is free to move on the surface and does not bind any binding sites on the surface. However, the picture of delocalized binding does not seem to be applicable to protons.

3.5.3. Macroscopic Description

Usually one distinguishes various species H_nA according to the number of protons bound $n = 0, 1, 2, \ldots, N$. The macroscopic description is equally applicable for localized as well as delocalized binding. This description does not distinguish the locations of the bound protons, but only the number of bound protons is relevant.

The notion of such macrospecies is perfectly well-defined on a statistical mechanical basis, but defined poorly from a molecular point of view. A macrospecies actually represents a mixture of various species in various protonation states (microspecies, see below), all of which have the same number of bound protons. The information about the locations of the protons has already been averaged out. The definition of a macrospecies is similar to a molecule which has different conformational states (see above). In this case, one usually does not explicitly refer to the various conformers, although the different conformational states may be well-defined, distinct species.[4,54,199]

The ionization problem can now be treated as one of a mixture of N different macrospecies H_nA. However, the concentrations (or probabilities) of these different species are not independent. While the number of solute molecules is fixed, the solvent represents a large reservoir of protons at a fixed chemical potential

$$\mu_H = \mu_H^{(0)} + kT \ln a_H \tag{59}$$

which is related to the solution pH. This situation is referred to as a *semi-grand canonical ensemble*. In the grand-canonical ensemble the chemical potential of all components is fixed, but in the present situation only the chemical potential of the protons is a fixed quantity.

The treatment simplifies considerably if a dilute solution is considered, and if macromolecule–macromolecule interactions can be ignored. Rigorous formulation calls for a statistical mechanical treatment, which was developed most clearly by Hill.[54] Here, we shall only give a few heuristic arguments; details can be found elsewhere.[54] The chemical equilibrium reaction reads

$$A + nH \rightleftharpoons H_nA \tag{60}$$

with the mass action law

$$\frac{c_{H_nA}}{c_A a_H^n} = \overline{K}_n = \frac{Q'_{H_nA}}{Q_H^n Q'_A} \tag{61}$$

where a_H is the activity of protons while c_{H_nA} and Q'_{H_nA} are the concentrations and canonical partition functions of species H_nA, respectively. The partition functions are evaluated in the presence of salt, the partition function of the protons Q_H without the salt. The probability of finding a species H_nA is proportional to its concentration. Based on Eq. (61) this probability is

$$P_n \propto c_{H_nA} \propto a_H^n \frac{Q'_{H_nA}}{Q_H^n Q'_A}. \tag{62}$$

We can introduce a formation free energy of the macrospecies H_nA by writing

$$P_n = \Xi^{-1} e^{-\beta F_n} \tag{63}$$

where Ξ is a normalization constant and the formation free energy is defined as

$$F_n = -kT \ln(Q'_{H_nA}/Q'_A) + nkT \ln(a_H/Q_H). \tag{64}$$

This free energy is defined with respect to the fully deprotonated species ($F_0 = 0$). The normalization constant Ξ can be interpreted as a partition function and is given by

$$\Xi = \sum_{n=0}^{N} e^{-\beta F_n} = \sum_{n=0}^{N} a_H^n \overline{K}_n \tag{65}$$

where $\overline{K}_n$ is the macroscopic association constant with $\overline{K}_0 = 1$. Equation (65) implies the relation $F_n = -kT \ln(a_H^n \overline{K}_n)$. The right hand side of Eq. (65) is known as the *binding polynomial*.[200] The titration curve, which is related to the thermal average of n, can be written as

$$\theta = \frac{\langle n \rangle}{N} = \frac{1}{N} \sum_{n=0}^{N} n P_n \tag{66}$$

where θ is the average degree of protonation. The analysis of Eqs. (64) and (66) shows that

$$\theta = \frac{a_H}{N} \frac{\partial \ln \Xi}{\partial a_H}. \tag{67}$$

The average degree of protonation can be written as the derivative of the partition function. By inserting Eq. (65) into Eq. (67) one obtains the classical expression for the titration curve of the polyprotic acid or base. This analysis leads to the known expressions for the macrostate probabilities[4]

$$P_n = \frac{\overline{K}_n a_H^n}{\sum_{n=0}^{N} \overline{K}_n a_H^n} \tag{68}$$

and for the overall titration curve

$$\theta = \frac{1}{N} \frac{\sum_{n=0}^{N} n \overline{K}_n a_H^n}{\sum_{n=0}^{N} \overline{K}_n a_H^n}. \tag{69}$$

3.5.4. Microscopic Description

While popular, the macroscopic description is not an intuitive way to describe binding of protons to macromolecules. As discussed above, proton binding is localized and thus one can specify the protonation state of each ionizable site. This observation is the basis of the microscopic description. Each of these protonation states represents a different species, the so-called *microspecies*.[4] The various protonation states can be specified as follows. Consider a molecule with N binding sites. The protonation state of site i ($i = 1, 2, \ldots, N$) can be characterized by a two-valued state variable s_i such that $s_i = 1$ if the site is protonated and $s_i = 0$ if the site is deprotonated. The protonation microstate of a molecule is then specified by the set of state variables $\{s_1, s_2, \ldots, s_N\}$ abbreviated as $\{s_i\}$. The number of such states is 2^N, a quantity which grows very rapidly as the number of sites increases.

The ionization problem is now reduced to the problem of treating a mixture of 2^N different species $A\{s_i\}$. Since the chemical potential of the protons is fixed, their concentrations (or probabilities) are not independent. Focussing on a dilute solution, we now repeat the analogous arguments, which were presented for the macroscopic description, for the microscopic association equilibrium of the species $A\{s_i\}$, where this notation suppresses the total number of bound protons. A more rigorous treatment requires a statistical mechanical formulation.[54] The chemical equilibrium reactions read

$$A\{0\} + nH \rightleftharpoons A\{s_i\} \tag{70}$$

where $A\{0\}$ represents the fully deprotonated species. The number of bound protons n is now already specified by

$$n = \sum_{i=1}^{N} s_i. \tag{71}$$

The mass action law of Eq. (70) is

$$\frac{c_{A\{s_i\}}}{c_{A\{0\}} a_H^n} = \frac{Q'_{A\{s_i\}}}{Q_H^n Q'_{A\{0\}}} \tag{72}$$

where $c_{A\{s_i\}}$ and $Q'_{A\{s_i\}}$ are the concentrations and canonical partition functions of species $A\{s_i\}$, respectively. The latter are evaluated in the presence of salt. The probability of a species $A\{s_i\}$ is proportional to its concentration. Based on Eq. (72) we find this probability to be

$$p(\{s_i\}) \propto c_{A\{s_i\}} \propto a_H^n \frac{Q'_{A\{s_i\}}}{Q_H^n Q'_{A\{0\}}}. \tag{73}$$

Again, a formation free energy of the microspecies $A\{s_i\}$ can be introduced,

$$p(\{s_i\}) = \Xi^{-1} e^{-\beta F(\{s_i\})} \tag{74}$$

where Ξ is the normalization constant introduced above and the formation free energy is defined as

$$F(\{s_i\}) = -kT \ln(Q'_{A\{s_i\}}/Q'_{A\{0\}}) + nkT \ln(a_H/Q_H). \tag{75}$$

As above, the fully deprotonated species is the reference point ($F(\{0\}) = 0$). The normalization constant must be now given by

$$\Xi = \sum_{\{s_i\}} e^{-\beta F(\{s_i\})} \tag{76}$$

which is similar to the expression given previously.

The macroscopic and microscopic pictures must, of course, be equivalent. The equivalence can be established by comparing Eqs. (65) and (76). Collecting coefficients we find

$$e^{-\beta F_n} = \sum_{\{s_i\}} e^{-\beta F(\{s_i\})} \, \delta_{n, \sum_j s_j} \tag{77}$$

where $\delta_{i,j}$ is the Kronecker delta ($\delta_{i,j} = 1$ for $i = j$ and $\delta_{i,j} = 0$ for $i \neq j$). The free energy of the macrospecies is related to the sum over all microspecies, which have the same number of bound protons. These relations were known to Simms[201] already a long time ago. One can show that the probability of a given microstate is

$$p(\{s_i\}) = \pi_n(\{s_i\})P_n(a_H) \tag{78}$$

where P_n is the macrostate probability and

$$\pi_n(\{s_i\}) = e^{-\beta F(\{s_i\})} \left[\sum_{\{s_i\}} e^{-\beta F(\{s_i\})} \, \delta_{n, \sum_j s_j} \right]^{-1} \tag{79}$$

is the probability to find a particular microstate within a given macrostate. This quantity is pH-independent and represents the mole fraction of a microspecies with the same number of protons bound.

Knowing the probabilities of various microspecies $A\{s_i\}$, all properties of the ionizing molecules can be evaluated. For example, the protonation curves of individual sites are given by

$$\theta_m = \sum_{\{s_i\}} s_m \, p(\{s_i\}) \tag{80}$$

which can be simplified using Eq. (78) to

$$\theta_m = \sum_{n=0}^{N} A_{mn} P_n(z) \tag{81}$$

where the pH-independent coefficients are given by

$$A_{mn} = \sum_{\{s_i\}} s_m \pi(\{s_i\}) \delta_{n, \sum_j s_j}. \tag{82}$$

The average of the protonation curves of individual sites leads to the average degree of protonation

$$\theta = \frac{1}{N} \sum_{m=1}^{N} \theta_m. \tag{83}$$

Introducing the formation free energy $F(\{s_i\})$ may not seem much of an advantage. This function basically represents a table of the free energies for all the possible microstates, a description which contains an enormous number of parameters for a larger number of sites. However, this free energy can be either approximated in a systematic way, or calculated directly.

A systematic approximation scheme for the free energy $F(\{s_i\})$, which is a function of discrete variables, can be obtained by expanding this quantity in a similar way as a function of continuous variables is expanded in a Taylor series. As shown in Appendix B, this free energy can be written in its general form as

$$F(\{s_i\}) = -\sum_i \mu_i s_i + \sum_{i<j} E_{ij} s_i s_j + \sum_{i<j<k} L_{ijk} s_i s_j s_k + \cdots \tag{84}$$

where the linear coefficients μ_i are effective chemical potentials, while the higher coefficients E_{ij} and L_{ijk} are pair and triplet interaction energies. All summation indices run over all sites $(i, j, k = 1, 2, \ldots, N)$. Only the interaction energies of order larger than N vanish, but in practice the series given in Eq. (84) often converges rapidly, such that the triplet and higher-order contributions can be neglected. Molecular symmetry permits one to reduce the number of coefficients entering Eq. (84). This expansion is also called the *cluster expansion*, since the corresponding terms correspond to clusters of sites. The first term represents the individual sites, the second term all pair of sites, the third triplets, etc.

The effective chemical potentials entering Eq. (84) can be expressed as

$$\frac{\beta \mu_i}{\ln 10} = p\hat{K}_i - pH \tag{85}$$

where $p\hat{K}_i$ are the microscopic pK values of the ionizable groups given that all other groups are deprotonated, while $pH = -\log_{10} a_H$ has its usual meaning. We also introduce dimensionless interaction parameters for pairs

$$\varepsilon_{ij} = \frac{\beta E_{ij}}{\ln 10}, \tag{86}$$

and triplets

$$\lambda_{ijk} = \frac{\beta L_{ijk}}{\ln 10}.$$ (87)

These coefficients are symmetrical, $E_{ij} = E_{ji}$, and without loss of generality one may set $E_{ii} = 0$. Similarly, $L_{ijk} = L_{jki} = L_{kij}$ and $L_{iij} = 0$.

An example of the usefulness of such a cluster expansion is the calculation of the *microscopic* pK values.[7] These ionization constants commonly refer to the protonation reaction, where one particular site is protonated. If we label this site with k, the association equilibrium can be written as

$$A\{s_i\} + H \rightleftharpoons A\{s'_i\}$$ (88)

where $s_i = s'_i$ for all $i \neq k$ but $s_k = 0$ and $s'_k = 1$. Using the free-energy Eq. (84), the microscopic pK value for Eq. (88) follows[202]

$$p\hat{K}_{A\{s_i\}} = p\hat{K}_i - \sum_j \varepsilon_{ij} s_j - \frac{1}{2} \sum_{j,k} \lambda_{ijk} s_j s_k - \ldots$$ (89)

Neglecting triplet contributions, this relation reflects the group additivity concept for the estimation of pK values, which is very popular in organic chemistry.[6] Once the interaction parameters are determined, all microscopic pK values can be calculated.

Another possibility to approach the ionization problem is to use a model to evaluate the free energy explicitly. For example, one may calculate the electrostatic free energy for each microstate and perform the necessary thermal averages using a Monte Carlo simulation procedure.[65,67,70,203] Taking the simple discrete charge model introduced above, the free energy can be written as

$$F(\{s_i\}) = -\sum_i \mu'_i s_i + F^{(el)}(\{s_i\})$$ (90)

where the effective chemical potential μ'_i refers to the fully uncharged state. Again the free energy of the macromolecule is written as a sum of two terms. The first term is a chemical contribution, which treats the sites as independent and considers only the short-range interactions. The second term is the electrostatic term, which corresponds to the electrostatic free energy of the macromolecule in the protonation state $\{s_i\}$.

On the DH level, the free energy of an assembly of point charges with charge of magnitude ξ_i is given by Eq. (39). If these sites are considered as the ionizable sites, the charge of site i is related to the state variable

$$\xi_i = s_i + Z_i$$ (91)

where Z_i is the charge of the site i in its deprotonated state. Inserting Eq. (91) into Eq. (39), we obtain

$$F(\{s_i\}) = -\sum_i \mu_i s_i + \sum_{i<j} E_{ij} s_i s_j \tag{92}$$

where the pair interaction energies are given by electrostatic energies

$$E_{ij} = W(\mathbf{r}_i, \mathbf{r}_j) \tag{93}$$

which is the interaction energy of two elementary charges i and j located at $\mathbf{r}_i$ and $\mathbf{r}_j$. The self-energy terms are absorbed in the chemical potential contributions, and one recovers Eq. (56).

Such problems can be treated analytically by direct enumeration, mean-field approximations, or numerically by Monte Carlo simulations. Direct enumeration yields exact results, but can only be applied to rather small systems (say $N <$ 30).[16,202] The Monte Carlo procedure starts with a particular microstate $\{s_i\}$. One picks a site at random, and then an attempt is made to change its protonation state. This move is accepted according to the Metropolis sampling scheme using the effective free energy $F(\{s_i\})$. Note that this free energy depends on the chemical potential of the proton (solution pH) and thus the number of bound protons fluctuates. This situation represents the semi-grand canonical ensemble. Monte Carlo simulation techniques can be used to calculate titration curves and other properties of the titratable molecules.[65,67,70,203]

3.5.5. Adding Internal Degrees of Freedom

The free energies (or canonical partition functions) for each microstate introduced above involve averaging over all internal degrees of freedom, which include, for example, all conformational degrees of freedom. If there is only one important conformational state, then only this particular state will contribute and the free energy can be evaluated for a fixed configuration from the electrostatic free energy to a good degree of approximation.

If there are different conformers possible, various conformers will contribute to each microstate. Let us label all the conformers with an index α. The free energy of a microstate can be decomposed as[140,199]

$$e^{-\beta F(\{s_i\})} = \sum_\alpha e^{-\beta F(\{s_i\} \mid \alpha)} \tag{94}$$

where $F(\{s_i\}|\alpha)$ is the free energy of the microstate $\{s_i\}$ and conformation α. The reference point for the free energy is taken for the fully deprotonated state and for the conformer with the lowest free energy. In analogy to Eq. (78), the probability to find a given microstate in a particular conformation can be written as

$$p(\{s_i\}|\alpha) = \pi_n(\{s_i\}|\alpha)P_n(a_{\mathrm{H}}) \tag{95}$$

where the relative probabilities to find a microstate in a particular conformation are pH-independent and read

$$\pi_n(\{s_i\} \mid \alpha) = e^{-\beta F(\{s_i\} \mid \alpha)} \left[\sum_{\{s_i\}} \sum_{\alpha} e^{-\beta F(\{s_i\} \mid \alpha)} \delta_{n, \sum_j s_j} \right]^{-1}. \qquad (96)$$

When these quantities are summed over all conformations

$$\pi_n(\{s_i\}) = \sum_{\alpha} \pi_n(\{s_i\} \mid \alpha) \qquad (97)$$

the relative probabilities of individual microstates are recovered. This approach can be used if the various conformers can be enumerated in a systematic fashion, and the free energy is specified accordingly.

Such problems can again be approached efficiently with the Monte Carlo simulation technique. The simulation is carried out as described above for the microstates, but now also the conformations must be moved at random and included in the free energy of the system. In such a simulation, any conformational degrees of freedom can be considered, for example, also the flexibility of polyelectrolyte chain. Such techniques were used either within the Flory framework of discrete rotational states[24] or within a fully flexible chain model.[203]

4. Small Molecules

As long as the number of ionizable sites of a molecule is small, its ionization behavior depends strongly on this number. The behavior of monoprotic acids and bases is discussed in all chemistry textbooks. Polyprotic acids and bases are usually discussed as well, but the essential distinction between the *macroscopic* and *microscopic* description is often ignored. We shall focus particularly on polyprotic molecules.

4.1. Monoprotic Acids and Bases

Let us start by briefly recalling the basic notions. Details are given in standard chemistry textbooks.[4,7,29,38,76]

4.1.1. Equilibrium Constants

The reaction between a monoprotic acid HA and the corresponding base A reads

$$H + A \rightleftharpoons HA \qquad (98)$$

and is described by the mass action law

$$K_{\text{ass}}^{(0)} = \frac{a_{\text{HA}}}{a_{\text{A}} a_{\text{H}}} \tag{99}$$

where $K_{\text{ass}}^{(0)}$ is the association or formation constant while a_{HA}, a_{A}, and a_{H} denote the activities of the acid, base, and protons in M (i.e., relative to a standard state of 1 M). The reference state is taken at infinite dilution in pure water. The corresponding dissociation reaction

$$\text{HA} \rightleftharpoons \text{H} + \text{A} \tag{100}$$

has a mass action law

$$K_{\text{diss}}^{(0)} = 1/K_{\text{ass}}^{(0)} = \frac{a_{\text{A}} a_{\text{H}}}{a_{\text{HA}}} \tag{101}$$

where $K_{\text{diss}}^{(0)}$ is the acid dissociation constant. The abbreviations

$$\text{pH} = - \log_{10} a_{\text{H}} \tag{102}$$

and

$$\text{p}K^{(0)} = - \log_{10} K_{\text{diss}}^{(0)} = + \log_{10} K_{\text{ass}}^{(0)} \tag{103}$$

are usually introduced. Whenever pH or pK is used, we always mean quantities defined according to the above equations. While pX is sometimes considered to be equivalent to $-\log_{10} X$, we shall simply define pH and pK exclusively according to Eqs. (102) and (103). This definition reflects the conventional meaning of "pH value" and "pK value."

For a dilute solution of the acid or base in an indifferent monovalent electrolyte, the overall ionic strength of the solution is determined by the concentration of the added salt. (We focus only on the high-salt regime in this section.) The activity of species i is $a_i = \gamma_i c_i$, where γ_i is the activity coefficient and c_i the concentration. As the activity coefficients are effectively constant for fixed ionic strength, they can be absorbed into a conditional (mixed) equilibrium constant, which can be written as

$$K_{\text{ass}} = K_{\text{ass}}^{(0)} \frac{\gamma_{\text{A}}}{\gamma_{\text{HA}}} = \frac{c_{\text{HA}}}{c_{\text{A}} a_{\text{H}}}. \tag{104}$$

The conditional association constant K_{ass} now depends on the ionic strength and is commonly expressed as

$$\text{p}K = - \log_{10} K_{\text{diss}} = + \log_{10} K_{\text{ass}} \tag{105}$$

where, as above, $K_{\text{diss}} = 1/K_{\text{ass}}$. In the following we shall only refer to association constants, and thus drop the index "ass" and always make use of the identity

$$\text{p}K = + \log_{10} K. \tag{106}$$

The activity coefficients are often estimated from the Debye–Hückel (DH) theory, or extensions thereof. Using the corresponding expression for the activity coefficient Eq. (33) in Eq. (104), we obtain the corrections of the pK value of a monoprotic acid or base due to a finite salt concentration

$$pK = pK^{(0)} + (1 + 2Z)\frac{\beta e^2}{8\pi\varepsilon_0 D_w \ln 10}\frac{\kappa}{1 + a\kappa} \qquad (107)$$

where Z is the charge of the corresponding base in units of elementary charge e, ε_0 the dielectric permittivity of vacuum, D_w the relative permittivity of water, a the ionic radius, and κ^{-1} the Debye length [cf. Eq. (21)]. Equation (107) predicts the pK value for a base ($Z = 0$) to increase with increasing ionic strength, while for an acid ($Z = -1$) the pK decreases with increasing ionic strength. The above equation can be also viewed as the ionic strength dependent term of the self-energy for a point charge in a dielectric sphere, and the result can be also obtained by inserting Eq. (37) into Eq. (57).

Equation (107) is correct at low salt concentrations (say <10 mM); at higher concentrations deviations can be observed. However, this relation is important for the determination of the ionization constant p$K^{(0)}$. Since the inverse Debye length κ^{-1} is proportional to the square root of the ionic strength [cf. Eq. (21)], one plots the experimental pK values as a function of the latter variable and extrapolates to zero.

4.1.2. Titration Behavior

As discussed in Section 2, in a potentiometric titration experiment one can directly measure the titration curve, or equivalently the proton adsorption isotherm. The degree of protonation θ is plotted as a function of pH (or the proton activity a_H). One may also consider various other quantities, such as the number of bound protons $N\theta$ or the total molecular charge $N\theta - Z_{tot}$, where Z_{tot} is the total charge of the fully deprotonated form. All these quantities are linear functions of the degree of protonation θ.

For a monoprotic acid or base, the degree of protonation is given by

$$\theta = \frac{c_{HA}}{c_A + c_{HA}} = \frac{Ka_H}{1 + Ka_H} \qquad (108)$$

which is equivalent to the Langmuir adsorption isotherm.[184] If θ is plotted as a function of pH one obtains the familiar sigmoidal curve shown in Fig. 29a. Often, Eq. (108) is written as the Henderson–Hasselbalch equation[204,205]

$$pH = pK + \log_{10}\frac{1 - \theta}{\theta} \qquad (109)$$

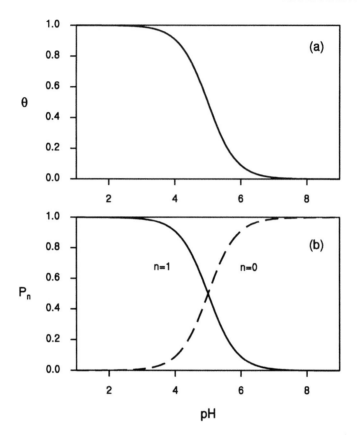

Figure 29.   Titration behavior of a monoprotic acid or base as a function of pH with $pK = 5$. (a) Titration curve represents the degree of protonation θ and (b) prevalence diagram the macrostate probabilities P_0 and P_1.

where we use the definition in Eq. (105). Also of interest are the relative mole fractions (or probabilities) of individual species. The species A binds no protons,

$$P_0 = \frac{c_A}{c_A + c_{HA}} = \frac{1}{1 + Ka_H} \qquad (110)$$

while the species HA binds one proton,

$$P_1 = \frac{c_{HA}}{c_A + c_{HA}} = \frac{Ka_H}{1 + Ka_H} \cdot \qquad (111)$$

These two quantities are shown in Fig. 29b. Clearly, $P_0 + P_1 = 1$ and P_1 is of course equivalent to the titration curve, as the concentration of the species HA corresponds to the total number of protons bound ($\theta = P_1$).

4.1.3. Experimental Data

A wealth of information on ionization behavior of monoprotic acids and bases has been assembled. This review does not attempt to cover this vast field, and we refer to comprehensive summaries in chemistry textbooks, and to detailed discussions and compilations.[6,7,38,206–213]

Conditional constants of monoprotic acids and bases are typically determined by potentiometric titration experiments. Such ionization constants depend on ionic strength and, at higher concentrations, on the type of salt used. A typical result of such an experiment is shown in Fig. 30. At low concentrations the simple DH law in Eq. (107) provides an adequate description. The pK value initially decreases with increasing ionic strengths, and the reverse trend is observed for a weak base. At higher concentrations substantial deviations become apparent, which further depend on the type of counterions. The DH law can be extended to higher concentrations, either in an empirical fashion such as the semiempirical Davies equation or by the more rigorous expressions due to Pitzer.[5,29,74,179,180] In Section 3 we have seen that a similar dependence is found from the primitive model, where the upturn

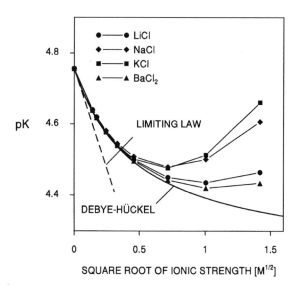

Figure 30. Dependence of pK values of acetic acid in different media on ionic strength.[74] The solid line is the Debye–Hückel (DH) result [cf. Eq. (107)] with an effective radius $a \approx 0.65$ nm; the dashed line represents the square-root limiting law.

is related to hard-sphere interactions (cf. Fig. 24). The ionic strength dependence of ionization constants was recently compared with these expressions.[214]

The ionization constants are usually reported at infinite dilution, but the ionic strength effects are not excessive. A summary of pK value ranges for common functional groups is given in Table 4. For more extensive tables the reader is referred to the literature.[8,207–212]

Table 4 also includes values outside the accessible window of pK values in water. This window is about 2–12. The pH scale can be extended by determination of pK values in solvents other than water.[75,206] Ionization constants can be determined in these solvents as well, for example by using spectroscopic methods, and various procedures were devised in order to transform these measurements back to the pH scale in water. Here we shall not discuss these developments; details are given in standard texts.[4,38,206,215]

A certain ambiguity in the definition of a *monoprotic* acid or base persists. Consider an acetate ion, for example, for which two protonation steps can be considered

$$CH_3 - COO^- + H^+ \rightleftharpoons CH_3 - COOH \quad pK \approx 5 \tag{112}$$

$$CH_3 - COOH + H^+ \rightleftharpoons CH_3 - COOH_2^+ \quad pK \approx -6. \tag{113}$$

While acetate could be considered as a diprotic acid, the pK values of these ionization reactions are very far apart (about 11 orders of magnitude in this case),

Table 4. *Summary of Representative pK Values of Common Organic Functional Groups*[a]

Reaction	pK
$R–COOH + H^+ \rightleftharpoons R–COOH_2^+$	–6
$R–O–R + H^+ \rightleftharpoons R–OH^+–R$	–4
$R–CH_2OH + H^+ \rightleftharpoons R–CH_2OH_2^+$	–2
$\phi–NH_2 + H^+ \rightleftharpoons \phi–NH_3^+$	3–5
$R–COO^- + H^+ \rightleftharpoons R–COOH$	4–5
$\phi–S^- + H^+ \rightleftharpoons \phi–SH$	6–8
$\phi–O^- + H^+ \rightleftharpoons \phi–OH$	8–11
$R–NH_2 + H^+ \rightleftharpoons R–NH_3^+$	10–11
$R–S^- + H^+ \rightleftharpoons R–SH$	10–11
$R–O^- + H^+ \rightleftharpoons R–OH$	18–19
$R–CONH^- + H^+ \rightleftharpoons R–CONH_2$	16–17

[a]An aliphatic carbon chain is denoted as R and an aromatic ring by ϕ. Taken from March.[206]

and one basically never faces the situation that the second protonation step becomes important. Therefore, acetate can be viewed as a monoprotic base.

This situation is generic for atoms in the first row of the periodic table; whenever a proton binds to a *single atom*, the second protonation or deprotonation step is unlikely and can be safely neglected. (When protons bind to *different atoms*, the situation is of course different and will be discussed later.)

The assumption of a single relevant protonation step might fail if protons bind to atoms in the second and higher rows of the periodic table, most notably, sulfur and selenium. For example, the splitting in the pK values for H_2S is about 3–5, and two-step protonation equilibria involving a single atom have to be considered. However, the splitting between pK values of $R-S^-$ (thiols) is also about 15.

4.2. Diprotic Acids and Bases

The simplest situation, where site–site interactions may become important, are diprotic acids and bases. Therefore, this situation is discussed in detail in the following.

4.2.1. Macroscopic Description

Polyprotic equilibria of small molecules are usually treated in the macroscopic picture. In the case of a diprotic acid or base, the description considers three *macrospecies* A, HA, and H_2A with the reaction equilibria[4]

$$A + H \rightleftharpoons HA \quad \overline{K}_1 \tag{114}$$

$$A + 2H \rightleftharpoons H_2A \quad \overline{K}_2 \tag{115}$$

or equivalently

$$A + H \rightleftharpoons HA \quad K_1 \tag{116}$$

$$HA + H \rightleftharpoons H_2A \quad K_2. \tag{117}$$

Thereby, the stepwise constants are related to the cumulative constants by $K_1 = \overline{K}_1$ and $K_2 = \overline{K}_2/\overline{K}_1$. Recall that p$K_1 = \log_{10} K_1$ and p$K_2 = \log_{10} K_2 = \log_{10} \overline{K}_2 - \log_{10} \overline{K}_1$. The relative concentration of the deprotonated species can be expressed in terms of the equilibrium constants

$$P_0 = \frac{c_A}{c_A + c_{HA} + c_{H_2A}} = \frac{1}{1 + \overline{K}_1 a_H + \overline{K}_2 a_H^2} \tag{118}$$

and similar equations apply to the other macrospecies

$$P_1 = \frac{\overline{K}_1 a_H}{1 + \overline{K}_1 a_H + \overline{K}_2 a_H^2} \tag{119}$$

$$P_2 = \frac{\overline{K}_2 a_H^2}{1 + \overline{K}_1 a_H + \overline{K}_2 a_H^2} . \tag{120}$$

These quantities are denoted as *macrostate probabilities*, and thus we have $P_0 + P_1 + P_2 = 1$. The overall titration curve is given by

$$\theta = \frac{1}{2} \frac{c_{HA} + 2c_{H_2A}}{c_A + c_{HA} + c_{H_2A}} \tag{121}$$

and in terms of the equilibrium constants

$$\theta = \frac{1}{2} \frac{\overline{K}_1 a_H + 2\overline{K}_2 a_H^2}{1 + \overline{K}_1 a_H + \overline{K}_2 a_H^2} . \tag{122}$$

The titration curve can be expressed as a weighted sum of the macrostate probabilities

$$\theta = \frac{1}{2} (P_1 + 2P_2). \tag{123}$$

These relations are illustrated graphically in Fig. 31. Figure 31a shows the overall titration curve, with its typical two-step behavior with an intermediate plateau at $\theta = 1/2$. The two protonation steps are located at pH $\approx pK_1$ and pH $\approx pK_2$. From such titration curves, the macroscopic ionization constants pK_1 and pK_2 can be determined. Figure 31b indicates the macrostate probabilities (also called the prevalence diagram). This diagram shows the relative probabilities of the individual macrospecies.

4.2.2. Microscopic Description

Typically, a diprotic acid or base has two *different* ionizable sites. Let us assume for the moment that these two sites are nonequivalent. Since each site can be either protonated or deprotonated, four different protonation microstates are then possible. These states can be easily enumerated using the occupation variables discussed in Section 3. For the site 1, introduce a variable s_1 such that $s_1 = 0$ if the site is deprotonated and $s_1 = 1$ if this site is protonated. An analogous variable s_2 is introduced for the second site. The protonation state of the molecule A can be thus specified by these two variables $\{s_1, s_2\}$. These four *microspecies* A$\{0,0\}$, A$\{1,0\}$, A$\{0,1\}$, and A$\{1,1\}$ are represented schematically in Fig. 32. Open circles represent a deprotonated site and filled circles protonated ones.[4,7,199]

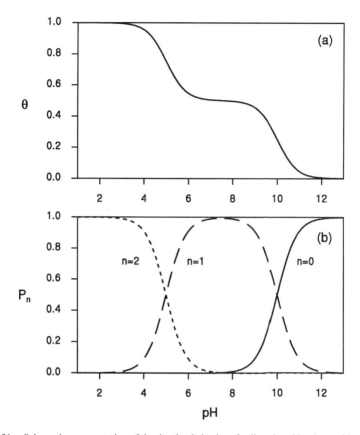

Figure 31. Schematic representation of the titration behavior of a diprotic acid or base with $pK_1 \simeq 10$ and $pK_2 \simeq 5$. (a) Overall titration curve and (b) prevalence diagram showing the macrostate probabilities.

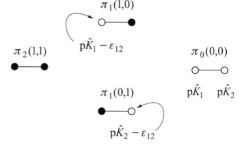

Figure 32. Microstates, microscopic ionization constants, and conditional microstate probabilities of a diprotic acid or base. Open circles represent a deprotonated site and filled circles protonated sites.

These four species are in chemical equilibrium and described by the following reactions. One proton can bind on two different sites

$$A\{0,0\} + H \rightleftharpoons A\{1,0\} \quad \hat{K}_1 \tag{124}$$

$$A\{0,0\} + H \rightleftharpoons A\{0,1\} \quad \hat{K}_2 \tag{125}$$

while there is only one possibility to bind two protons

$$A\{0,0\} + 2H \rightleftharpoons A\{1,1\} \quad \hat{K}_A = \hat{K}_1 \hat{K}_2 u_{12}. \tag{126}$$

As already recognized by Bjerrum,[2] the proper way to parametrize the latter equilibrium constant is by the dimensionless ratio

$$u_{12} = \frac{\hat{K}_A}{\hat{K}_1 \hat{K}_2} \tag{127}$$

which is nothing but the *pair interaction parameter*

$$\varepsilon_{12} = -\log_{10} u_{12} \tag{128}$$

introduced in Section 3. This parameter was also abbreviated as $\Delta \log k_{1-2}$ and referred to as the interactivity parameter.[199]

The equilibrium constants entering Eqs. (124)–(126) are microscopic association constants. The first two equations involve just a single proton, and their equilibrium constants can be interpreted as microscopic pK values. In particular, $\text{p}\hat{K}_1 = \log_{10}\hat{K}_1$ and $\text{p}\hat{K}_2 = \log_{10}\hat{K}_2$ are the ionization constants of sites 1 and 2, respectively, given the other site is deprotonated. The other two microscopic pK values of interest are related to the equilibrium constants of the reaction equilibria

$$A\{1,0\} + H \rightleftharpoons A\{1,1\} \quad \hat{K}_1' = \hat{K}_1 u_{12} \tag{129}$$

$$A\{0,1\} + H \rightleftharpoons A\{1,1\} \quad \hat{K}_2' = \hat{K}_2 u_{12} \tag{130}$$

and the corresponding microscopic pK values are given by

$$\text{p}\hat{K}_1' = \text{p}\hat{K}_1 - \varepsilon_{12} \tag{131}$$

$$\text{p}\hat{K}_2' = \text{p}\hat{K}_2 - \varepsilon_{12}. \tag{132}$$

One can interpret $\text{p}\hat{K}_1'$ and $\text{p}\hat{K}_2'$ as the microscopic pK values of the sites 1 and 2, respectively, given the other site is protonated. A summary of these ionization constants is given in Fig. 32.

Let us relate the *microscopic* and *macroscopic* pictures. We note that A is identical to A$\{0,0\}$ and H$_2$A is identical to A$\{1,1\}$. However, the macroscopic species HA represents actually two different microscopic ones, namely A{0,1} and A{1,0}. The corresponding relations between the concentrations are[4,199]

$$c_A = c_{A\{0,0\}} \tag{133}$$

$$c_{HA} = c_{A\{0,1\}} + c_{A\{1,0\}} \tag{134}$$

$$c_{H_2A} = c_{A\{1,1\}} \tag{135}$$

Comparing the mass action laws we find [see also Eq. (77)]

$$\overline{K}_1 = \hat{K}_1 + \hat{K}_2 \tag{136}$$

$$\overline{K}_2 = \hat{K}_1 \hat{K}_2 u_{12}. \tag{137}$$

The macroscopic model is fully specified by two association constants $\overline{K}_1$ and $\overline{K}_2$, while for the microscopic description of the same ionization problem three parameters, $\hat{K}_1$, $\hat{K}_2$, and ε_{12}, are needed.

The relative mole fractions (or probabilities) of the individual microscopic species $A\{0,0\}$, $A\{1,0\}$, $A\{0,1\}$, and $A\{1,1\}$ can now be evaluated. The probability to find $A\{0,0\}$ among all other microspecies is

$$p(0,0) = \frac{c_{A\{0,0\}}}{c_{A\{0,0\}} + c_{A\{1,0\}} + c_{A\{0,1\}} + c_{A\{1,1\}}} \tag{138}$$

and analogous relations hold for the remaining species. From the appropriate mass action laws, one observes that these mole fractions are proportional to the mole fractions of the corresponding macrospecies. One obtains the following relations [see also Eq. (78)]

$$p(0,0) = \pi_0(0,0)\, P_0 \tag{139}$$

$$p(1,0) = \pi_1(1,0)\, P_1 \tag{140}$$

$$p(0,1) = \pi_1(0,1)\, P_1 \tag{141}$$

$$p(1,1) = \pi_2(1,1)\, P_2 \tag{142}$$

where P_0, P_1, and P_2 are the macrostate probabilities, which determine the entire pH dependence, and $\pi_0(0,0)$, $\pi_1(0,1)$, $\pi_1(1,0)$, and $\pi_2(1,1)$ are constant conditional microstate probabilities. The latter quantity represents the probability (or mole fraction) to find a given microspecies among all possible ones with the same total number of protons bound. For example, $\pi_1(0,1)$ denotes the probability to find the microspecies A{0,1} among all possible microspecies with one bound proton. There is just a single microscopic state for the fully deprotonated and protonated state, and the corresponding conditional probabilities are unity. For the singly protonated state, there are two possible microstates. The results read

$$\pi_0(0,0) = 1 \tag{143}$$

$$\pi_1(1,0) = \frac{\hat{K}_1}{\hat{K}_1 + \hat{K}_2} \tag{144}$$

$$\pi_1(0,1) = \frac{\hat{K}_2}{\hat{K}_1 + \hat{K}_2} \tag{145}$$

$$\pi_2(1,1) = 1. \tag{146}$$

Note that these probabilities are constants and do not depend on the proton activity. In other words, the ratio of the concentrations of species $A\{0,1\}$ and $A\{1,0\}$ always remains constant.

The key quantity to experimentally probe microscopic equilibria are titration curves of individual sites. In favorable cases, such site-specific titration curves can be directly measured by spectroscopic methods. The quantity of interest is therefore the relative concentration of all species where a given site is protonated. The degree of protonation of a given site is the sum of the probabilities of those microspecies where this particular site is protonated. These probabilities can be expressed in terms of the microscopic association constants and the proton activity. We find that

$$\theta_1 = p(1,0) + p(1,1) = \frac{\hat{K}_1}{\hat{K}_1 + \hat{K}_2} P_1 + P_2 \tag{147}$$

$$\theta_2 = p(0,1) + p(1,1) = \frac{\hat{K}_2}{\hat{K}_1 + \hat{K}_2} P_1 + P_2. \tag{148}$$

Note that the pH dependence of these quantities is given by P_1 and P_2, but their relative weights depend on the relative concentrations of the two different singly protonated microspecies.

The relations are illustrated schematically in Fig. 33. We use the microscopic ionization constants $p\hat{K}_1 = 10$ and $p\hat{K}_2 = 9$ together with the interaction parameter $\varepsilon_{12} = 4$. The macroscopic pK values become $pK_1 \approx 10.04$ and $pK_2 \approx 4.95$. For these parameters, the overall titration curve and macrostate probabilities were already shown in Fig. 31. Upon protonation with one proton, the corresponding ionization constants are $p\hat{K}_1' = 6$ and $p\hat{K}_2' = 5$. The conditional microstate probabilities are $\pi_1(1,0) \approx 0.91$ and $\pi_1(0,1) \approx 0.09$.

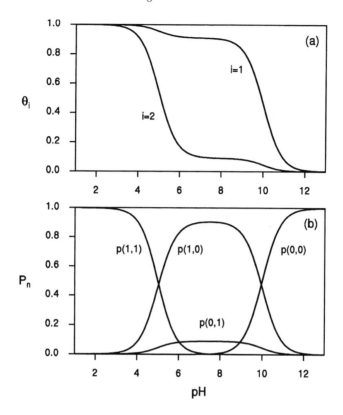

Figure 33.   Schematic representation of microscopic protonation equilibria for a diprotic acid or base. (a) Titration curves of individual sites. (b) Probabilities of individual microstates. Parameters used are $p\hat{K}_1 = 10$, $p\hat{K}_2 = 9$, and $\varepsilon_{12} = 4$.

Figure 33a shows the titration curves of the individual sites. The intermediate plateaus just reflect the conditional microstate probabilities $\pi_1(1,0)$ and $\pi_1(0,1)$. From such site titration curves, which can be probed by nuclear magnetic resonance (NMR) and other spectroscopic techniques, the microscopic constants can be derived. Note that these curves are symmetric, but only for $N = 2$. Figure 33b shows the actual microstate probabilities as a function of pH. While such a presentation is often used in the literature, it is redundant and difficult to read for polyprotic molecules. Since the conditional microstate probabilities are constant, the pH dependence of any microspecies is dictated by the pH dependence of the corresponding macrospecies. Its relative proportion is given by the constant conditional microstate probabilities.

Experimental determination of microscopic ionization constants for diprotic molecules was discussed frequently and we shall not attempt to cover this topic here extensively. A detailed review on this topic was recently published.[199]

4.2.3. *Conformational Degrees of Freedom*

The case of a diprotic molecule is useful to illustrate how conformational degrees of freedom can be considered explicitly. In the above discussion, these degrees of freedom were already considered implicitly, but were averaged out. Let us now distinguish a number of well-defined conformers, numbered according to the index α. Each of these conformers can be characterized by three parameters: two microscopic constants $\hat{K}_1^{(\alpha)}$ and $\hat{K}_2^{(\alpha)}$ and the corresponding interaction parameter $\varepsilon_{12}^{(\alpha)}$. The corresponding quantities for each conformer α are given by the same equations as above, for example the macroscopic association constants by Eqs. (136) and (137), and the conditional microstate probabilities by Eqs. (143)–(146). The macroscopic constants of the molecule involve an average over all conformational states[199]

$$\overline{K}_n = \sum_\alpha x_\alpha \overline{K}_n^{(\alpha)} \tag{149}$$

where x_α is the probability of finding a conformer α in the deprotonated state. The probabilities of individual conformational states can now be expressed as[199]

$$p^{(\alpha)}(0,0) = \pi_0^{(\alpha)}(0,0)\rho_0^{(\alpha)}P_0 \tag{150}$$

$$p^{(\alpha)}(1,0) = \pi_1^{(\alpha)}(1,0)\rho_1^{(\alpha)}P_1 \tag{151}$$

$$p^{(\alpha)}(0,1) = \pi_1^{(\alpha)}(0,1)\rho_1^{(\alpha)}P_1 \tag{152}$$

$$p^{(\alpha)}(1,1) = \pi_2^{(\alpha)}(1,1)\rho_2^{(\alpha)}P_2 \tag{153}$$

where $\rho_n^{(\alpha)}$ is the conditional probability to find conformation α within the macrostate n and is given by

$$\rho_n^{(\alpha)} = \frac{x_\alpha \overline{K}_n^{(\alpha)}}{\sum_\alpha x_\alpha \overline{K}_n^{(\alpha)}} \tag{154}$$

Note that $\rho_0^{(\alpha)} = x_\alpha$. From these quantities various measurable properties can be derived. For example, the overall probability of a given conformational state α is given by

$$\rho^{(\alpha)} = \rho_0^{(\alpha)}P_0 + \rho_1^{(\alpha)}P_1 + \rho_2^{(\alpha)}P_2. \tag{155}$$

Again, the entire pH dependence is given by the macrostate probabilities P_n. Applications of these relations will be discussed later.

4.2.4. Equivalent Sites

For equivalent ionizable sites (symmetric molecule) the description simplifies substantially. In this case, all corresponding parameters become equal and indices can be dropped. We shall use the abbreviations $p\hat{K}_1 = p\hat{K}_2 = p\hat{K}$ and $\varepsilon_{12} = \varepsilon$. The equations for the macroscopic ionization constants simplify to

$$pK_1 = p\hat{K} + \log_{10} 2 \tag{156}$$

$$pK_2 = p\hat{K} - \log_{10} 2 - \varepsilon \tag{157}$$

In contrast to the case of nonequivalent sites, for equivalent sites the microscopic parameters can be deduced from the macroscopic constants. For example, the splitting of the pK values of a symmetric diprotic acid is directly related to the

Table 5. Experimental Macroscopic pK_n Values of Various Carboxylic Acids[211] with Different Number of Atoms between the Acidic Protons at Ionic Strengths 1.0 M and Extrapolated to Infinite Dilution[a]

	L	$I = 1\,M$			$I \to 0$		
		pK_n	$p\hat{K}, p\hat{K}'$	ε	pK_n	$p\hat{K}, p\hat{K}'$	ε
Oxalic acid	4	3.56	3.26	1.88	4.27	3.97	2.42
HOOC–COOH		1.08	1.38		1.25	1.55	
Malonic acid	5	5.03	4.73	1.84	5.70	5.40	2.25
HOOC–CH_2–COOH		2.59	2.89		2.85	3.15	
Succinic acid	6	5.11	4.81	0.54	5.64	5.34	0.83
HOOC–$[CH_2]_2$–COOH		3.97	4.27		4.21	4.51	
Glutaric acid	7	4.90	4.60	0.17	5.42	5.12	0.49
HOOC–$[CH_2]_3$–COOH		4.13	4.43		4.33	4.63	
Adipic acid	8	4.94	4.64	0.08	5.42	5.12	0.40
HOOC–$[CH_2]_4$–COOH		4.26	4.56		4.42	4.72	
Pimelic acid	9	4.94	4.64	0.07	5.43	5.13	0.34
HOOC–$[CH_2]_5$–COOH		4.27	4.57		4.49	4.79	
Fumaric acid (trans)	6	3.90	3.60	0.51	4.48	4.18	0.86
HOOC–$[CH]_2$–COOH		2.79	3.09		3.02	3.32	
Maleic acid (cis)	6	5.60	5.30	3.35	6.27	5.97	3.75
HOOC–$[CH]_2$–COOH		1.65	1.95		1.92	2.22	

[a]Microscopic $p\hat{K}$ and pair interaction parameter ε are calculated from Eqs. (156) and (157). The number of atoms in between the ones accepting the protons is denoted by L.

Table 6.   Experimental pK_n Values of Linear Aliphatic Amines and Related
Compounds[211] with Different Number of Atoms between the Acidic
Protons at an Ionic Strength 1.0 M[a]

	L	pK_n	$p\hat{K}$, $p\hat{K}'$	ε
Hydrazine[b]	2	8.18	7.88	>7
H_2NNH_2		<1	<1	
Ethylenediamine (en)	4	10.20	9.90	2.13
$H_2N(CH_2)_2NH_2$		7.47	7.77	
Trimethylenediamine	5	10.76	10.46	1.06
$H_2N(CH_2)_3NH_2$		9.10	9.40	
Tetramethylenediamine	6	11.08	10.78	0.57
$H_2N(CH_2)_4NH_2$		9.91	10.21	
Pentamethylenediamine	7	11.39	11.09	0.20
$H_2N(CH_2)_5NH_2$		10.59	10.89	
Hexamethylenediamine[b]	8	11.02	10.72	0.18
$H_2N(CH_2)_6NH_2$		10.24	10.54	

[a]Microscopic $p\hat{K}$ and pair interaction parameter ε are calculated from Eqs. (156) and (157). The number of atoms in
between the ones accepting the protons is denoted as L.
[b]Ionic strength 0.5 M.

interaction parameter ε. These parameters are listed for a number of carboxylic
acids and linear amines in Tables 5 and 6.

One observes that the interaction parameter ε decreases with increasing distance between the ionizable sites. This decrease was probably first correctly
interpreted by Bjerrum[2] as originating from the Coulomb interaction between the
charged sites. We shall return to this central point later.

4.3.   Oligoprotic Acids and Bases

Analogous arguments, as discussed for the diprotic molecule above, can now
be put forward for polyprotic molecules.

4.3.1.   Macroscopic Description

The cumulative macroscopic association equilibria for oligoprotic molecules
are

$$A + nH \rightleftharpoons H_nA \quad \overline{K}_n \qquad (158)$$

for $n = 1, \dots, N$, when $\overline{K}_n$ are the cumulative association constants. The corresponding stepwise equilibria read

$$H_{n-1}A + H \rightleftharpoons H_nA \quad K_n \tag{159}$$

where the stepwise association constants K_n define the macroscopic pK values through $\mathrm{p}K_n = \log_{10} K_n$. The stepwise constants K_n can be expressed in terms of the cumulative constants as

$$K_n = \overline{K}_n / \overline{K}_{n-1} \quad \text{for} \quad n = 1, \dots, N \tag{160}$$

where $\overline{K}_0 = 1$. The macrostate probabilities are given by

$$P_n = \frac{c_{H_nA}}{c_A + c_{HA} + c_{H_2A} + c_{H_3A} + \cdots} \quad \text{for} \quad n = 0, \dots, N \tag{161}$$

and recall that $\sum_{n=1}^{N} P_n = 1$. These quantities can be again expressed in terms of the cumulative association constants as

$$P_n = \frac{\overline{K}_n a_H^n}{\sum_{n=0}^{N} \overline{K}_n a_H^n}. \tag{162}$$

The average degree of protonation (titration curve) is

$$\theta = \frac{1}{N} \frac{c_{HA} + 2c_{H_2A} + 3c_{H_3A} + \cdots}{c_A + c_{HA} + c_{H_2A} + c_{H_3A} + \cdots} \tag{163}$$

which can be also expressed in terms of equilibrium constants as

$$\theta = \frac{1}{N} \frac{\sum_{n=0}^{N} n\overline{K}_n a_H^n}{\sum_{n=0}^{N} \overline{K}_n a_H^n}. \tag{164}$$

This quantity can be also written in terms of the macrostate probabilities

$$\theta = \frac{1}{N} \sum_{n=0}^{N} nP_n. \tag{165}$$

In potentiometric titrations, one directly measures θ as a function of pH and, in principle, one has access to all macroscopic ionization constants. The best approach is a direct least-squares fit of the titration curve with Eq. (164). In practice, however, depending on the accuracy of the data and spacing of the ionization constants, this procedure is applicable only up to N within the range of 5–12. Titration curves contain information on the macroscopic constants only; additional information is needed if the microscopic constants have to be determined.

4.3.2. Microscopic Description for Triprotic Acids and Bases

Microscopic description of the ionization equilibria of a triprotic molecule appears already quite complicated, and it becomes even more so for a larger number of sites. This complication arises as there is a large number of different microstates and a corresponding large number of microscopic constants. Nevertheless, the problem becomes easily tractable with the cluster expansion method presented in Section 3, which automatically provides the proper parametrization of the problem. This parametrization is not easily recognized from chemical equilibrium considerations. Let us first summarize the results for a triprotic acid ($N = 3$) and then discuss various applications of the cluster method for higher oligoprotic acids ($N > 3$).

For a triprotic acid or base there are 8 different species at equilibrium (see Fig. 34). To the fully deprotonated species, one proton can bind in three different ways

$$A\{0,0,0\} + H \rightleftharpoons A\{1,0,0\} \qquad \hat{K}_1 \qquad (166)$$

$$A\{0,0,0\} + H \rightleftharpoons A\{0,1,0\} \qquad \hat{K}_2 \qquad (167)$$

$$A\{0,0,0\} + H \rightleftharpoons A\{0,0,1\} \qquad \hat{K}_3 \qquad (168)$$

two protons bind also in three different ways

$$A\{0,0,0\} + 2H \rightleftharpoons A\{1,1,0\} \qquad \hat{K}_1\hat{K}_2 u_{12} \qquad (169)$$

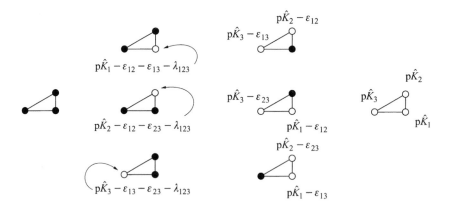

Figure 34. Microstates and all microscopic pK values of a triprotic acid or base. The fundamental parameters are the microscopic ionization constants p$\hat{K}_i$, and the interaction parameters for pairs ε_{ij} and triplets λ_{ijk} with $i, j, k = 1, 2, 3$.

$$A\{0,0,0\} + 2H \rightleftharpoons A\{1,0,1\} \qquad \hat{K}_1\hat{K}_3u_{13} \tag{170}$$

$$A\{0,0,0\} + 2H \rightleftharpoons A\{0,1,1\} \qquad \hat{K}_2\hat{K}_3u_{23} \tag{171}$$

but there is only one way to bind three protons

$$A\{0,0,0\} + 3H \rightleftharpoons A\{1,1,1\} \qquad \hat{K}_1\hat{K}_2\hat{K}_3u_{12}u_{13}u_{23}v_{123}. \tag{172}$$

The equilibria are characterized by 7 microscopic constants, that are already expressed in terms of 7 appropriate parameters. We introduce 3 microscopic ionization constants $\hat{K}_1$, $\hat{K}_2$, and $\hat{K}_3$, and 4 dimensionless ratios u_{12}, u_{13}, u_{23}, and v_{123}. Equivalent representation is given in terms of the microscopic pK values defined as p$\hat{K}_i = \log_{10} \hat{K}_i$, and the interaction parameters for pairs $\varepsilon_{ij} = -\log_{10} u_{ij}$ and triplets $\lambda_{ijk} = -\log_{10} v_{ijk}$ with $i, j, k = 1, 2, 3$.

These 7 parameters are used to yield the corresponding association constants for the reactions given above. For the 8 different microscopic species, 12 possible microscopic ionization constants can be considered. These constants are summarized in Fig. 34. The macroscopic association constants can also be expressed in terms of these microscopic constants,[201]

$$\overline{K}_1 = \hat{K}_1 + \hat{K}_2 + \hat{K}_3 \tag{173}$$

$$\overline{K}_2 = \hat{K}_1\hat{K}_2u_{12} + \hat{K}_1\hat{K}_3u_{13} + \hat{K}_2\hat{K}_3u_{23} \tag{174}$$

$$\overline{K}_3 = \hat{K}_1\hat{K}_2\hat{K}_3u_{12}u_{13}u_{23}v_{123}. \tag{175}$$

The various microstate probabilities can be also evaluated. For example, the conditional probability to find the species A{100} among all singly protonated species ($n = 1$) is [see also Eq. (79)]

$$\pi_1(1,0,0) = \frac{\hat{K}_1}{\hat{K}_1 + \hat{K}_2 + \hat{K}_3} \tag{176}$$

while the probability of A{110} among all doubly protonated species ($n = 2$) is

$$\pi_2(1,1,0) = \frac{\hat{K}_1\hat{K}_2u_{12}}{\hat{K}_1\hat{K}_2u_{12} + \hat{K}_1\hat{K}_3u_{13} + \hat{K}_2\hat{K}_3u_{23}}. \tag{177}$$

Analogous expressions can be constructed for all other microstates. We have a single fully deprotonated and protonated microstate and thus $\pi_0(0,0,0) = \pi_3(1,1,1) = 1$. With these conditional probabilities at hand, we obtain the site titration curves. The degree of protonation of site 1 is given by [see also Eq. (80)]

$$\theta_1 = p(1,0,0) + p(1,1,0) + p(1,0,1) + p(1,1,1)$$

$$= \pi_1(1,0,0)P_1 + [\pi_2(1,1,0) + \pi_2(1,0,1)]P_2 + P_3. \tag{178}$$

For the other sites similar expressions follow.

In the literature,[4,111,199] such problems are usually approached within a chemical equilibrium framework by introducing microscopic constants such as, for example, $\hat{K}_2^1 = \hat{K}_1 u_{12}$ and $\hat{K}_1^2 = \hat{K}_2 u_{21}$. The difficulty with this approach is that $\hat{K}_1^2 \neq \hat{K}_2^1$, and one does not easily recognize that these two constants are not independent, but linked with a single parameter $u_{12} = u_{21}$. In other words, the symmetry of the pair interaction matrix $\varepsilon_{ij} = \varepsilon_{ji}$ is usually not taken into account. Analogous symmetries arise for the triplet interaction parameters and read $\lambda_{ijk} = \lambda_{jki} = \lambda_{kij}$.

As an example of this kind of analysis, Fig. 35 shows the site protonation curves of D-*myo*-inositol 1,2,6-tri(phosphate) in 0.1 M Et$_4$NBr at 25 °C determined by ^{31}P NMR.[111] The structure of this molecule and the labeling of the three ionizable sites is shown in Fig. 36 (top). The NMR data were interpreted with a microscopic equilibrium model for a triprotic acid with a simultaneous least-squares fit of all site titration curves.[111] The analysis leads to the following parameters: the microscopic ionization constants $\hat{p}K_1 \approx 9.44$, $\hat{p}K_2 \approx 7.32$, and $\hat{p}K_2 \approx 8.80$, the pair interaction parameters $\varepsilon_{12} \approx 0.53$, $\varepsilon_{13} \approx 1.64$, and $\varepsilon_{23} \approx 0.13$, and the triplet interaction parameter $\lambda_{123} \approx 0.56$.

Figure 35a shows the site protonation curves calculated using these parameters together with the experimental data, while Fig. 35b indicates the corresponding macrostate probabilities. The microscopic constants and microspecies probabilities can be calculated from these parameters as well. A summary of these results is given in Fig. 36.

We should point out that the calculated site protonation curves in Fig. 35 and the microscopic values given in Fig. 36 are very close to their counterparts in Ref. 111, but not exactly the same. Small discrepancies arise from the fact that Mernissi *et al.*[111] have analyzed the data without the symmetry constraints $\varepsilon_{ij} = \varepsilon_{ji}$ and $\lambda_{ijk} = \lambda_{jki} = \lambda_{kij}$. They have correctly recovered these symmetries from their analysis, but the different estimates were subject to an experimental error. The data for the parameters reported represent mean values of these different estimates, and thus lead to the minor differences between the model presented here and in Ref. 111. A better description of the experimental results shown in Fig. 35 could be probably achieved within the discussed symmetric model by slight parameter adjustment.

Such detailed analysis is seldomly carried out. Other examples, where the microscopic ionization equilibria of asymmetric triprotic molecules were fully resolved, include various bioligands such as amino acids and dihydroxyphenylalanine (DOPA).[216] These aspects were reviewed recently.[199] Complete site protona-

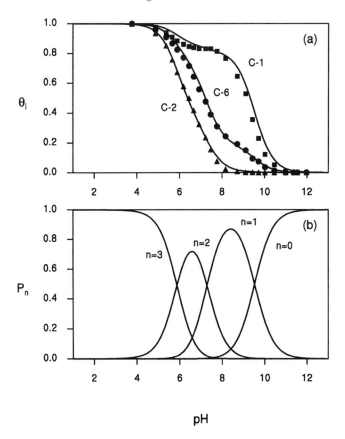

Figure 35.   Ionization behavior of D-*myo*-inositol 1,2,6-tri(phosphate) in 0.1 M Et$_4$NBr at 25 °C. The structural formula is given in Fig. 36. (a) Comparison of calculations and experimental data obtained by ^{31}P NMR measurements,[111] (b) calculated macrostate probabilities. Solid lines are calculated with the parameters $p\hat{K}_1 \simeq 9.44$, $p\hat{K}_2 \simeq 7.32$, and $p\hat{K}_2 \simeq 8.80$; $\varepsilon_{12} \simeq 0.53$, $\varepsilon_{13} \simeq 1.64$; $\varepsilon_{23} \simeq 0.13$, and $\lambda_{123} \simeq 0.56$.

tion curves are often not available, and one usually estimates the microscopic pK values from similar molecules, which can be used to mimic the ionization behavior of individual groups to good approximation.

Tetraprotic acids or bases ($N = 4$) involve 15 parameters (taking all the symmetries into account), and the problem of the determination of all microconstants appears prohibitively complicated. However, the situation is easily handled with the cluster approach (see Section 3). Before we discuss a few examples, we shall focus on some archetypal situations.

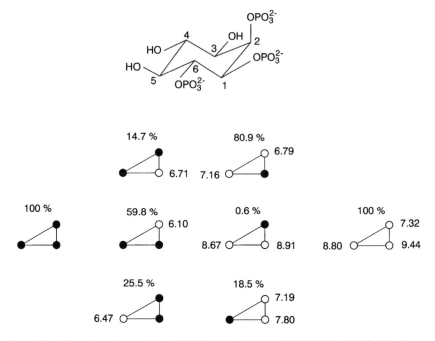

Figure 36.   Microstates of D-*myo*-inositol 1,2,6-tri(phosphate) in 0.1 M Et$_4$NBr at 25 °C. The sites are numbered according to the carbons in the ring as C–1, C–2, and C–6. The corresponding microscopic pK values and conditional microstate probabilities are given. Parameters are given in the caption of Fig. 35. Adapted from Ref. 111.

4.3.3.   *Equivalent Sites*

In a fully-symmetric molecule all sites are equivalent, and each site interacts with all other sites in the same fashion. The ionization constants of all sites are equal and given by p$\hat{K}$, and so are all interaction parameters. For the sake of illustration, let us first assume that only the pair interactions are nonzero and given by a common interaction parameter ε. The formation free energy of a microstate is

$$F(\{s_i\}) = -\mu n + \frac{n(n-1)}{2} E \qquad (179)$$

where n is the number of protons bounds, and the chemical potential μ and the pair interaction energy E are defined by $\beta\mu/\ln 10 = p\hat{K} - pH$ and $\beta E/\ln 10 = \varepsilon$. The factor $n(n-1)/2$ represents the number of possible pairs of occupied sites. We shall see later that this model is precisely equivalent to the mean-field approximation (see Section 5). Since this free energy depends only on the number of sites occupied, the number of configurations with given n is just the number of ways to arrange n protons among N sites. The cumulative association constants are thus given by[217]

$$\overline{K}_n = \binom{N}{n} \hat{K}^n u^{n(n-1)/2} \tag{180}$$

where $\varepsilon = -\log_{10} u$ and the binomial coefficient is

$$\binom{N}{n} = \frac{N!}{n!(N-n)!}. \tag{181}$$

The corresponding macroscopic pK values become

$$pK_n = p\hat{K} + \log_{10} \frac{N+1-n}{n} - \varepsilon(n-1). \tag{182}$$

Within this model the steps between successive polyprotic pK values are approximately the same.

All microstates are equally likely, and all sites titrate in the same fashion. For this reason there is only one possible microscopic pK for a given degree of protonation n which is

$$p\hat{K}_n = p\hat{K} - \varepsilon(n-1). \tag{183}$$

A few applications of this model are summarized in Tables 7 and 8. We note that this simple two-parameter model is able to capture the main trends in the macroscopic pK values for several types of molecules. Apparently, the higher-order interactions are not that important. One can thus estimate microscopic ionization constants and interaction parameters for larger oligoprotic molecules from macroscopic constants alone.

The present arguments are easily generalized for interaction parameters to arbitrarily high order. Including triplet interactions we have

$$pK_n = p\hat{K} + \log_{10} \frac{N+1-n}{n} - (n-1)\varepsilon - \frac{(n-1)(n-2)}{2}\lambda - \dots \tag{184}$$

Note that the above equation is exact for a fully symmetric triprotic molecule.

4.3.4. Linear Molecules

The second archetypal situation is a linear chain. This situation is more complex, since sites within a finite chain are not all equivalent. Let us assume, for simplicity, that all sites have a common microscopic ionization constant p$\hat{K}$, and that the ionizable sites are sufficiently far apart, such that nearest-neighbor pair interactions are the only relevant ones (high-salt regime). If we number the sites along the chain in a consecutive fashion, we have

$$\varepsilon_{ij} = \begin{cases} \varepsilon & \text{for } j = i \pm 1 \\ 0 & \text{elsewhere} \end{cases} \tag{185}$$

Table 7. Experimental pK_n Values for Symmetric Organic Molecules[211,212] with Different Number of Atoms between the Acidic Protons at an Ionic Strength 0.1 M[a]

	N	L	pK_n		p$\hat{K}$	ε
			exp.	calc.		
Benzene-1,3,5-tricarboxylic acid	3	5	4.49	4.48	4.00	0.26
$C_6H_{15}COOH_3$			3.71	3.74		
			3.01	3.00		
Tris(aminomethyl)methane	3	5	10.39	10.44	9.96	1.50
$HC(CH_2NH_2)_3$			8.56	8.56		
			6.44	6.48		
Tris(2-aminoethyl)amine[b]	3	5	10.14	10.24	9.64	0.27
$N(CH_2CH_2NH_2)_3$			9.68	9.49		
			8.74	8.74		
Tetrakis(aminomethyl)methane	4	5	9.89	10.16	9.56	1.91
$C(CH_2NH_2)_4$			8.17	7.83		
			5.67	5.56		
			3.03	3.23		

[a]Microscopic p$\hat{K}$ and pair interaction parameter ε from the mean field model [cf. Eq. (182)]. We denote by N the number of ionizable sites, by L the number of atoms in between the sites.

[b]Ionic strength 0.5 M. For this molecule, the protonation step of the teriary amine is neglected as it occurs at very low pH only.

where $j = 2, \ldots, N - 1$ and ε is the nearest-neighbor pair interaction parameter. The effective free energy Eq. (84) can be written as

$$F(\{s_i\}) = -\sum_i \mu s_i + E \sum_i s_i s_{i+1}$$

$$= -n\mu + Em \tag{186}$$

where m denotes the number of nearest-neighbor pairs in the chain. Again we use $\varepsilon = \beta E/\ln 10$ and n denotes the number of occupied sites. The analysis proceeds analogously to arguments presented above; the main difference is that the free energy no longer depends only on the number of occupied sites n, but also on the number of nearest-neighbor pairs m. Now we have to evaluate the number of possible ways of arranging n sites and m nearest-neighbor pairs on a chain with total N sites. This number can be deduced from combinatorial arguments or by direct evaluation and reads[21,25]

Table 8. Experimental pK_n Values of Various Inorganic Acids and Bases[5,211] with Different Number of Atoms between the Acidic Protons at an Ionic Strength 0.5 M[a]

	N	L	pK_n		$p\hat{K}$	ε
			exp.	calc.		
Phosphorous acid	2	3	6.14	6.14	5.84	4.44
$OPH(OH)_2$			1.10	1.10		
Phosphoric acid[b]	3	3	11.80	11.82	11.34	4.48
$OP(OH)_3$			6.90	6.86		
			1.88	1.90		
Silicic acid	4	3	—		18.24	2.72
$Si(OH)_4$			—			
			12.62	12.62		
			9.47	9.47		
Iron(III)–aquoion[b]	6	3	—		13.18	2.15
$[Fe(H_2O)_6]^{3+}$			—			
			9.6	9.0		
			6.3	6.6		
			3.5	4.2		
			2.2	1.6		
Yttrium(III)–aquoion[b]	6	3	—		11.63	0.63
$[Y(H_2O)_6]^{3+}$			—			
			10.5	10.5		
			9.6	9.6		
			8.7	8.7		
			7.7	7.7		

[a]Microscopic $p\hat{K}$ and pair interaction parameter ε from the mean field model. We denote by N the number of ionizable sites, by L the number of atoms in between the sites.
[b]Extrapolated to vanishing ionic strength.

$$\binom{n-1}{m}\binom{N-n-1}{N-m} \tag{187}$$

The cumulative association constants for a chain with a total of N sites thus become[202]

$$\overline{K}_n = \hat{K}^n \sum_{m=\max(0,2n-N-1)}^{n-1} \binom{n-1}{m}\binom{N-n-1}{N-m} u^m \tag{188}$$

where u is defined by $\varepsilon = - \log_{10} u$. Here, no simple analytical expression for the pK values can be given, but from Eq. (188) their values are readily obtained. While most other properties, like microscopic pK values, microstate probabilities, site protonation curves, etc., can be evaluated analytically, the more straightforward approach is to evaluate these quantities by direct enumeration on the computer. This approach is not limited to a linear chain, but can be used to evaluate the ionization properties of arbitrary molecules with specified microscopic ionization constants and interaction parameters.

For illustration, Table 9 summarizes the resulting macroscopic pK values. For $N = 2$ their splitting is determined by the pair interaction parameter, up to the statistical factor $\log_{10} 4$. For $N = 3$ the first two pK values are close to the microscopic pK. Two protons bind independently. However, the last protonation step has a pK value which is the lowest; the proton has to overcome two pair interactions. A similar pattern is observed for $N = 4$. Two protons bind first, the next protonation step must overcome one pair interaction, while for the last protonation step two pair interactions must be overcome.

The resulting protonation pattern can be now better understood by investigating the microscopic constants. There are three possible microscopic pK values, namely 10 when the site has no nearest neighbors protonated, 8 when one neighbor is protonated, and 6 when both are protonated. The first proton binds to any site with equal probability, but when two protons are bound the microspecies with both terminal sites protonated is favored. The last proton must bind on this site, which has a microscopic pK of 6.

Table 9.   Calculated Macroscopic pK_n
Values for the Linear Chain with
Nearest-Neighbor Interactions with $p\hat{K} = 10$
and $\varepsilon = 2$ for Molecules with Different
Number of Sites N

N	n	pK_n
1	1	10.00
2	1	10.30
	2	7.70
3	1	10.48
	2	9.53
	3	5.99
4	1	10.60
	2	9.88
	3	7.82
	4	5.69

The applicability of the nearest-neighbor pair interaction model can be demonstrated by considering homologous series of linear aliphatic amines. Consider linear polypropylene amines whose ionization constants are given in Table 10 at an ionic strength of 0.1 M. For the interpretation one must observe that primary and secondary amine groups may have different proton affinities. The corresponding microscopic ionization constants will be denoted as $p\hat{K}^{(I)}$ and $p\hat{K}^{(II)}$, respectively. However, we do assume that the microscopic pK values of all (nonequivalent) secondary amine groups and that all nearest-neighbor pair interaction parameters, denoted by ε, are the same. We neglect next-nearest pair interactions and triplet interactions. This model has 3 adjustable parameters. These parameters are determined with a least-squares fit of all macroscopic pK values for the homologous series given in Table 10. The model accounts for all 9 experimental pK values to a reasonable accuracy. The resulting parameters are $p\hat{K}^{(I)} \approx 10.07$, $p\hat{K}^{(II)} \approx 9.70$, and $\varepsilon \approx 1.05$.

The same approach can be used to interpret site protonation curves. Such information can be obtained on polyamines with natural abundance ^{15}N NMR, or less directly with ^{13}C NMR.[108,112,114,115] Consider an example from the recent study by Koper *et al.*[108] Natural abundance ^{15}N NMR measurements were carried out on various polyamines as a function of pH. Figure 37 shows site protonation curves of 1,5,9-triazanonane with the formula $H_2N(CH_2)_3NH(CH_2)_3NH_2$. Due to the low concentrations of the ^{15}N isotope, however, these experiments must be carried out at rather high amine concentrations (50% by weight), and medium effects lead to substantial pK shifts. Thus any results obtained in this medium will

Table 10. *Comparison of Experimental Macroscopic* pK *Values of Linear Polypropyl Amines at Ionic Strength 0.1* M[211,212] *with Calculated Values Obtained from the Linear Chain Model*[202]a

Molecule		pK_n	
	N	Exp.	Calc.
Trimethylenediamine	2	10.52	10.37
$H_2N(CH_2)_3NH_2$		8.74	8.72
1,5,9-Triazanonane	3	10.65	10.45
$H_2N(CH_2)_3NH(CH_2)_3NH_2$		9.57	9.72
		7.69	7.57
1,5,9,13-Tetrazatridecane	4	10.46	10.53
$H_2N[(CH_2)_3NH]_2(CH_2)_3NH_2$		9.82	9.90
		8.54	8.68
		7.21	7.28

aWe denote by N the number of ionizable sites. The parameters $p\hat{K}^{(I)} \approx 10.07$, $p\hat{K}^{(II)} \approx 9.70$, and $\varepsilon \approx 1.05$ are used.

be only in qualitative, but not in quantitative agreement with the values quoted in Table 10.

The model used to explain these site protonation curves is analogous to the model discussed above. We take into account that the primary and secondary amine groups have different ionization constants and incorporate nearest-neighbor pair interaction. In this medium, next-nearest-neighbor pair interactions seem to be important as well. This simple model neglects triplet interactions, but is able to describe the site protonation curves shown in Fig. 37 reasonably well. The resulting microscopic ionization constants are $p\hat{K}^{(I)} \simeq 11.8$ and $p\hat{K}^{(II)} \simeq 11.5$, while the pair interaction parameters are $\varepsilon \simeq 1.55$ for nearest neighbors and $\varepsilon' \simeq 1.20$ for next-nearest neighbors. The corresponding macrostate probabilities are also shown in

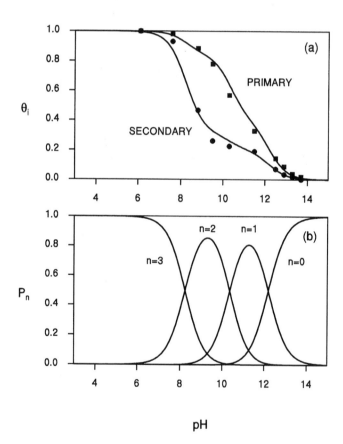

Figure 37. Protonation of 1,5,9-triazanonane, $H_2N(CH_2)_3NH(CH_2)_3NH_2$, in water at 50% by weight. (a) Site titration curves where the points are measured by ^{15}N NMR spectroscopy[108] and the solid line is the model with parameters $p\hat{K}^{(I)} \simeq 11.8$ (primary amine), $p\hat{K}^{(II)} \simeq 11.5$ (tertiary amine), $\varepsilon \simeq 1.55$ (nearest neighbors), $\varepsilon' \simeq 1.20$ (next-nearest neighbors). (b) The macrostate probabilities.

Fig. 37b. Figure 38 shows all resulting microscopic ionization constants and microstate probabilities. In the first protonation step the primary amine group protonates more readily, and in the second protonation step mainly both primary groups are protonated due to strong nearest-neighbor pair repulsion.

In Table 11 we further report on linear polyethylene amines with two carbon atoms between nitrogen atoms. In this case, we expect stronger interactions and possibly some interactions of higher order. The next-nearest pair interactions turn out to be unimportant but the nearest-neighbor triplet interaction parameter λ is not negligible. We determine the four model parameters from 9 experimental pK values of en, dien, and trien, and obtain p$\hat{K}^{(I)} \approx 9.42$, p$\hat{K}^{(II)} \approx 8.44$, $\varepsilon \approx 1.97$, and $\lambda \approx 0.42$.

The model can be solidly tested by predicting the pK values for tetren. Even though these corresponding pK values were not used in the fitting procedure, the predictive capabilities of the model are evident (see Table 11). Having this description at hand, one can also calculate all microconstants and microstate probabilities. The results for tetren are summarized in Fig. 39. In the first two steps, mainly the two primary amine groups protonate. In the third step, the middle secondary group protonates, resulting in a very stable microspecies where every second site is protonated. The remaining sites are much more acidic and protonate only at lower pH. Note that this *microscopic* information was solely derived on the basis of *macroscopic* pK values of a homologous series of molecules.

The cluster expansion method can be applied to molecules with an arbitrary number of sites. If one is interested in resolving the microscopic equilibria, one should typically restrict oneself to molecules with a limited number of sites (say $N < 15$), not because of difficulties with the cluster expansion approach, but simply because the number of microstates and possible microconstants become overwhelming, and they can be no longer visualized. Two examples will be considered.

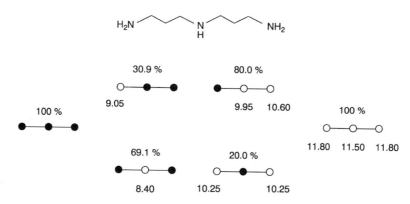

Figure 38. Ionization behavior of 1,5,9-triazanonane, $H_2N(CH_2)_3NH(CH_2)_3NH_2$, in water at 50% by weight derived from ^{15}N NMR spectroscopy.[108] Microstates, microscopic pK values, and conditional microstate probabilities.

*Table 11.   Comparison of Experimental Macroscopic pK Values of
Linear Short-Chain Polyethylene Amines[211,212] with Calculated
Values Obtained from the Linear Chain Model[202]a*

Molecule	N	pK_n	
		Exp.	Calc.
Ethylenediamine (en)	2	9.89	9.72
$H_2N(CH_2)_2NH_2$		7.08	7.15
1,4,7-Triazaheptane (dien)	3	9.84	9.74
$H_2N(CH_2)_2NH(CH_2)_2NH_2$		9.02	9.10
		4.23	4.08
1,4,7,10-Tetrazadecane (trien)	4	9.74	9.76
$H_2N[(CH_2)_2NH]_2(CH_2)_2NH_2$		9.07	9.16
		6.59	6.69
		3.27	3.36
1,4,7,10,13-Pentazatridecane (tetren)	5	9.74	9.78
$H_2N[(CH_2)_2NH]_3(CH_2)_2NH_2$		9.14	9.21
		8.05	8.30
		4.70	4.73
		2.97	2.99

*a*We denote by N the number of ionizable sites. We use the parameters $p\hat{K}^{(I)} \simeq 9.42$, $p\hat{K}^{(II)} \simeq 8.44$, $\varepsilon \simeq 1.97$, and $\lambda \simeq 0.42$.

The first application of the cluster expansion method is the estimation of the necessary interaction parameters (and thus all microscopic constants) from the macroscopic ones. This procedure is only applicable if the necessary molecular structure is sufficiently simple, and the pattern interaction parameter matrix is known for the molecular class under investigation.

As an example for this approach consider the potentiometric titration of tetramethylenedinitrilotetrakis(4-butylamine) in 0.5 M KCl.[83] The structural formula is given in Fig. 40. The potentiometric titration curve of this molecule was discussed in Section 2 (see Fig. 9). This molecule has 6 ionizable sites. For similar aliphatic polyamines, we know that nearest-neighbor pair interactions are the only important ones. With this simplification, one can devise a model with only 4 parameters: two microscopic ionization constants for the primary and tertiary amine group $p\hat{K}^{(I)}$ and $p\hat{K}^{(III)}$, and two interaction parameters ε and ε' for the butyl and propyl bond, respectively. The titration curve is directly fitted to this model and we obtain for the ionization constants $p\hat{K}^{(I)} \simeq 10.41$, $p\hat{K}^{(III)} \simeq 9.41$, and interaction parameters $\varepsilon \simeq 1.03$ and $\varepsilon' \simeq 0.59$. As a test of the model, one can compare the macroscopic ionization constants obtained from this model and those obtained from

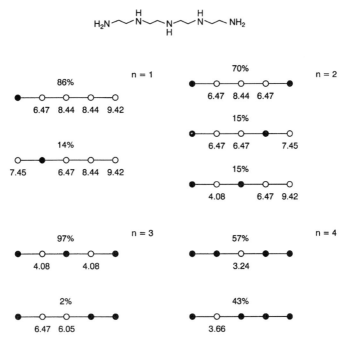

Figure 39. Microstates, microscopic pK values, and conditional microstate probabilities for 1,4,7,10,13-pentazatridecane (tetren). See also Table 11.

a direct fit of the titration curve. It can be shown that these two parameter sets agree rather well.[83] With this description at hand, all microscopic ionization constants as well as the microstate and macrostate probabilities can be calculated. The properties of the microstates are summarized in Fig. 40. A typical protonation pattern emerges: first the secondary amine groups protonate forming a very stable microspecies with all four primary amine groups protonated. The two remaining tertiary amine groups protonate at much lower pH.

A second example serves to illustrate the potential of the cluster expansion approach. As an example, consider a molecule with 14 ionizable groups. Figure 41 shows the site protonation curves obtained by ^{15}N NMR for a second-generation poly(propylene imine) dendrimer, also abbreviated as DAB-*dendr*-(NH$_2$)$_8$. (The corresponding NMR spectra were shown in Section 2 as an example.) The structural formula is given in Fig. 42 (top). The nearest-neighbor pair interaction model, which contains only 5 parameters, can interpret these results in a satisfactory fashion. The model parameters are as follows: microscopic ionization constants for primary amine groups p$\hat{K}^{(I)}$ and two different outer and inner tertiary amine groups p$\hat{K}^{(II)}$ and p$\hat{K}^{(III)'}$, respectively, and two nearest-neighbor pair interaction parameters, for the butyl bond ε and for the propyl bond ε'. The best fit with this model is

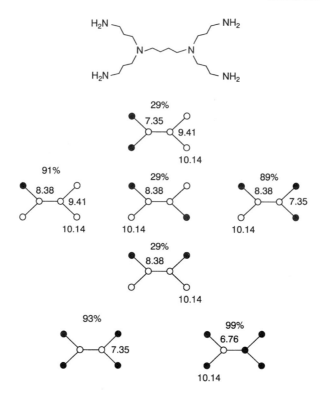

Figure 40.   Most abundant microstates with conditional probabilities and microscopic pK values for tetramethylenedinitriloterakis(4-butylamine) in 0.5 M KCl.[83] Parameters are p$\hat{K}^{(I)} \simeq 10.14$, p$\hat{K}^{(III)} \simeq$ 9.41, $\varepsilon \simeq 1.03$, and$\varepsilon' \simeq 0.59$.

shown in Fig. 41, and is based on the obtained ionization constants p$\hat{K}^{(I)} \simeq 10.70$, p$\hat{K}^{(III)} \simeq 10.35$, p$\hat{K}^{(III)'} \simeq 10.40$, and the interaction parameters $\varepsilon \simeq$ 1.30 and $\varepsilon' \simeq 0.75$. The appropriateness of this description is demonstrated by the fact that the same set of parameters can be also used to successfully model the site titration curves for other dendrimer generations, such as DAB-*dendr*-(NH$_2$)$_4$ and DAB-*dendr*-(NH$_2$)$_{16}$.[108]

These parameters are sufficient to derive a complete set of microconstants and conditional microstate probabilities. The most abundant microspecies are summarized in Fig. 42, having a similar protonation pattern as above. Four primary amine groups protonate first. The remaining four primary amine groups protonate only after one tertiary amine group in the center of the molecule is protonated. The second tertiary group in the center protonates only after all primary amine groups are protonated. The resulting microspecies is very stable, as it minimizes the nearest-neighbor interactions by protonating the outermost and innermost shell of the dendrimer.

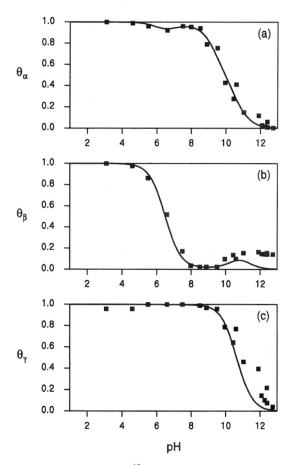

Figure 41. Site protonation curves obtained by ^{15}N NMR for second-generation poly(propylene imine) dendrimer DAB-dendr-(NH$_2$)$_8$ in a 50% aqueous solution.[108] The structural formula is given in Fig. 42.

4.4. Noninteracting Sites

An illustrative special case arises when there is no interaction between the sites. This situation occurs if all ionizable sites are far apart on a large molecule. As will be discussed in Section 5, the titration behavior of proteins can be often modelled within this approximation rather well.

4.4.1. Microscopic versus Macroscopic Picture

Without site–site interactions, the protonation of a given site is not influenced by the protonation state of the neighboring sites. Each site has then one particular

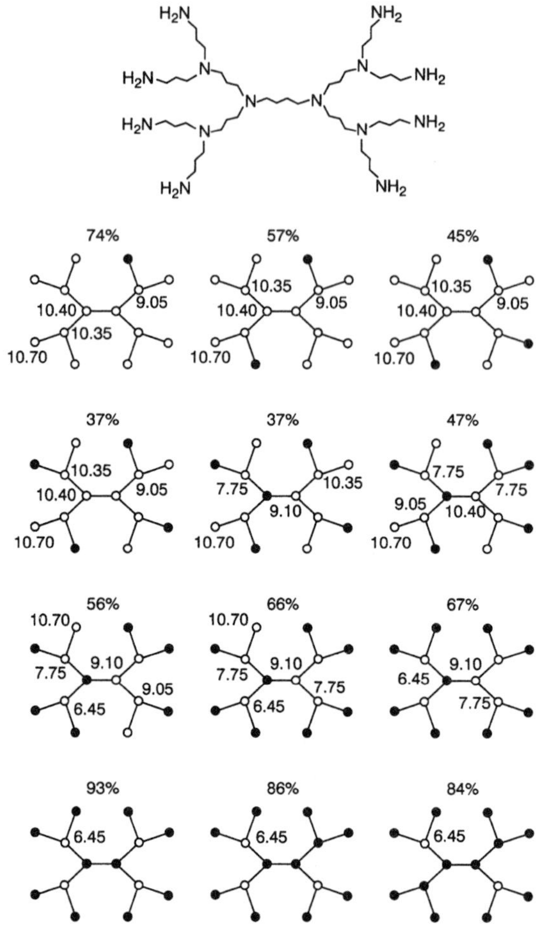

Figure 42.   Selected microstates, microscopic pK values, and conditional probabilities for the second-generation poly(propylene imine) dendrimer DAB-dendr-(NH$_2$)$_8$ in a 50% aqueous solution.[108]

microscopic ionization constant, which does not depend on the protonation state of neighboring sites. The ionizable groups protonate independently, and the site-specific titration curves conform to the simple case of a monoprotic acid or base

$$\theta_i = \frac{\hat{K}_i a_H}{1 + \hat{K}_i a_H} \tag{189}$$

for $i = 1, \ldots, N$. The overall titration curve follows immediately, as it is always given by the average of the site-specific titration curves [cf. Eq. (83)]

$$\theta = \frac{1}{N} \sum_{i=1}^{N} \theta_i = \frac{1}{N} \sum_{i=1}^{N} \frac{\hat{K}_i a_\mathrm{H}}{1 + \hat{K}_i a_\mathrm{H}} \tag{190}$$

Clearly, noninteracting sites titrate as a dilute mixture of monoprotic acids and bases. The corresponding microscopic association constants $\hat{K}_i$ do not depend on the protonation of the neighboring sites; a given site has always the same microscopic pK value.

Since all sites can be thought as being part of a single macromolecule, one must be able to interpret this titration curve as the one of a polyprotic acid. Comparing Eqs. (190) and (164), one finds the corresponding macroscopic association constants for a diprotic acid ($N = 2$)

$$\overline{K}_1 = \hat{K}_1 + \hat{K}_2 \tag{191}$$

$$\overline{K}_2 = \hat{K}_1 \hat{K}_2, \tag{192}$$

and for a triprotic acid ($N = 3$)

$$\overline{K}_1 = \hat{K}_1 + \hat{K}_2 + \hat{K}_3 \tag{193}$$

$$\overline{K}_2 = \hat{K}_1 \hat{K}_2 + \hat{K}_2 \hat{K}_3 + \hat{K}_1 \hat{K}_3 \tag{194}$$

$$\overline{K}_3 = \hat{K}_1 \hat{K}_2 \hat{K}_3. \tag{195}$$

For $N > 3$ the appropriate relations can be derived as well and were known for some time.[4] The equivalence between Eqs. (190) and (164) can be established much more directly by considering the semi-grand partition function (also referred to as the binding polynomial).[200] In the case of independent sites, the partition function simply reads

$$\Xi = \prod_{i=1}^{N} (1 + \hat{K}_i a_\mathrm{H}). \tag{196}$$

Expanding this polynomial and comparing with Eq. (65) directly yields the above relations.

For identical and noninteracting sites with a microscopic affinity constant $\hat{K}$, all site titration curves are the same and the overall titration curve is

$$\theta = \frac{\hat{K} a_\mathrm{H}}{1 + \hat{K} a_\mathrm{H}} \tag{197}$$

which is again identical to the titration curve of a monoprotic acid or base. However, the results can be again interpreted as originating from a polyprotic acid. The partition function now simplifies to

$$\Xi = (1 + \hat{K}a_{\mathrm{H}})^N.$$

(198)

Application of the binomial theorem and comparing with Eq. (65) leads directly to the macroscopic association constants

$$\overline{K}_n = \hat{K}^n \binom{N}{n}.$$

(199)

The corresponding macroscopic pK values become[4]

$$\mathrm{p}K_n = \mathrm{p}\hat{K} + \log_{10} \frac{N+1-n}{n}.$$

(200)

The macrostate probabilities are given by

$$P_n(z) = \binom{N}{n} \frac{(\hat{K}a_{\mathrm{H}})^n}{(1 + \hat{K}a_{\mathrm{H}})^N}.$$

(201)

The resulting behavior is shown in Fig. 43, and a few examples are given in Table 12.

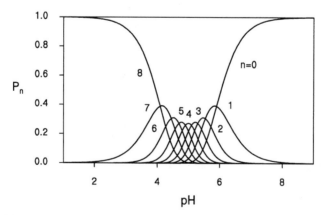

Figure 43. Macrostate probabilities for a polyprotic acid or base with identical noninteracting sites with p$\hat{K} = 5$ ($N = 8$).

Table 12. *Molecules with Noninteracting, Identical Ionizable Sites*[a]

Molecule	N	I[M]	pK_n		$p\hat{K}$
			Exp.	Calc.	
Azelaic acid	2	3.0	5.31	5.29	4.99
$HOOC-[CH_2]_7-COOH$			4.66	4.69	
cis,cis-cyclohexane-1,3,5-tricarboxylic acid	3	0.5	4.96	4.80	4.33
$C_6H_9(COOH)_3$			4.33	4.33	
			3.75	3.85	

[a]Macroscopic pK values are compared with predictions of Eq. (200).

4.4.2. Affinity Distributions

A rather simple approach has become popular for the interpretation of acid–base titration curves of complex molecules, particularly with a large number of ionizable sites ($N \gg 1$). Usually, one can represent the titration curve of a complex molecule as originating from a dilute mixture of monoprotic acids or bases. Its acid–base titration curve is nothing but a simple linear combination of individual titration curves of independent monoprotic acids or bases and can be written as [cf. Eq. (190)]

$$\theta = \sum_i \frac{\tilde{K}_i a_H}{1 + \tilde{K}_i a_H} \rho_i \tag{202}$$

where ρ_i represents the mole fraction of the monoprotic acid of type i with an association constant $\tilde{K}_i$. For a large number of individual acids and bases one may introduce a continuous representation

$$\theta(a_H) = \int_0^\infty \frac{K a_H}{1 + K a_H} P(K) dK \tag{203}$$

where $P(K)$ is the affinity distribution which is a function of the association constant K. This function can also be reported as a pK distribution (pK spectrum) which is considered as a function of $pK = \log_{10} K$ [cf. Eq. (106)]. If properly normalized, the latter function can be expressed as $KP(K) \ln 10$.

The idea to view the titration behavior of a complex mixture in terms of a statistical distribution of independent sites is rather old.[43,44] These techniques have become popular in biochemistry and environmental chemistry for the interpretation of acid–base titration curves of complex materials.[45–47,218,219] This problem can be approached in three different ways.

One possibility is to assume a certain form of the affinity distribution and evaluate the integral analytically. The distribution $P(K) \propto K^{-1}$ in the interval $K_1 \le$

$K \leq K_2$ (and zero elsewhere) can be often used to interpret titration curves. This distribution corresponds to a constant pK spectrum in a finite interval (box distribution). The evaluation of the integral yields the simple expression, also referred to as the Temkin isotherm,

$$\theta(a_H) = \left(\ln \frac{K_1}{K_2} \right)^{-1} \ln \frac{1 + a_H K_1}{1 + a_H K_2} . \tag{204}$$

Clearly, the corresponding isotherms can be derived for other forms of the affinity distribution as well.

A second approach is to start with a certain functional form of the titration curve and then evaluate the corresponding affinity distribution. This approach was introduced by Sips[220] who obtained an explicit inversion relation of Eq. (203), but entirely analogous methods were developed in the analysis of dielectric spectra.[221] For continuous distributions, the equation proposed by Sips can be simplified to[222,223]

$$P(K) = \frac{1}{\pi K} \left| \Im m \; \theta \left(-\frac{1}{K} \right) \right| \tag{205}$$

where $\Im m$ represents the imaginary part. A classical example for this procedure is the Langmuir–Freundlich isotherm

$$\theta(a_H) = \frac{(a_H \overline{K})^\nu}{1 + (a_H \overline{K})^\nu} \tag{206}$$

where ν is the Freundlich exponent ($0 < \nu < 1$). This equation is frequently written as

$$pH = pK + \nu^{-1} \log_{10} \frac{1 - \theta}{\theta} \tag{207}$$

where ν is now referred to as the Hill exponent.[13] Inserting Eq. (206) into Eq. (205) leads to the Sips distribution

$$P(K) = \frac{1}{\pi K} \frac{(K/\overline{K})^\nu}{1 + 2(K/\overline{K})^\nu \cos(\pi\nu) + (K/\overline{K})^{2\nu}} . \tag{208}$$

For illustration, the two mentioned isotherms and the corresponding affinity distributions are shown in Fig. 44. The box distribution corresponding to the Temkin isotherm decays sharply to zero, and leads to a rapid approach to the plateau and a Henry regime $\theta \propto a_H$ for $a_H \to 0$. On the other hand, the Sips distribution decays only algebraically and the Henry regime is never attained. For $a_H \to 0$ the isotherm decays slowly as $\theta \propto a_H^\nu$.

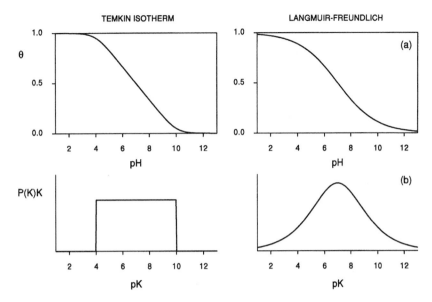

Figure 44. Illustrative isotherms and corresponding affinity distributions for the Temkin and Langmuir–Freundlich isotherms. (a) Titration curve and (b) affinity distribution. For the Temkin isotherm we use $K_1 = 10^4$ M and $K_2 = 10^{10}$ M, and for the Langmuir–Freundlich isotherm $\overline{K} = 10^5$ M and $\nu = 0.3$.

The third approach is to obtain the affinity distribution directly from an experimentally observed titration curve. One must realize, however, that extracting an affinity distribution from experimental data is a nontrivial task, which belongs to the class of so-called *ill-posed* problems. One must apply special regularization techniques in order to avoid serious numerical instabilities.[47,224] These regularization techniques rely on a specific *a priori* hypothesis about the structure of the distribution. For example, such a hypothesis could be that the distribution is a smooth function or that it consists of a small number of discrete peaks.[47] Once such a hypothesis is decided upon, various computer algorithms are available to calculate the corresponding affinity distribution. More details on such techniques can be found in Appendix C. Various approximate methods were also developed earlier to obtain affinity distributions from experimental titration curves. Since the regularization methods discussed here can be easily run on any personal computer, such approximate methods now become largely obsolete.

The concept of the affinity distributions appears intuitive in the case of fulvic acids, for example, which are complex mixtures of low-molecular-weight organic molecules occurring in natural environments such as soils or surface waters. In fulvic acids, a substantial fraction of the ionizable groups is probably in the form of monoprotic acids or on effectively isolated sites, having negligible interactions with neighboring sites. Figure 45a shows an experimental titration curve of a fulvic acid.[225] One observes that this titration curve differs markedly from a titration curve of a monoprotoic acid or base. In this system, a meaningful interpretation can be given in terms of a linear combination of independent ionizable sites.

The corresponding affinity distributions were calculated from the titration curve shown in Fig. 45 using the mentioned regularization methods. A distribution, which is derived on the basis of the *smoothness hypothesis*, is shown in Fig. 45b, while in Fig. 45c a distribution regularized for a *small number of discrete peaks* is given. Both distributions lead to an excellent fit of the experimental data. In the present case of the fulvic acid, the smoothness hypothesis may in fact represent a reasonable approximation of the true distribution of the ionization constants of the monoprotic acids present.[226]

In many other instances, the same affinity distribution approach was applied to systems such as proteins, humics, or oxides.[134,219,226–228] Clearly, these substances are no simple mixtures of monoprotic acids or bases, yet the affinity distribution description seem to be compatible with experimental data. In systems where the ionizable groups cannot be considered as independent, the appropriateness of the procedure and the significance of the affinity distribution is far from clear. However,

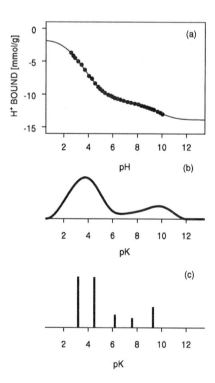

Figure 45. Acid–base properties of a fulvic acid in 1.0 M KCl.[225] (a) Experimental titration curve. Affinity distributions derived from numerical inversion techniques with regularization for (b) smoothness and (c) small number of sites. Fits of the titration curves derived from affinity distributions (b) and (c) are essentially identical; the curve calculated from smooth spectrum (b) is shown in (a). Reprinted from Ref. 226 with permission from Elsevier Science.

it turns out that the affinity distribution is usually well defined even when the sites are interacting, although its interpretation is no longer straightforward. Appendix C discusses some of these aspects in more detail.

From the practical point of view, an affinity distribution constitutes an empirical representation of a titration curve and may, indeed, be a useful fingerprint of the ionization behavior of a complex system. However, this interpretation provides little insight into the actual molecular mechanisms.

4.5. Interpretation and Prediction of Ionization Constants

The ability to predict ionization properties of small molecules may not seem essential—a wide variety of such molecules is covered in the extensive compilations of pK values.[207–213] Predictive methods are important, however, since data of this kind might not be available for systems involving a larger number of ionizable sites, such as proteins, polyelectrolytes, and ionizable interfaces. For this reason, the prediction of ionization properties of small molecules represents an invaluable testing ground in order to assess the reliability of the predictive methods.

Empirical techniques are available for the prediction of pK values, even for rather complex compounds. These techniques often rely on fully empirical correlations; in other cases, they are based on semiempirical schemes and use various molecular properties as input.

4.5.1. Empirical Methods

For organic molecules, empirical methods are rather powerful and well established,[6,229] especially for the estimation of the ionization constants of monoprotic acids and bases, but are equally applicable to polyprotic molecules, leading to microscopic constants. Thus, interaction parameters can be estimated as well.

Since an excellent introduction into the subject is available,[6] these methods will not be discussed here in much detail. The basic tenets are linear free-energy relationships. To a first approximation, the contribution of a substituent is additive in the free energy, and leads to a corresponding change in the pK. This observation provides the basis of Hammett and Taft equations, which are the most widely used methods of pK prediction.[75] The starting point is a residue with an ionization constant pK_0 which is taken as a reference state. Usually, the reference residue is bound to an alphatic chain. The pK value of the same group in different molecule environment is then expressed as

$$pK = pK_0 - \rho \sum_i \sigma_i \tag{209}$$

where ρ is a group specific constant, and σ_i are the contributions of the individual neighboring substituents, which are independent of the chemical nature of the group. Tables of these constants are available.[6]

Consider glycine H_2N-CH_2-COOH as an example. In the available tables one finds that the carboxylic acid residue bound to an aliphatic chain has $pK_0 \simeq 4.8$ at infinite dilution. An amino group attached to the α-carbon has a contribution of $\rho\sigma \simeq 0.5$ and $pK \simeq 4.3$. If the amino group is ionized, then $\rho\sigma \simeq 2.4$, which leads to $pK \simeq 2.4$. This procedure leads to both microscopic ionization constants; their difference gives the pair interaction parameter $\varepsilon \simeq 2.9$.

This example provides just a simple illustration of this rather well-developed method. Benzene rings and other aromatic compounds can be also handled with good confidence, and computer programs for this kind of analysis are available. Although these methods are applicable to complex molecules, they are best suited for molecules with a small number of ionizable sites (typically $N \leq 3$). For polyprotic molecules (already $N > 3$), these methods become cumbersome quickly.

While these techniques yield only microscopic ionization constants, in principle, interaction parameters are also accessible, but the resulting values are usually much less accurate. From linear free-energy relationships one will necessarily derive only interaction parameters between pairs of sites; triplet and higher-order contributions are always neglected. Interaction parameters can be also estimated from empirical correlations to a good degree of accuracy. Figure 46 shows pair interaction parameters for linear amines as a function of the number of atoms in between the ones accepting the protons.[202]

Methods for estimating ionization constants for inorganic compounds are less well established. In such systems, oxygens usually represent the ionizable sites. Depending on the substituent, the hydroxyl group will act as either a base or an acid, but never simultaneously in both ways. Examples of this behavior are given in Fig. 47.

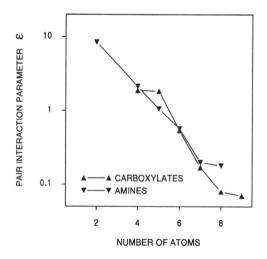

Figure 46.   Interaction parameter ε as a function of the number of atoms L in between the acidic protons for diprotic aliphatic carboxylic acids and amines at ionic strength of 1 M (see also Tables 5 and 6).

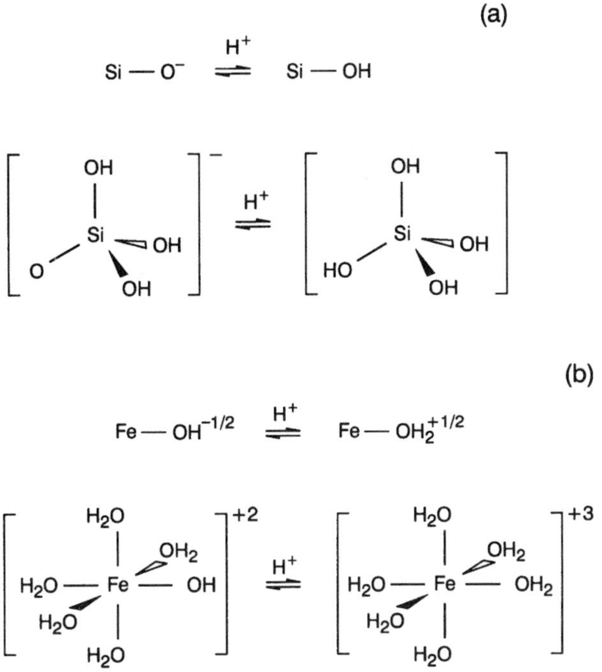

Figure 47. Oxygen as the typical ionizable group in inorganic complexes. (a) Silicic acid has four ionizable sites and hydroxyl groups which act as acids, and (b) iron(III)–hexaquocomplex with six ionizable sites where the hydroxyl groups acts as bases.

One of the oldest methods to estimate pK values for inorganic compounds is due to Pauling.[40,230] One simply counts the number of singly and doubly bonded oxygens to the central atom. This model was motivated by an electrostatic treatment[231] and later Ricci[232] pointed out that essentially equivalent predictions of the ionizations constants can be achieved by counting the bound oxygens. More recently, semiempirical methods for estimation of ionization constants were developed, which rely on valence bond considerations. The simplest definition of the bond valence is given by the charge of the cation divided by the coordination number.[40,230,233] Consider the examples shown in Fig. 47. For Si(IV) the valency of the Si–O bond is 1, and together with the charge of −2 of the oxygen atom one obtains the charge of −1. For Fe(III) the Fe–O bond has a valency of 1/2, and with the charge of −1 of the OH group it leads to the charge of −1/2.

A simple application of this valence bond principle was recently described by Bleam.[233] One can establish a correlation between the intrinsic ionization constants (i.e., ionization constants of the neutral ion) and the valency for various types of inorganic species. These considerations have been refined by using a more precise definition of the bond valence and hydrogen bonding. This definition of the actual bond valence assumes an exponential dependence between the length of the oxygen bond and the bond valence, and has been particularly successful in predicting structural and crystallographic properties of various inorganic compounds.[230,234] Ionization constants of a variety of inorganic compounds correlate also to this definition of valence rather well (see Fig. 48).

4.5.2. Methods Based on First Principles

Ab initio calculation of pK values of small molecules is a difficult undertaking. The main problem is that the free-energy difference of individual species, determining this equilibrium constant, originates from the difference of two large numbers. Very accurate free-energy estimates are therefore needed. Such quantities are commonly calculated from the scheme shown in Fig. 49. For the association reaction in water

$$A^- + H^+ \rightleftharpoons HA \tag{210}$$

the free-energy difference is obtained from

$$\Delta F^{(aq)} = \Delta F^{(gas)} + \Delta F_{HA}^{(hydr)} - \Delta F_{H^+}^{(hydr)} - \Delta F_{A^-}^{(hydr)} \tag{211}$$

where $\Delta F^{(gas)}$ and $\Delta F^{(aq)}$ refer to the reaction free energies in the gas phase (gas-phase acidities) and in water (pK values), respectively. The hydration free energies for species i are denoted by $\Delta F_i^{(hydr)}$.

With quantum mechanical techniques one calculates the free energy of proton dissociation in the gas phase (gas-phase acidities). This calculation requires accu-

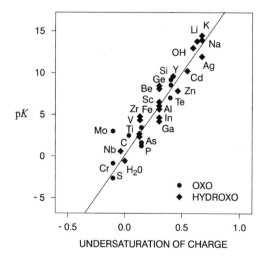

Figure 48. Correlation between pK values of various inorganic oxo and hydroxo complexes and the actual bond valence.[333]

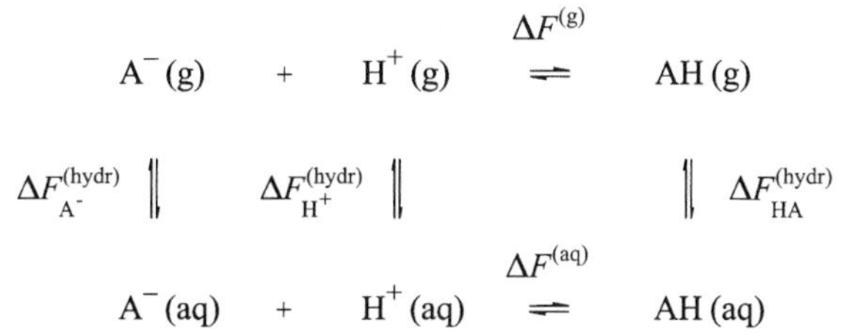

$$A^-(g) \quad + \quad H^+(g) \quad \overset{\Delta F^{(g)}}{\rightleftharpoons} \quad AH(g)$$

$$\Delta F_{A^-}^{(hydr)} \Big\updownarrow \qquad \Delta F_{H^+}^{(hydr)} \Big\updownarrow \qquad\qquad\qquad \Big\updownarrow \Delta F_{HA}^{(hydr)}$$

$$A^-(aq) \quad + \quad H^+(aq) \quad \overset{\Delta F^{(aq)}}{\rightleftharpoons} \quad AH(aq)$$

Figure 49. Thermodynamic cycle used for *ab initio* pK calculations. The free-energy differences for the ionization reaction are $\Delta F^{(gas)}$ in gas phase and $\Delta F^{(aq)}$ in water. The hydration free-energy differences are denoted as $\Delta F_i^{(hydr)}$.

rate energy hypersurfaces, which are then used to evaluate the partition function for individual species. For small molecules in gas phase, such methods are well established.[145–147,235–237] The next step requires the free solvation energies of the individual species. This problem is usually approached with dielectric cavity models or molecular dynamics simulations.[10,145–147,238] The difficult part of this calculation is to calculate free energies to sufficient accuracy. To overcome these problems, solvation free-energy differences can be directly evaluated along a fictitious thermodynamic path, where one solute is mutated into another. Rustad *et al.*[238] have used this technique to estimate the pK values of Fe(III)–hexaquocomplex. Similarly, differences between pK values of various small organic molecules in water were successfully estimated.[10,146,147]

Gas-phase acidities have been approximately calculated by evaluating total energies using quantum mechanical approaches for a number of simple compounds.[235–237,239] In many cases, these results were also correlated to pK values in solution. However, due to the large solvation contributions, the appropriateness of such procedures remains questionable.

A better approach is to calculate the solvation energies from a dielectric model. If coupled to *ab initio* calculations, such methods can yield the correct range of pK values of simple molecules in water. Table 13 illustrates these trends by comparing pK values calculated according to different models with experiment. Chen *et al.*[146] use various density functional methods coupled to continuum dielectric models, while Kallies *et al.*[147] employ an extended basis set as well as density functional methods for the quantum mechanical calculations together with polarized continuum models. One observes that the range of the pK values can be reasonably well estimated, but the accuracy is still limited to about 1–2 pK units due to the inherent inaccuracies of the total energies.

The calculation of the pair interaction parameters in symmetric diprotic acids and bases has a long history. Already Bjerrum[2] recognized that the interaction parameter decreases with increasing distance between the ionizable groups and suggested that this interaction originates from the Coulomb repulsion energy

Table 13.   *Ab initio pK Calculations of a Few Simple Compounds in Water*[a]

Base form	Chen *et al.*[146]	Kallies *et al.*[147]	Experiment
Aniline $C_6H_5-NH_2$		2.6 – 4.4	4.6
Acetate ion CH_3-COO^-	4.1 – 6.9		4.7
Imidazole $C_3N_2H_4$	7.6 – 9.8	7.1 – 8.6	7.1
Ammonia NH_3		6.9 – 8.6	9.3
Methylamine CH_3-NH_2	11.2 – 13.7	10.0 – 10.2	10.6
Hydroxide ion OH^-	14.0 – 16.9		15.7

[a]Chen *et al.*[146] used density functional methods coupled to continuum dielectric models, while Kallies *et al.*[147] use an extended basis set for the quantum mechanical calculations together with polarized continuum models. Ranges of the pK values calculated according to different schemes are compared with experiment.

$$W = \frac{e^2}{4\pi\varepsilon_0 D_w} \frac{1}{r} \tag{212}$$

where r is the distance between the charged groups. The interaction parameter ε_{12} is simply related to this energy by $\varepsilon_{12} = \beta W/\ln 10$. This observation was a cornerstone in the development of a molecular picture of ionization reactions for larger molecules.

Figure 50 compares experimental data for symmetric dicarboxylic acids extrapolated to infinite dilution with the Bjerrum model [cf. Eq. (212)]. The distances correspond to *trans* conformers, as obtained from molecular force field calculations, and are in good agreement with X-ray data of solids.[240] The simple model does extremely well at distances >0.6 nm; the disagreement at shorter distances was already interpreted by Bjerrum as originating from the low relative permittivity of the hydrocarbon chain in between the ionizable residues. Kirkwood and Westheimer[3] have analyzed this idea quantitatively by estimating the electrostatic free energy of two point charges in spherical and ellipsoidal cavities of low relative permittivity.[241,242] For large eccentricities of the ellipsoid, this model reduces to the original Bjerrum model. Much more recently, such calculations were made with a refined shape of the dielectric cavities, which reflect the proper molecular geometry, and using more realistic charge distributions within the molecules.[240,243] The results of these calculations using a relative permittivity of 2 inside the cavity by Rajasekaran *et al.*[243] compare very well with experimental data (see Fig. 50).

In all these models the salt and the solvent are represented as a structureless continuum. Recently, Figueirido *et al.*[140] have compared such continuum models for succinic acid with a molecular dynamics study. The interaction parameters were estimated from the free-energy differences of the various ionization states for each conformer. Succinic acid has three rotational conformers: *trans* (t) and two equiva-

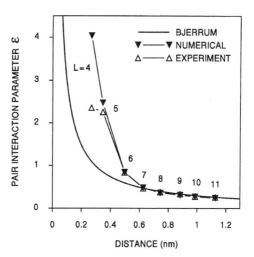

Figure 50. Interaction parameter ε as a function of distance between ionizable groups for symmetric dicarboxylic acids.[240] Experimental data are compared with finite difference solutions of the Poisson equation for realistic molecular geometries and with the Bjerrum model [cf. Eq. (212)]. The distances between the ionizable groups are obtained from molecular force field calculations. The number of atoms in between the acidic protons L corresponds to the entries in Table 5.

lent *gauche* (g) conformers (see Fig. 28). The resulting interaction parameters obtained from molecular dynamics are summarized in Table 14. For comparison, corresponding results from dielectric continuum models using the actual shape of the molecules are also given.

Since the ionizable groups are more distant in the *trans* conformer, the interactions between the groups are weaker and the *trans* conformer shows a smaller splitting than the *gauche* conformer. This observation is also in line with structural analogs. At infinite dilution, the interaction parameter for fumaric (*trans*) and maleic (*cis*) acids are 0.8 and 3.7, respectively. These values are comparable to those of the appropriate conformers given in Table 14.

Each of these conformers will have its own ionization constants and experimentally one observes their thermal average. The experimentally evaluated formation constants $\overline{K}_1$ and $\overline{K}_2$ and the corresponding interaction parameter ε are obtained

Table 14. Comparison of the Interaction Parameters for Succinic Acid $\varepsilon = pK_2 - pK_1 - \log_{10}4$ from Experimental Data and Calculated by Various Methods

	trans	*gauche*	Average
Experimental[7]			0.8
Molecular dynamics[140]a	0.5	3.2	1.0
Dielectric[140]	1.0	2.3	1.5
Dielectric[243]	0.8		1.3[b]
Dielectric[240]	1.8		2.3[b]

[a]Error bar is about 0.3 pK units.

[b]The calculated value from Ref. 140 was used to evaluate the thermal averages. Adapted from Ref. 140.

from these thermal averages [cf. Eq. (149)]. In order to do so, Figueirido *et al.*[140] have estimated that $\rho_2^{(t)} \simeq \rho_1^{(t)} \simeq 0.35$ and concluded that the intrinsic ionization constants of the carboxylic groups in both conformers must be comparable $(p\hat{K}^{(t)} \simeq p\hat{K}^{(g)} \simeq 4.5)$. The resulting interaction parameter is compared with experiment in Table 14, showing the molecular dynamics results to agree rather well with the experimental values. The electrostatic models appear to overestimate the spitting slightly.[140] The resulting microscopic pK values and conformer populations are summarized in Fig. 51, suggesting that succinic acid will undergo a conformational transition with pH. At low pH, *trans* and *gauche* conformers are both present, while at higher pH one finds only the *trans* conformer. According to the interpretation of Figueirido *et al.*,[140] the excellent agreement between the experiment and continuum calculations for succinic acid as reported by Rajasekaran *et al.*[243] is due to a fortuitous cancellation of errors (cf. Table 14). More work along these lines is needed before the accuracy of continuum calculations and the importance of conformational states can be assessed unambiguously.

This example suggests the importance of various conformational states, an issue which has not been addressed in much detail in the literature. As discussed in Section 3, a given microspecies must be regarded as a mixture of various conformers. In some cases, one might be able to identify the most likely conformer and thus a preferred geometry, but as a rule one must expect that several conformers will contribute significantly. For the calculation of ionization constants and interaction parameters with fixed geometry models, the problematic assumption is probably not the one of fixed geometry, but rather that only one particular conformer is being considered. If one attempts to link the ionization behavior of larger molecules to their structure, conformational degrees of freedom require more detailed considerations.

The role of conformation degrees of freedom on ionization equilibria has been addressed in rather detailed fashion from the experimental point of view. Various authors have attempted to estimate the contributions of different conformers in ionization equilibria.[199,244] Aspartic acid was studied in quite some detail. In its

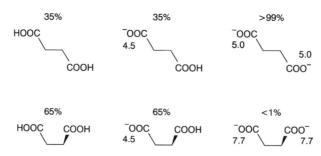

Figure 51. Microscopic ionization constants and conditional conformer probabilities for succinic acid.[140]

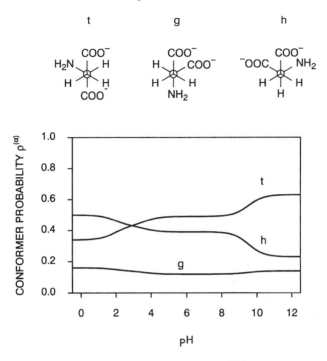

Figure 52. Conformational populations of aspartic acid[244] as a function of pH.

fully deprotonated state, aspartate $^-OOC-CH_2-HC(-NH_2)-COO^-$ is a three protic base with three possible conformational states (see Fig. 52). By combining potentiometry, NMR, and results on similar molecules, Noszál *et al.*[244] have proposed a set of microconstants for all three different conformers. Table 15 summarizes the conditional conformer probabilities $\rho_n^{(\alpha)}$, while Fig. 52 shows that the probabilities of various conformers are strongly pH-dependent, and pH variations cause substantial changes in conformational populations.

Table 15. *Conditional Conformer Probabilities of Various Macrospecies n of Aspartic Acid in 2 M NaCl*[244]a

n	pK_n	Conformer probability		
		t	g	h
0	—	0.36	0.14	0.23
1	9.53	0.49	0.12	0.39
2	3.65	0.43	0.14	0.43
3	1.96	0.34	0.16	0.50

aThe corresponding macroscopic pK values are also given.

5. Proteins

All proteins contain ionizable residues. These groups are mostly attached to amino acid side chains and largely involve amines and carboxylic groups, but phenolic and imidazole groups are important as well (see Fig. 5). As a first approximation, one may use the pK values of individual amino acids to estimate the pK values of the ionizable residues. If, however, the dielectric environment of the residues within the protein is different from the one for the small molecule in solution, the pK values of the ionizable residues will be shifted from the values of the dissolved amino acids. Moreover, the individual ionizable groups in a protein will interact mutually. If these interactions are weak the individual residues will titrate independently, but stronger interactions may substantially modify the titration behavior of individual groups. The final complication is related to the number of ionizable sites in a protein, which is typically on the order of 10–100. This number is too small to invoke some averaging procedure, but rather large if each site is to be treated individually.

Many proteins have been investigated experimentally with respect to their ionization behavior and it is beyond the scope of the present article to offer an exhaustive review. Potentiometric titration and optical spectroscopy techniques were heavily used in the past.[139,245–247] The importance of these techniques has declined due to the excellent resolution of nuclear magnetic resonance (NMR). As discussed in Section 2, NMR can quantitatively monitor the ionization of individual sites in a complex protein. This method was used to study various proteins, including different lysozymes,[248,249] ribonuclease,[118,119] ovomucoid,[113,250] calbindin,[68,109,116] calmodulin,[251] myoglobin,[15] typsin inhibitor,[117] papain,[252] human growth hormone,[126] thioredoxin,[253–256] xylanase,[257] subtilisin,[258] and barnase.[259] The data vary in quality; in some cases the most elementary experimental parameters, such as protein concentration, electrolyte type, or its concentration, are missing.[260] The data interpretation is not always straightforward. In order to measure the site-specific titration curves with confidence, it is essential to simultaneously monitor the chemical shift of various nuclei in the vicinity of the ionizable group[68,120]; protein titration curves based on the measurements of single nuclei cannot be always trusted (see Section 2).

Ionization behavior of proteins has been interpreted on different levels of sophistication (see Fig. 22). Initial approaches treated the protein as a sphere immersed in a structureless solvent and assumed a uniform charge density or a certain charge distribution. More recently, models have been put forward which treat the protein as well as the solvent in certain atomistic detail. The microscopic modeling is, of course, appealing as it avoids all the possible pitfalls associated with a continuum treatment. Even with the present day's supercomputers, the microscopic approach still lacks numerical reliability. In spite of the availability of various schemes to treat strongly interacting systems, it is important to realize that ionizable groups in proteins interact only weakly. The simplest model, which considers all sites to be independent and assumes a common ionization constant for a given type of amino acid, often leads to quite reliable predictions of the ionization behavior of proteins. The titration behavior of proteins has also been recently reviewed by others.[17,66,260–262]

5.1. The Null Model

The ionizable groups in proteins are the chain terminals and the amino acid side groups. Recall, for example, that the amino acid lysine has three ionizable groups. During polymerization to poly(lysine) the adjacent amine and carboxyl groups form the amide bond which is part of the polymer backbone. Thus, only the primary amine groups in the side chains of the amino acids and the terminal groups of the protein backbone will ionize within the protein.

Amines and carboxylates represent the most important types of ionizable groups. Amines occur in side chains of lysine or on the N-terminal and dissociate according to the reaction

$$-NH_3^+ \rightleftharpoons -NH_2 + H^+ \tag{213}$$

The amine group is neutral when deprotonated ($Z_i = 0$) and positively charged in its protonated state, with pK around 10–11. Carboxylates represent the reactive groups for aspartic acid, glutamic acid, and the C-terminal according to the reaction

$$-COOH \rightleftharpoons -COO^- + H^+ \tag{214}$$

with pK around 4–5. These groups are negatively charged in the deprotonated state ($Z_i = -1$). Among other ionizable groups that occur in proteins the most important are guanidyl in arginine, imidazol in histidine, phenyl in tyrosine, and thiol in cystein (see Table 16 and Fig. 53).

Table 16. Summary of the p$\hat{K}$ Values for the Null Model, where the Ionizable Sites are Assumed to Protonate Independently^a

Residue	Abbr.	Group	Charge	Null model[206]	Improved null model[263]
N-terminal	N-ter	amine	0	7.5	8.2
Glutamic acid	Glu	carboxyl	−1	4.4	4.0
Aspartic acid	Asp	carboxyl	−1	4.0	2.7
Tyrosine	Tyr	phenyl	−1	9.6	10.7
Lysine	Lys	amine	0	10.4	10.1
Arginine	Arg	guanidyl	0	12.0	—
Histidine	His	imidazol	0	6.3	6.9
Cystein	Cys	thiol	−1	8.3	—
C-terminal	C-ter	carboxyl	−1	3.8	2.7

^{a}The original null model as summarized by Antosiewicz *et al.*,[260] and the improved null model[263] result from a statistical analysis of various proteins. The values refer to 0.01–0.1 M monovalent electrolyte solutions.

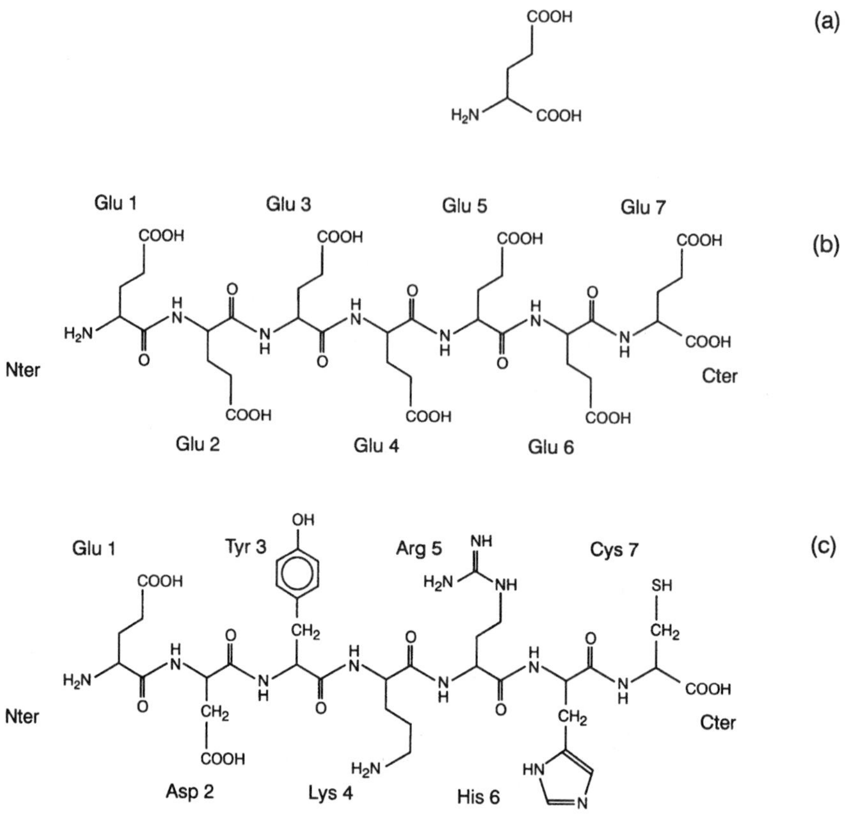

Figure 53. The comparison of individual amino acid molecules and the corresponding polypeptides. (a) Glutamic acid, (b) poly(glutamic acid), (c) oligopeptide Glu-Asp-Tyr-Lys-Arg-His-Cys, which shows the most important ionizable groups in a protein.

Ionizable residues in a protein molecule are often situated far apart and usually interact only weakly. For this reason, the simplest (and often still the most success-ful) model of the ionization of the individual residues is the so-called *null model*. This model neglects site–site interactions altogether, and any individual site ionizes as if no other sites were present, as already discussed in Section 4. The degree of ionization of any individual site is given by

$$pH = p\hat{K}_i + \log_{10} \frac{1 - \theta_i}{\theta_i} \tag{215}$$

where θ_i is the degree of protonation of site i and $p\hat{K}_i$ is the corresponding microscopic ionization constant. The overall degree of protonation (macroscopic titration curve) follows from the average of the site-specific titration curves [cf. Eq. (190)]

$$\theta = \frac{1}{N} \sum_i \theta_i \qquad (216)$$

where N is the total number of ionizable sites. Table 16 summarizes different estimates for the pK values for common ionizable residues. The original null model by Antosiewicz *et al.*[260] is based on older literature data, while the improved null model[263] is based on a statistical analysis of site-specific titration curves of various proteins.

Nozaki and Tanford carried out potentiometric titrations of ribonuclease in 6 M guanidine hydrochloride (GuHCl).[139] The medium was chosen to eliminate the site–site interactions, and one can assume that the sites ionize independently to a reasonable degree of approximation. The resulting potentiometric titration curve is represented in Fig. 54 together with the corresponding distribution of pK values. The actual values are given in Table 17. The numbers of individual groups are derived from the amino acid sequence, and all carboxyl and tyrosine side chain groups were originally modeled by distributing them equally among the two extreme pK values shown in Table 17.

The weak interaction between the ionizable residues is best demonstrated by inspecting a site-specific titration curve. Nozaki and Tanford[139] have investigated the ionization of the tyrosines side chains through UV/VIS spectroscopy. Their

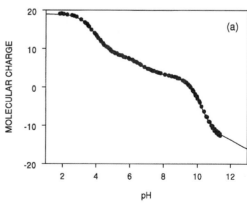

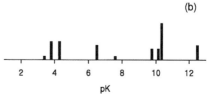

Figure 54. (a) Potentiometric titration of ribonuclease A in 6 M guanidine hydrochloride (GuHCl) by Nozaki and Tanford.[139] The solid line is calculated from the affinity distribution shown in (b). For individual pK values see Table 17.

Table 17.   A Summary of p$\hat{K}$ *Values to Interpret the Macroscopic Titration Data of Ribonuclease*[a]

Residue	Number	Nozaki *et al.*[139]	Tanford *et al.*[245]
N-ter	1	7.6	7.8
Glu and Asp	10	3.8–4.3	4.0–4.7
Tyr	6	9.8–10.2	10.0
Lys	10	10.4	10.2
Arg	4	>12.5	>12.0
His	4	6.5	6.5
C-ter	1	3.4	3.8

[a]The media employed are 6 M guanidine hydrochloride (GuHCl) used by Nozaki and Tanford,[139] and KCl used by Tanford and Hauenstein.[245]

result is shown in Fig. 55 and can be very well described by the protonation of a single, noninteracting site. The slight broadening of the curve can be attributed to minor differences between the pK values of the different tyrosine residues.

While the high ionic strength medium eliminates to a large extent the site–site interactions, it will be shown later that even at lower ionic strengths the effects of site–site interactions are usually small. For that reason the simplistic null model currently represents the most powerful framework to understand ionization properties of proteins, and often beats various sophisticated computational schemes.

An early application of the null model was the prediction of the point of zero charge (PZC) of a protein. This quantity typically coincides with the so-called

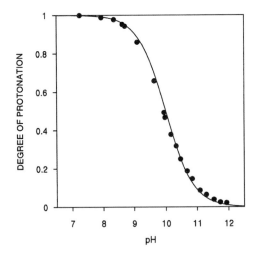

Figure 55.   Protonation of all tyrosine residues in ribonuclease A in 6 M guanidine hydrochloride (GuHCl) as measured by Nozaki and Tanford.[139] Experimental data points were obtained by UV/VIS spectroscopy and the solid line is the null model with p$\hat{K}$ = 10.

Table 18.   Comparison of Experimental Isoionic Points with Predictions of the Null
Model for Various Proteins, including Hen Egg White Lysozyme (HEWL)[a]

Protein	Null model	Experiment
Ribonuclease A	9.4	9.6
Ribonuclease T_1	4.1	3.8
Barnase	8.8	9.0
HEWL	10.6	11.2
Lysozyme T4	9.9	10.0

[a]Adapted from Ref. 260.

isoionic point, which is the pH of a protein solution without any added acid or base. Table 18 compares such isoionic points calculated for various proteins on the basis of the null model and those observed experimentally. Given the simplicity of the model, the quality of these predictions is surprisingly good.

5.2.   The Smeared-Out Charge Model

A simple way to approximate the effects of the interactions between ionizable sites is by treating the protein molecule as a uniformly charged object.[(12)] The charge originates from the ionized groups on its surface. Let us first assume, for simplicity, that all ionizable groups are identical and have the same ionization constant. We shall remove this restriction shortly. For such a system, the free energy of a microscopic configuration is approximated by

$$F(\{s_i\}) = F^{(\text{ch})} + F^{(\text{el})} \qquad (217)$$

where $F^{(\text{ch})}$ and $F^{(\text{el})}$ are the chemical and electrostatic free energy of the configuration. As discussed in Section 3 the chemical contribution is simply

$$F^{(\text{ch})} = -\mu n \qquad (218)$$

where n is the number of protons bound to the proteins and $\beta\mu/\ln 10 = p\hat{K}^{(\text{int})} - \text{pH}$. The constant $p\hat{K}^{(\text{int})}$ refers to the uncharged molecule (intrinsic constant).

The titration curve can be obtained by minimizing the free energy of the system with respect to the number n of bound protons. Thereby, one minimizes the free energy of the macrostate F_n, which also includes a mixing entropy term. The latter originates from the various ways of arranging n protons between N sites and leads to the binomial factor in Eq. (180). For $N \gg 1$ the Stirling approximation $N! = N \ln N - N$ can be used, and one obtains the free energy per site (in units of thermal energy kT)

$$\frac{\beta F_n}{N} = \frac{\beta F(\{s_i\})}{N} + \theta \ln \theta + (1 - \theta) \ln (1 - \theta) \tag{219}$$

where $\theta = n/N$ is the degree of protonation and $\beta = 1/(kT)$. We now minimize the free energy by setting $\partial F_n/\partial n = 0$. The derivative $\partial F^{(el)}/\partial n$ is related to the electrostatic surface potential since [cf. Eq. (25)]

$$\psi_0 = \frac{\partial F^{(el)}}{\partial Q} \tag{220}$$

where Q is the actual charge of the system

$$Q = e(n + Z) \tag{221}$$

and Z is the total charge of the protein in its deprotonated state (in units of elementary charge e). The result of this analysis leads to the relationship

$$pH = p\hat{K}^{(int)} - \beta e \psi_0/\ln 10 + \log_{10} \frac{1 - \theta}{\theta} \tag{222}$$

where the electrostatic potential ψ_0 must be evaluated for the charge resulting at the actual degree of protonation θ. Another way to write this equation is

$$pH = pK_{eff} + \log_{10} \frac{1 - \theta}{\theta} \tag{223}$$

where one defines an effective association constant by

$$pK_{eff} = p\hat{K}^{(int)} - \beta e \psi_0/\ln 10. \tag{224}$$

This quantity is just the microscopic constant of the ionizable group at a given electrostatic potential ψ_0. For this reason, pK values of ionizable groups can be viewed as probes of the electrostatic potential. For sufficiently small potentials, one can generally write [cf. Eq. (28)]

$$\psi_0 = \frac{Q}{C} \tag{225}$$

where C is the total capacitance of the system. Inserting this relation in Eq. (222) and combining with Eq. (221), one obtains a closed-form expression for the isotherm

$$pH = p\hat{K} - \bar{\epsilon}\theta + \log_{10} \frac{1 - \theta}{\theta} \cdot \tag{226}$$

The average pair interaction parameter per site is introduced as

$$\bar{\varepsilon} = \frac{\beta e^2}{\ln 10}\frac{N}{C} \tag{227}$$

and the microscopic ionization constant becomes

$$p\hat{K} = p\hat{K}^{(int)} - Z\bar{\varepsilon}. \tag{228}$$

The microscopic constant $p\hat{K}$ refers to the fully deprotonated molecule, while $p\hat{K}^{(int)}$ refers to the uncharged molecule. Returning to the notation in Eq. (224), one obtains

$$pK_{eff} = p\hat{K} - \bar{\varepsilon}\theta. \tag{229}$$

This model was originally proposed in 1924 by Linderstrøm-Lang, and is equivalent to the constant capacitance model,[50,51] the Frumkin isotherm,[184] or the Bragg–Williams approximation.[59,60]

In the protein literature it is customary to use this equation in order to define an apparent ionization constant $pK_{1/2}$ through the midpoint of the curve (i.e., pK_{eff} at $\theta = 1/2$). While the information on the detailed shape of the titration curve is lost within this description, the parameter $pK_{1/2}$ still represents the most important characteristics of the site-specific titration curve in a weakly interacting system. It should be also noted that the Hill isotherm [cf. Eq. (207)] resembles the isotherm resulting from the mean-field treatment [cf. Eq. (226)], and the effect of site–site interactions can be mimicked by adjusting the Hill exponent $0 < \nu < 1$. In protein literature, the Hill isotherm is frequently used to quantify the broadening of titration curves in an empirical fashion.[109,250]

The present analysis of the large system limit ($N \to \infty$) complements our discussion of the same model for finite N in Section 4. There we have treated a model with equal site–site interactions with a pair interaction parameter ε, which is related to the average pair interaction per site by

$$\bar{\varepsilon} = \frac{\varepsilon}{N} \tag{230}$$

Figure 56 illustrates the transition between the system with finite and infinite N with the parameters $\bar{\varepsilon} = 4$ and $p\hat{K} = 10$. For $N \to \infty$ the isotherm given in Eq. (226) is compared with the corresponding results discussed for finite N [cf. Eq. (180)], see Table 19 for details.

The main effect of mean-field type interactions is to broaden the titration curve with respect to the isolated site (null model). For a small number of ionizable sites the macroscopic pK values split up, but the larger this number, the oscillations smooth out, and in the large system limit a broad titration curve results. Figure 56 illustrates this effect based on an interaction parameter that is somewhat too large for a protein; typical values of $\bar{\varepsilon}$ are usually rather small, and hardly exceed 2.

Table 19. Calculated Macroscopic pK_n Values for
the Mean-Field Model with $p\hat{K} = 10$ and $\bar{\varepsilon} = 4$ for
Molecules with Different Number of Sites N
[cf. Eq. (180)][a]

N	n	ε	pK_n
1	1	—	10.00
2	1	2	10.30
	2		7.70
3	1	4/3	10.48
	2		8.67
	3		6.86
4	1	1	10.60
	2		9.18
	3		7.82
	4		6.39

[a]Note that the interaction parameter ε decreases with increasing N.

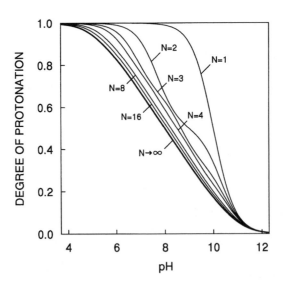

Figure 56.   Titration curves for the mean-field model with different number of sites N. For finite N the titration curves are calculated from polyprotic pK values given in Eq. (182) while for $N \to \infty$, Eq. (183) is used. The parameters are $\bar{\varepsilon} = 4$ and $p\hat{K} = 10$.

The present model is simply generalized to the case of nonequivalent ionizable groups. However, an analytic expression for the titration curve can no longer be given. Assume that the sites have ionization constants $p\hat{K}_i$ and charges Z_i in the deprotonated state. The titration curve is now given by the coupled set of equations

$$\text{pH} = p\hat{K}_i - \bar{\varepsilon}\theta + \log_{10} \frac{1 - \theta_i}{\theta_i} \qquad (231)$$

where θ_i is the degree of protonation of site i and θ the average degree of protonation of the entire molecule given by Eq. (216). The microscopic constants are

$$p\hat{K}_i = p\hat{K}_i^{(\text{int})} - Z_i\bar{\varepsilon} \qquad (232)$$

which is the same relation as in the case of identical groups [cf. Eq. (228)].

This simple model can very well describe titration of proteins. Figure 57 shows the classical titration data for ribonuclease by Tanford and Hauenstein,[245] which are compared with this mean-field model using appropriate numbers of amino acids from the protein sequence and the ionization constants given in Table 17. The mean

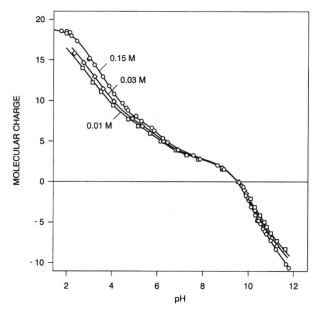

Figure 57.   Potentiometric titration data for ribonuclease in KCl by Tanford and Hauenstein.[245] The experimental data points are compared with a mean-field model where appropriate numbers of amino acids from the protein sequence enter. The ionization constants are given in Table 17. The mean interaction parameter is $\bar{\varepsilon} \approx 1.5$ and 0.01 M, $\bar{\varepsilon} \approx 2.3$ and 0.03 M, and $\bar{\varepsilon} \approx 2.8$ and 0.15 M.

interaction parameter is ionic strength dependent, and the values taken were $\bar{\varepsilon} \approx 1.5$ at 0.01 M, $\bar{\varepsilon} \approx 2.3$ at 0.03 M, and $\bar{\varepsilon} \approx 2.8$ at 0.15 M.

So far, no assumptions concerning the geometry of the protein were made. By assuming a linear charge–potential relationship, however, we have restricted ourselves to the linear Debye–Hückel (DH) regime. In the DH framework, the capacitance can be evaluated for various geometries. For a sphere, the diffuse layer capacitance per unit volume reads [cf. Eq. (28)]

$$\tilde{C} = \tilde{C}_d = \frac{3\varepsilon_0 D_w}{a^2}(1 + \kappa a) \tag{233}$$

where a is the radius of the sphere. The interaction parameter is now given by [cf. Eq. (227)]

$$\bar{\varepsilon} = \frac{\beta e^2 \rho}{\ln 10 \tilde{C}} \tag{234}$$

where $\rho = N/V$ is the site density per unit volume of the protein and $\tilde{C} = C/V$ the specific capacitance per unit volume (V being the volume of the protein). Other geometries and extensions to the nonlinear Poisson–Boltzmann (PB) approximation will be discussed in Sections 6 and 7.

5.3. The Tanford–Kirkwood Model

An early attempt to treat the effect of discrete charges on the titration behavior of a protein is due to Tanford and Kirkwood.[61] They assumed the protein to be a low dielectric sphere with relative permittivity D_d immersed in an electrolyte solution of screening length κ^{-1} and relative permittivity D_w (see Fig. 58). Ionized groups are modeled as point charges within the dielectric sphere. The free energy of all possible configurations are calculated by solving the DH equation for this

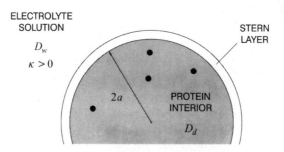

Figure 58.   Illustration of the Tanford–Kirkwood model. The protein is modeled by a dielectric sphere of a relative permittivity D_d which is immersed into an electrolyte solution with screening length κ^{-1} and relative permittivity D_w. Ionizable residues are represented by point charges and can be located anywhere within the sphere.

system (see Appendix A). As discussed in Section 3, the titration curve follows from thermal averaging at a fixed chemical potential of the protons over all configurations.

The Tanford–Kirkwood model can be formulated by introducing the state variables s_i. Recall that $s_i = 0$ if ionizable site i is deprotonated and $s_i = 1$ if it is protonated. As discussed in Section 3, in the linear regime the free energy of an ionizing protein has the quadratic form

$$F = -\sum_i \mu_i s_i + \frac{1}{2} \sum_{ij} E_{ij} s_i s_j \qquad (235)$$

where $\beta\mu_i / \ln 10 = \mathrm{pH} - p\hat{K}_i$ and $E_{ij} = W(\mathbf{r}_i, \mathbf{r}_j)$ is the electrostatic interaction energy of two point charges at locations $\mathbf{r}_i$ and $\mathbf{r}_j$. Analytic expressions for these interaction energies due to Kirkwood[11] are summarized in Appendix A. In protein literature, one often considers the free energy

$$F = -\sum_i \mu_i' \xi_i + \frac{1}{2} \sum_{ij} E_{ij} \xi_i \xi_j \qquad (236)$$

where $\beta\mu_i'/\ln 10 = p\hat{K}_i^{(\mathrm{int})} - \mathrm{pH}$ and ξ_i is the charge of residue i given by

$$\xi_i = s_i + Z_i. \qquad (237)$$

Equations (235) and (236) are identical, since

$$p\hat{K}_i = p\hat{K}_i^{(\mathrm{int})} - \sum_j \varepsilon_{ij} Z_j \qquad (238)$$

where $\varepsilon_{ij} = \beta E_{ij}/\ln 10$ as usual. Recall that $p\hat{K}_i$ refers to the microscopic ionization constant when all other sites are deprotonated, while $p\hat{K}_i^{(\mathrm{int})}$ is the intrinsic ionization constant. The latter also corresponds to a microscopic ionization constant, but given that all other sites have no charge ($\xi_i = 0$).

While the expressions for the pair interactions are analytical (see Appendix A), they must be evaluated numerically. For short distances ($r \to 0$), the presence of the electrolyte can be neglected and the charges interact via a simple Coulomb law in the dielectric

$$W(r) = \frac{e^2}{4\pi\varepsilon_0 D_d} \frac{1}{r}. \qquad (239)$$

An example of this interaction potential is illustrated in Fig. 59. Two elementary charges are placed 0.1 nm from the surface of a sphere of 1.5 nm radius and dielectric permittivity $D_d = 3$, which is immersed in a 0.01 M electrolyte solution. By variation of the dihedral angle between both charges, their mutual distance can be varied up to 2.8 nm. Figure 59 shows that the corresponding interaction potential falls off much faster than the simple Coulomb law in the dielectric. In fact, its

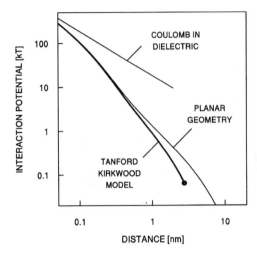

Figure 59. Site–site interaction potential for the Tanford–Kirkwood model. The sphere of radius 1.5 nm and of dielectric permittivity $D_d = 3$ is immersed in an electrolyte solution of ionic strength 0.01 M. The charges are located 0.1 nm below the surface and their distance is varied by changing the corresponding dihedral angle. No Stern layer is assumed. The Coulomb law and the interaction potential in the corresponding planar geometry are also shown.

distance dependence is approximated reasonably well with the planar geometry (see Section 7 and Appendix A).

The assumption of a low dielectric permittivity for the protein interior makes the free energies of the Tanford–Kirkwood model quite sensitive to the positions of the titrating groups. A more robust model can be constructed by assuming a uniform dielectric permittivity which is comparable to the value of water; in this case the mutual interaction energies are substantially weaker.

5.3.1. Solution Techniques of the Ionization Problem

While everything can be evaluated analytically so far, the difficult part of the calculation is to obtain the titration curves. As long as the number of sites is not too large, this problem is best solved by *direct enumeration* of states.[202] In practice, this approach is feasible up to roughly $N < 30$, but for larger N it becomes prohibitively time-consuming, even for present supercomputers (recall that for $N = 30$ the number of states to be enumerated is $2^N \simeq 10^9$). Many proteins have several hundreds of titratable sites, and in such a case this method is not applicable.

An extremely effective and accurate method to treat a larger number of sites is *Monte Carlo simulation*. One generates random configurations of the protonation microstates; they are weighted according to the Metropolis algorithm with the free energy Eq. (235) (see Section 3). The method is reasonably fast and is exact in principle, although subject to random sampling errors. These errors can be controlled through the length of the simulation run.

Titration curves can also be calculated by means of a *mean-field approximation*. However, one must note that such methods are approximate and their predictions should be always checked against exact results, such as direct enumeration or Monte

Carlo simulations. We have already discussed one approximation of this kind, namely Eq. (231). In the Tanford–Kirkwood model, the average interaction parameter is now given by

$$\bar{\varepsilon} = \frac{1}{N} \sum_{ij} \varepsilon_{ij}. \tag{240}$$

While the resulting expression is approximate, it represents a very useful tool for estimating the effects of site–site interactions in a protein. Another possibility of a mean-field approximation is the so-called Tanford–Roxby method.[63] The spirit of this approximation is to replace in Eq. (235) the value of each state variable s_i by its expectation value $\langle s_i \rangle = \theta_i$. Minimization of the free energy yields the equation

$$\text{pH} = p\hat{K}_i - \sum_{j} \varepsilon_{ij}\,\theta_j + \log_{10} \frac{1 - \theta_i}{\theta_i} \tag{241}$$

which must be solved numerically.

Another useful approximation scheme is the *mean-field cluster expansion*, which was originally developed for the treatment of magnets a long time ago.[264] Recently, these techniques have been used to treat protein titrations with success.[265] More detailed discussion of such mean-field techniques is given in Appendix B.

Figure 60 illustrates the ionization behavior within the Tanford–Kirkwood model. Figure 60a displays the degree of protonation θ as a function of pH, while Fig. 60b shows the alternative representation of pK_{eff} as a function of θ, whereby pK_{eff} is defined according to Eq. (223). We consider identical basic ionizable groups with a $p\hat{K} = 10$, which are arranged in a dodecahedron and are placed 0.1 nm from the surface of a sphere of 1.5 nm radius and dielectric permittivity $D_d = 3$. This sphere is immersed in a 0.01 M electrolyte solution. (The interaction potential for this situation is shown in Fig. 59.) The titration curve was evaluated exactly by direct enumeration of states and is shown to compare favorably with the mean-field model [cf. Eqs. (231) and (240)]. The mean-field interaction parameter $\bar{\varepsilon} \simeq 2.4$ would be slightly underestimated by the smeared-out charge double-layer model, which predicts in the same setting $\bar{\varepsilon}_{dl} \simeq 2.7$ [cf. Eq. (234)]. This model system with identical sites shows a broadening of the titration curve to an extent which is uncommon in actual proteins. The Tanford–Kirkwood model was originally applied to proteins with such low relative permittivities D_d for the protein interior. As will be discussed later, for proteins these values appear substantially higher, and results in weaker site–site interactions in accordance with experiment.

5.3.2. Shifts in Ionization Constants

With the Tanford–Kirkwood model one can also estimate intrinsic microscopic constants $p\hat{K}_i^{(\text{int})}$ of the ionizable residues within the protein. The pK value of an amino acid inside a protein is not necessarily the same as the corresponding value

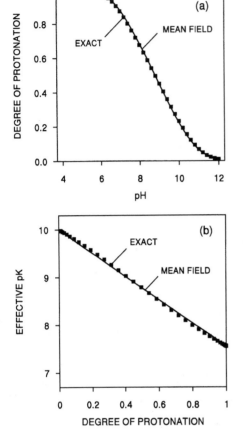

Figure 60. Titration curves for a model system according to the Tanford–Kirkwood model. The sphere of radius 1.5 nm and dielectric permittivity $D_d = 3$ is immersed in an electrolyte solution of ionic strength 0.01 M. Basic ionizable groups of $p\hat{K} = 10$ are located 0.1 nm below the surface and arranged in a dodecahedral geometry ($N = 20$). No Stern layer is assumed. (a) Titration curve and (b) pK_{eff} as a function of the degree of protonation.

of the same amino acid in solution. While the immediate chemical neighborhood is similar, the dielectric environment is different. It was shown in Section 3 that the free energy of a point charge depends on its dielectric environment and one thus expects a shift in the pK between an amino acid in solution and in a protein. The appropriate relation reads [cf. Eq. (58)]

$$p\hat{K}_i^{(int,pr)} = p\hat{K}_i^{(int,aq)} - (1 + 2Z_i) \frac{\beta}{\ln 10} (W_i^{(s,pr)} - W_i^{(s,aq)}) \qquad (242)$$

where $W_i^{(s,pr)}$ and $W_i^{(s,aq)}$ refer to the self-energy contributions for the site i in the protein and in water, respectively (see Section 3).

In the Tanford–Kirkwood model, the self-energy contributions must be evaluated from the expressions given in Appendix A, but approximations can be invoked here. The self-energy term for a site, which is buried deep within the protein core, can be approximated by the self-energy of a point charge in a uniform dielectric. As discussed in Section 3, this state is considered as the reference state for the self-energies and one simply has $W_i^{(s,pr)} = 0$. The shift of the intrinsic ionization constants at zero ionic strength is [cf. Eqs. (58) and (36)]

$$p\hat{K}_i^{(0,pr)} = p\hat{K}_i^{(0,aq)} + (1 + 2Z_i)\frac{\beta e^2}{8\pi\varepsilon_0 a \ln 10}\left(\frac{1}{D_w} - \frac{1}{D_d}\right) \qquad (243)$$

where a is the molecular size of the isolated amino acid. Note that the shift vanishes for $D_d = D_w$, while the shift becomes huge (5–20 pK units) in the case of $D_d \ll D_w$ and a realistic molecular radius a.

This analysis suggests that an amine group ($z = 0$) within the protein should have a somewhat smaller $p\hat{K}^{(int)}$ than the corresponding constant of the amino acid at zero ionic strength, and be hardly ionic strength dependent. The reverse shift is expected for a carboxylic group ($z = -1$). These trends were indeed observed for acidic and basic indicators in nonionic micelles. However, the magnitude of the shifts is moderate, typically only one pK unit (see Section 7 and Ref. 266). In the case of proteins these shifts are difficult to establish, since it is usually impossible to study the ionization of a given group in a neutral protein molecule, even with site-directed mutants. For example, carboxylic groups in a protein typically ionize when the protein is positively charged, and under such circumstances one expects the apparent pK to be smaller than $p\hat{K}_i^{(int)}$ (see Eq. (224))].

5.4. Recent Developments in Dielectric Continuum Models

While the Tanford–Kirkwood model introduces several drastic simplifications, conceptually it reflects many important features of modern dielectric continuum models. With the availability of accurate three-dimensional protein structures from X-ray scattering and NMR, it has become apparent that the geometrical shape of a protein is often poorly modeled by a sphere, while nowadays one can routinely obtain such solutions for irregularly shaped bodies. An early study of titrating groups in a protein is due to Bashford *et al.*[267] They applied the PB equation to histidine residues in azurin assuming a nonuniform dielectric permittivity. Such computational approaches are becoming increasingly popular, and we shall summarize some of these developments now.

5.4.1. General Methodology

The problem is typically approached from a three-dimensional structure established by X-ray scattering, which yields coordinates of all atoms of a protein in a crystal. Protons represent an exception due to their weak X-ray scattering power and no direct information about their coordinates can be obtained. These coordinates are instead generated through a computational procedure and various schemes are available for this purpose.[17,263] The delicate part of this analysis is that the crystallographic structure may represent a poor model of the structure in solution. To circumvent this problem, the X-ray geometries are sometimes optimized by minimizing its free energy, or by using structures obtained from a molecular dynamics run.[268,269] The appropriateness of both techniques relies on the availability of an accurate force field. More recently, rather detailed information about the three-dimensional structures in solution have been obtained from NMR, and such structures are now often used as a starting point in titration calculations.[250,263]

By assigning an appropriate van der Waals radius to each atom, a three-dimensional body is generated which represents the protein molecule. Some authors incorporate a Stern layer of thickness a few Å, obtained by rolling a sphere on the surface of the protein. In the interior of the protein body, point charges are defined according to the atomic coordinates of the charged groups. From the available molecular force fields one uses the partial charges as appropriate to the neutral state of each group. The ionization process can be modeled in various ways. The simplest approach is to change the partial charge on the atom being ionized by a unit elementary charge (single-site model).[260] In the case of carboxylic groups, the charge is often distributed symmetrically among both oxygens. A more detailed approach replaces the partial charges appropriate to the neutral state with a completely different set of partial charges appropriate to the ionized state.[270–272]

The value of the relative dielectric permittivity in the protein represents a very controversial issue. This problem has been intensively debated,[16,260,270,272–274] but in our opinion the experimental evidence for a dielectric discontinuity is unreliable. Based on the work on small molecules, initial studies have assumed a low relative permittivity (around 2–4), but the models are plagued by unrealistically large shifts in some of the resulting pK values and enormous sensitivity on the exact location of the partial charges, particularly if located close to the dielectric boundary. As will be exemplified below, a number of reports have recently confirmed that a much better description of the pK shifts can be obtained with substantially higher values of relative permittivity, around 20–40.[260,263,271,272] Some authors have even obtained good agreement with experiment by assuming the same permittivity for the protein interior as for water.[68,109,272,275] Molecular dynamics simulations with an explicit representation of all protein atoms and the solvent indicate that the dielectric constant of a protein is fairly high[276–279] and lies in the range of 10–30. Similar conclusions were reached by Svensson et al.[197] by studying calcium binding to proteins. Their results indicate that the protein should have a very high dielectric response and, if a substantially lower value than 80 is assumed, the agreement with experiment is lost.

The solution of either the DH or PB equation is obtained by numerical methods.[151,280] A cartesian grid is generated and the partial differential equation is solved numerically by standard finite difference methods. Various focussing methods have been devised in order to generate accurate solutions in spite of local regions of high curvature and with substantial variation of the electrostatic potential.

The nonlinear PB equation requires considerably more computer effort and it has become standard practice to use the linearized DH version. The numerical technique has achieved widespread acceptance and is available as a commercial software package. Juffer *et al.*[272] have recently proposed a boundary element method to solve the DH approximation for an irregularly shaped body. Two coupled integral equations are solved on the dielectric boundary, which can be discretized using an appropriate triangular mesh. A different grid procedure was used by Fushiki *et al.*,[281] where spherical polar coordinates were used to solve DH or PB equations.

The free energies are calculated from the electrostatic potential in a straightforward fashion, and the former serve as input for the titration calculation. Equal microscopic pK values are typically assigned to the same-type amino acids (see Table 17). Some authors have also evaluated the appropriate self-energies along similar lines as discussed for the Tanford–Kirkwood model and included the appropriate pK shifts in the calculation. The statistical problem of the interacting titrating sites is then approached by direct enumeration, Monte Carlo simulation, or mean-field techniques as discussed above. More recently, some authors have also incorporated conformational degrees of freedom in the titration calculations. Beroza *et al.*[269] have treated the side chain flexibility by allowing for two different conformational states. Monte Carlo sampling then includes thermal averaging over the different ionization and conformational states of the molecule (see Section 3). The effects of bound water molecules[17] or dielectric saturation[282] have also been addressed.

5.4.2. Case Studies

As a first example consider hen egg white lysozyme (HEWL) (see Fig. 5). Numerous computational studies of its ionization behavior have been carried out.[16,65,260,263,271,272,283] As evident from Fig. 61, the site-titration curves obtained

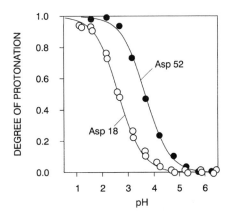

Figure 61. Representative site-titration curves for hen egg white lysozyme (HEWL) in 0.1 M NaCl solution. The points are obtained from NMR[248] while the solid lines represent the best fit with the null model [cf. Eq. (215)].

by Bartik *et al.*[248] by NMR conform to the null model. There is hardly any broadening of the site-specific titration curves and the effect of site–site interactions is weak. The apparent $pK_{1/2}$ values are obtained from such measurements and summarized in Table 20. (The arginine side chains are not reported in this table.) However, not all residues could be monitored by the NMR and several pK values extracted from the macroscopic titration curve[284] are somewhat uncertain. Since the experimental pK values compare rather well with those from the null model, one concludes that the pK shifts are also rather moderate. However, some of the carboxylic groups are shifted to the acidic region and Glu 35 is substantially shifted to the basic region, as well established by the NMR.[248]

The results of two early studies based on a dielectric model are summarized in Table 20. Bashford and Karplus[16] used the DH model based on crystallographic

Table 20.   Apparent pK Values for Hen Egg White Lysozyme (HEWL) in 0.1 M, Calculated according to Bashford and Karplus (BK) and Yang and Honig (YH) using the Triclinic Structure[16,283]a

Residue	BK[16]	YH[283]	Antosiewicz et al.[263]		Juffer et al.[272]		Null	Exp.
			$D_d = 4$	$D_d = 20$	$D_d = 4$	$D_d = 30$		
N-ter	6.1	5.4	—	—	−3.8	7.0	8.2	7.9
Lys 1	9.6	10.2	9.0	10.1	8.9	10.4	10.1	10.6
Glu 7	2.1	3.4	2.9	2.9	2.9	3.0	4.0	2.9
Lys 13	11.6	12.0	10.5	10.6	17.3	11.4	10.1	10.3
His 15	4.0	6.7	0.3	4.7	5.5	5.9	6.9	5.4
Asp 18	3.1	4.0	2.6	2.7	—	2.7	2.7	2.7
Tyr 20	14.0	—	15.4	10.2	12.2	9.0	10.7	10.3
Tyr 23	11.7	—	11.7	9.5	14.1	9.4	10.7	9.8
Lys 33	9.6	10.2	10.9	10.5	13.6	10.1	10.1	10.4
Glu 35	6.3	5.6	5.8	4.3	9.0	3.8	4.0	6.2
Asp 48	1.0	1.6	4.5	3.3	5.2	3.2	2.7	1.6
Asp 52	7.0	5.2	3.4	3.3	7.0	3.5	2.7	3.7
Tyr 53	20.8	—	22.3	11.0	>20	12.5	10.7	12.1
Asp 66	1.7	3.1	0.1	1.8	−0.6	4.0	2.7	0.9
Asp 87	1.2	1.6	3.3	2.5	0.0	4.0	2.7	2.1
Lys 96	10.4	10.5	10.6	11.2	15.3	10.9	10.1	10.7
Lys 97	10.6	11.4	10.8	11.1	13.3	11.2	10.1	10.1
Asp 101	7.9	6.4	5.3	3.5	9.8	4.9	2.7	4.1
Lys 116	9.9	10.6	8.8	10.0	11.9	10.3	10.1	10.2
Asp 119	3.2	3.4	3.8	3.1	3.5	3.9	2.7	3.2
C-ter	2.3	2.6	2.8	2.7	2.7	4.1	2.7	2.8
rms	2.5	1.2	3.0	0.8	4.4	1.2	1.0	
max	8.7	2.5	10.2	1.9	11.7	3.1	2.2	

*a*Protein dielectric permittivity $D_d = 4$ and 20 was used together with X-ray structures and the single-site model.[263] Improved null model according to Ref. 260. Experimental data as quoted in Ref. 263; the value for N-ter is taken from Ref. 272. Some of the experimental values are not well established; see also Refs. 248 and 272. Root-mean-square (rms) and maximum (max) deviations are given.

coordinates with a relative permittivity of 4 and a Stern layer of 2 Å thickness. The ionization process is modeled by changing the partial charge of the ionizable group by a unit elementary charge (single-site model) and by evaluating the shifts of $pK^{(int)}$ values through appropriate self-energies. Their results were obtained with an approximate scheme to treat the site–site interactions between ionizable sites. Beroza *et al.*[65] used exact Monte Carlo treatment of site–site interactions, while Yang and Honig[283] improved this analysis by using a molecular dynamic technique to obtain more realistic geometric structures, with a better description of the charge distribution. But these improvements only lead to marginal changes in the predictions. Table 20 indicates that the simple null model still performs much better than any of these approaches.

This point is also evident in Fig. 62, which displays a scatter plot of the calculated versus experimental pK values. A comparison of the performance of the

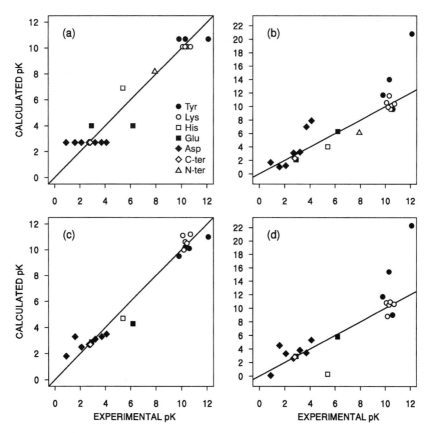

Figure 62. Graphical comparison of calculated and experimental apparent pK values for all ionizable residues of HEWL. (a) Null model, (b) results from Bashford and Karplus for triclinic structure.[16] Poisson–Boltzmann (PB) models from Antosiewicz *et al.*[263] with different interior permittivities (c) $D_d = 20$ and (d) $D_d = 4$. Points taken from Table 24.

null model (cf. Fig. 62a) with the low dielectric permittivity model (cf. Fig. 62b) shows the latter model to produce unrealistically large shifts in the pK values. This feature is a typical anomaly of such models, as was recently concluded from pK calculations where the dielectric permittivity was varied systematically.[260,263,272] Table 20 compares some of these results with experiment, whereby relative permittivities of $D_d = 4$ and 20 were used by Antosiewicz *et al.*[263] and $D_d = 4$ and 30 by Juffer *et al.*[272] Both authors observe a significant improvement in model performance for a larger dielectric permittivity, as is clearly evident in the scatter plots shown in Fig. 62. The high dielectric constant model compared rather well with the experimental values (cf. Fig. 62c), while the low dielectric model again performs poorly (cf. Fig. 62d).

The scatter plot showing all types of residues (as shown in Fig. 62) is always suggestive of good agreement between the model and experiments—even the trivial null model compares with experiment reasonably well. In order to decide whether a given model has indeed predictive capabilities or not, one has to focus on the root-mean-square deviations (as indicated in Table 20), or on scatter plots

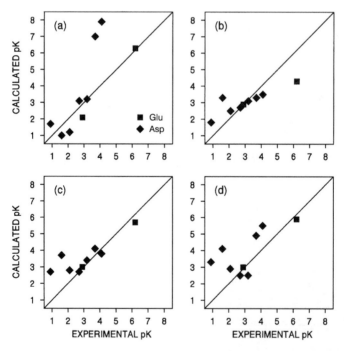

Figure 63.   Graphical comparison of calculated and experimental apparent pK values for carboxylic groups of HEWL. (a) Results from Bashford and Karplus for triclinic structure.[16] (b) High permittivity model from Antosiewicz *et al.*[263] (c) Calculations including conformational degrees of freedom based on global conformational search.[286] (d) Results of a microscopic approach.[270] Points are taken from Table 24.

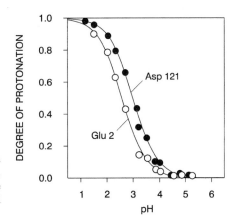

Figure 64. Representative site-titration curves for ribonuclease A at ionic strengths < 0.06 M. The points are obtained from NMR[248] while the solid lines represent the best fit with the null model [cf. Eq. (215)].

Table 21. *Comparison of Calculated and Experimental Apparent* $pK_{1/2}$ *Values with Two Different Values of the Relative Interior Permittivity* D_d *for Ribonuclease* $A^{(263)a}$

Residue	$D_d = 4$	$D_d = 20$	Null	Exp.
N-ter	6.7	7.0	8.2	7.6
Glu 2	1.9	2.4	4.0	2.8
Glu 9	4.8	4.0	4.0	4.0
His 12	−0.1	4.6	6.9	5.8
Asp 14	7.7	1.9	2.7	<2.0
Asp 38	2.8	2.7	2.7	3.1
His 48	−4.7	6.5	6.3	6.3
Glu 49	6.7	4.6	4.0	4.7
Asp 53	3.2	3.6	2.7	3.9
Asp 83	3.3	2.0	2.7	3.5
Glu 86	4.9	3.8	4.4	4.1
His 105	4.3	6.2	6.9	6.6
Glu 111	4.5	3.7	4.4	3.5
His 119	6.0	5.7	6.9	6.1
Asp 121	1.9	2.1	2.7	3.1
C-ter	1.9	2.3	3.8	2.4
rms	3.6	0.6	0.8	
max	11.0	1.5	1.4	

[a]Results are quoted for crystal structure B and the tautomer denoted as NE2-H at 0.2 M.[263] Results of the improved null model from Antosiewicz *et al.*[263] are shown for comparison. Experimental data are taken from the same source. Root-mean-square (rms) and maximum (max) deviations are given, whereby the limiting values are used.

for a single type of residues. Figure 63 shows this type of presentation for the carboxylic residues in HEWL, and shows that the low dielectric permittivity model performs poorly (see Fig. 63a) while the high dielectric model has a decisive predictive power (see Fig. 63b). Antosiewicz *et al.*[263] compared results based on the single-site model and the X-ray structure with various more sophisticated PB models, such as using structures based on NMR or other parameter sets for the charging process. They concluded that all improvements yielded only marginal changes in the predicted pK values for the model with the higher relative permittivity of $D_d = 20$, but that the results were always significantly better than the null model. With a lower relative permittivity of $D_d = 4$, these additions yielded substantial improvements but still the model did not attain the accuracy of the simple null model. Very similar conclusions concerning the choice of the interior dielectric permittivity were reached by Juffer *et al.*[272] and Karplus and coworkers.[271]

Ribonuclease A is another protein where dielectric continuum models have been extensively applied. The titration behavior of this protein was already studied

Table 22. Comparison of Apparent p$K_{1/2}$ Values for Turkey Ovomucoid Third Domain (OMTKY3)[a]

Residue	Forsyth *et al.*[250]		Juffer *et al.*[272]			Null	Exp.
	NMR	X-ray	$D_d = 4$	$D_d = 20$	$D_d = 80$		
N-term	7.5	7.1	3.2	6.8	7.2	8.2	8.0
Asp 7	3.3	2.9	2.5	2.3	3.2	2.7	<2.7
Glu 10	3.5	3.4	0.7	4.1	3.8	4.0	4.2
Tyr 11	10.0	9.8	10.1	10.0	9.2	10.7	10.2
Lys 13	11.2	12.3	16.3	13.7	11.7	10.1	9.9
Glu 19	2.8	3.2	4.2	4.1	3.7	4.0	3.2
Tyr 20	9.9	9.8	7.9	9.3	9.1	10.7	11.1
Asp 27	3.4	4.0	6.5	4.6	3.5	3.5	<2.3
Lys 29	12.1	11.5	12.0	11.8	11.4	10.1	11.1
Tyr 31	11.2	12.9	>20	14.4	10.8	10.7	>12.5
Lys 34	11.7	11.7	13.6	13.0	11.5	10.1	10.1
Glu 43	4.4	4.3	4.8	5.1	4.4	4.0	4.8
His 52	6.2	6.3	9.3	7.5	6.7	6.9	7.5
Lys 55	11.3	11.1	13.5	11.7	11.1	10.1	11.1
Cter 56	2.7	2.7	2.3	3.1	3.1	2.7	<2.5
rms	0.9	1.1	3.5	1.6	1.1	0.8	
max	1.6	2.4	7.5	3.8	2.0	1.8	

[a]Experimental values and PB model calculations based on X-ray and NMR structures are taken from Forsyth *et al.*[250] DH model calculations for different values of interior relative permittivity D_d by Juffer *et al.*[272] The arginine residue was omitted. Results of the improved null model from Antosiewicz *et al.*[263] are shown for comparison. Root-mean-square (rms) and maximum (max) deviations are given, whereby the limiting values are used.

in pioneering work by Tanford and coworkers[139,245] and, more recently, by NMR.[119,285] The individual sites again titrate as independent sites to a very good degree of approximation, as illustrated in Fig. 64. The experimentally determined $pK_{1/2}$ values are summarized in Table 21 and compared with results obtained from dielectric models.

Similar trends are observed as discussed for HEWL above, namely the model with the low dielectric permittivity performs rather poorly, while assuming a higher dielectric constant in the protein leads to better agreement with experiment than for the null model. Ribonuclease A represents an interesting system due to its ability to bind phosphate. It was established that binding of phosphate shifts the apparent pK values for His 12 and His 119 from 5.8 and 6.1 to 7.2 and 7.6, respectively. The model with high dielectric permittivity is able to predict these shifts reasonably well.[263]

The titration behavior of turkey ovomucoid third domain (OMTKY3), which contains 16 ionizable groups as summarized in Table 22, was recently studied in detail both experimentally and computationally. This protein is stable over a wide range of solution conditions and X-ray as well as NMR structures are available.[113,250] Forsyth *et al.*[250] have carried out an extensive experimental and computational study of the titration behavior of this protein. Site-specific titration

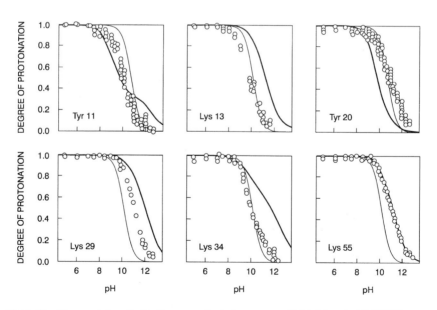

Figure 65. Representative site-titration curves for turkey ovomucoid third domain (OMTKY3) at an ionic strength of 0.015 M. Points are obtained from NMR while the thick solid line is calculated with a discrete-site Poisson–Boltzmann (PB) model based on the NMR structure.[250] The thin lines are calculations based on the improved null model from Antosiewicz *et al.*[263]

curves were determined by two-dimensional NMR methods at two different ionic strengths. The calculations were based on a PB model similar to the one used by Antosiewicz *et al.*[260,263] The relative permittivity of the protein interior was assumed to be $D_d = 20$. A few representative results are shown in Fig. 65. The points are experimental data, while solid lines represent results of PB calculations. A comparison with the null model (thin lines) shows that for this protein site–site interactions are sufficiently strong to make the titration curves deviate significantly from the predictions of the null model. The results for all residues are summarized in Table 22, which also includes recent DH boundary element method results by Juffer *et al.*[272] These results indicate that a high permittivity model is able to predict reasonably well the range of pK values, while the low dielectric permittivity model yields unreasonably large pK shifts.

Figure 66 compares the model predictions and experiment for the carboxylic residues in OMTKY3. With the exception of a few residues, one observes that all high permittivity models are able to reproduce experimental values

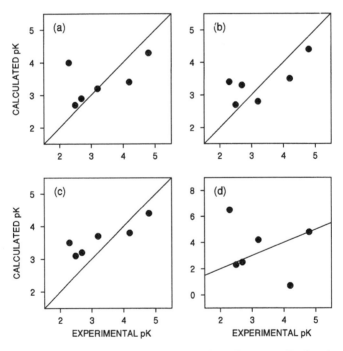

Figure 66. Graphical comparison of calculated and experimental apparent pK values for carboxylic groups for turkey ovomucoid third domain (OMTKY3). Poisson–Boltzmann (PB) calculations by Forsyth *et al.*[250] are based on geometrical structures derived from (a) X-ray and (b) NMR. Debye–Hückel (DH) calculations by Juffer *et al.*[272] for two different relative permittivities of the protein interior. (c) $D_d = 80$ and (d) $D_d = 4$. Points are taken from Table 22.

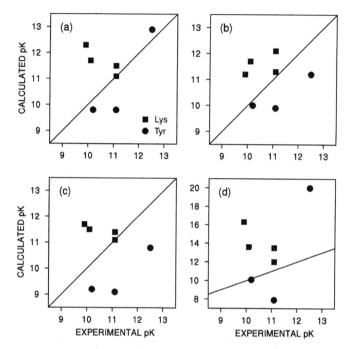

Figure 67. Graphical comparison of calculated and experimental apparent pK values for tyrosine and lysine groups of the turkey ovomucoid third domain (OMTKY3). Poisson–Boltzmann (PB) calculations by Forsyth *et al.*[250] based on geometrical structures derived by (a) X-ray and (b) NMR. Debye–Hückel (DH) calculations by Juffer *et al.*[272] with different relative permittivities of the protein interior, (c) $D_d = 80$ and (d) $D_d = 4$. Points taken from Table 22.

reasonably well. The low permittivity model leads to unrealistically large shifts. No significant differences are observed between the X-ray or the NMR derived structures. Figure 67 summarizes the analogous data for tyrosines and lysines, and shows similar results regarding the low and high dielectric permittivity models.

In spite of some of these difficulties, Antosiewicz *et al.*[263] have shown that the dielectric PB model with relative permittivity $D_d = 20$ explains the shifts in apparent pK values upon changes in ionic strengths (see Table 23). While the PB model might have difficulties in yielding the actual value of the apparent pK value, it is capable of predicting pK shifts upon changes in the ionic strength. Similar observations have been made by Kesvatera *et al.*[68]

5.4.3. Side Chain Flexibility

The role of side chain flexibility was recently studied within the framework of a dielectric model.[269,286] Beroza *et al.*[269] included in the averaging procedure not only the various ionization states, but also conformational degrees of freedom of

Table 23. *Comparison of Shifts in Apparent pK Values upon Changes in Ionic Strength
for Turkey Ovomucoid Third Domain (OMTKY3)[a]*

Residue	Calculated			Exp.
	0.01 M	1.0 M	ΔpK	ΔpK
Asp 7	2.92	3.50	0.58	0.57
Glu 10	3.37	3.91	0.54	0.25
Asp 27	3.60	4.03	0.43	0.39
Glu 43	4.35	4.49	0.14	0.30

[a]Results from PB model calculations with $D_d = 20$ are compared with experiment.[263]

the side chains. The authors used a global minimization of the free energy, which involved a search in the torsion space of the side chains.[286] This approach was computationally rather expensive, so they presented a simpler scheme, which involved two discrete conformational states for the side chain. The authors used X-ray as well as Monte Carlo generated structures, which maximized the solvent accessibility. The free energies were estimated by a continuum treatment with a low dielectric permittivity for the protein interior. Some of the results are summarized in Table 24, and the results of the global minimization scheme are shown in Fig. 63c. One can see that including the side chain flexibility substantially improves the quality of the model; the predictions are in fact comparable with those of a model with a high dielectric permittivity. Apparently, the conformational flexibility has a similar effect as a larger dielectric response of the protein interior.

5.5. Beyond Dielectric Continuum Models

Dielectric continuum models were harshly criticized by Warshel for being "fundamentally incorrect" in their treatment of buried charges.[273,274] Warshel and coworkers argue, probably quite correctly, that internal charges can only exist within a protein if stabilized by local protein dipoles or other charges. Such a stabilization mechanism is altogether absent in a macroscopic model. They suggest that the most important contribution to the apparent pK value of an ionizable group in a protein is mainly given by the pK shift of the intrinsic constant due to the self-energy contribution, and is rather weakly influenced by the site–site interactions. They further argue that this shift cannot be accurately estimated from a dielectric continuum model, since it depends on the local environment of the ionizable residue.

Indeed, these authors propose a semiempirical treatment of these effects, which they call the protein dipole Langevin dipoles model.[270,273,274] The latter explicitly includes interactions between the charge of the ionized group with fixed and induced dipoles within the protein and water. They also compare this approach with the linear response approximation.[140] These techniques were recently applied to predict the apparent pK values of the carboxylic side chains in lysozyme.[270] As shown in Table 24, the Langevin dipoles approach indeed leads to reasonable agreement with experiment. In particular, the substantial pK shift of the side chain Glu 35 is predicted correctly.

An alternative to PB calculations is to use Monte Carlo simulations on the primitive model level. Finite protein concentrations can be efficiently handled with the *cell model* (see Section 3). The protein, which is modeled similarly to the dielectric continuum models, is placed in a spherical cell.[287] The cell radius is determined by the protein concentration. Counterions and any extra salt are added to the cell and these species are allowed to be mobile, while the protein atoms are fixed in the Monte Carlo procedure. Free energy is calculated with the Widom technique[196] and titrating residues are handled by means of the grand canonical Monte Carlo method. The insertion or deletion of a proton in the protein is accompanied by a random insertion/deletion of a salt ion in order to maintain electroneutrality.[287]

Such Monte Carlo techniques were extensively used by Kesvatera *et al.*[68,109] to interpret the ionization process of calbindin D_{9k}. Calbindin is a highly charged protein and its ionization behavior has been studied in detail by various NMR

Table 24. Apparent p$K_{1/2}$ Values of Carboxylic Side Chains of Hen Egg White Lysozyme (HEWL)[a]

Residue	BK[16]	$D_d = 20$[263]	Null	Two-State	Global	Micro.[270]	Exp.
				Beroza *et al.*[269]			
Glu 7	2.1	2.9	4.0	2.6	3.0	3.0	2.9
Asp 18	3.1	2.7	2.7	3.0	2.7	2.5	2.7
Glu 35	6.3	4.3	4.0	4.7	5.7	5.9	6.2
Asp 48	1.0	3.3	2.7	3.3	3.7	4.1	1.6
Asp 52	7.0	3.3	2.7	5.7	4.1	4.9	3.7
Asp 66	1.7	1.8	2.7	2.7	2.7	3.3	0.9
Asp 87	1.2	2.5	2.7	3.2	2.8	2.9	2.1
Asp 101	7.9	3.5	2.7	3.4	3.8	5.5	4.1
Asp 119	3.2	3.1	2.7	3.4	3.4	2.5	3.2
rms	1.8	0.9	1.3	1.3	1.0	1.4	
max	3.8	1.9	2.2	2.0	2.1	2.5	

[a]Results for triclinic structure from Barshford and Karplus (BK),[16] and for X-ray structure and single-site model with relative dielectric permittivity $D_d = 20$.[263] Calculations based on extended X-ray structures based on Monte Carlo energy minimization including two discrete conformational states and global conformational from Beroza *et al.*[269] Results of a microscopic approach by Sham *et al.*[270] Results of the improved null model from Antosiewicz *et al.*[263] are shown for comparison. Experimental data from Table 20. Root-mean-square (rms) and maximum (max) deviations are given.

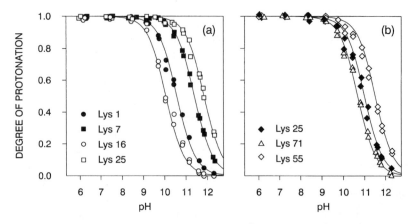

Figure 68.   Site-specific titration curves of lysines in the Ca^{2+} form of calbindin at a protein concentration of 0.5 mM and no salt added.[68] Points are from ^1H NMR and ^{13}C NMR data and thin lines are best fits according to the null model. (a) Lys 1, Lys 7, Lys 16, and Lys 25, (b) Lys 29, Lys 55, and Lys 71. The residues Lys 12, Lys 41, and Lys 72 are not shown for clarity, but they have a very similar appearance. The apparent pK values are summarized in Table 25.

techniques. Figure 68 displays the titration curves for lysines together with best fits with the null model. By monitoring ^1H NMR and ^{13}C NMR simultaneously, Kesvatera et al.[68] have carefully checked if the chemical shifts indeed reflect the ionization of the amino acid residue in question. One observes that all lysine side

Table 25.   *Apparent* p$K_{1/2}$ *Values for Lysine Residues of Calbindin* D_{9k} *in the* Ca^{2+} *Form at Protein Concentration of 0.5* mM *and No Added Salt*[a]

Residue	MC	$D_d = 4$	$D_d \approx 20$	$D_d \approx 80$	Null	Exp.
Lys 1	12.0	13.7	11.2	11.1	10.1	10.6
Lys 7	12.3	11.3	10.9	11.1	10.1	11.4
Lys 12	11.8	12.0	10.7	10.9	10.1	11.0
Lys 16	11.4	11.0	9.9	10.7	10.1	10.1
Lys 25	12.8	18.5	12.7	11.6	10.1	11.8
Lys 29	11.5	11.6	10.7	10.8	10.1	11.0
Lys 41	10.9	10.3	10.3	10.5	10.1	10.9
Lys 55	12.1	13.5	11.1	11.3	10.1	11.4
Lys 71	11.1	10.1	10.1	10.5	10.1	10.7
Lys 72	11.8	10.6	10.6	10.9	10.1	11.0
rms	0.9	2.5	0.5	0.3	1.0	
max	1.4	6.7	0.9	0.6	1.7	

[a]Experimental and Monte Carlo (MC) simulation data on the primitive model level are from Kesvatera et al.[68] Calculations based on the DH approximation by Juffer et al.[272] at 0.1 M ionic strength are labeled with the respective relative perimittivities D_d of the protein interior. Results of the improved null model from Antosiewicz et al.[263] are shown for comparison. Root-mean-square (rms) and maximum (max) deviations are given.

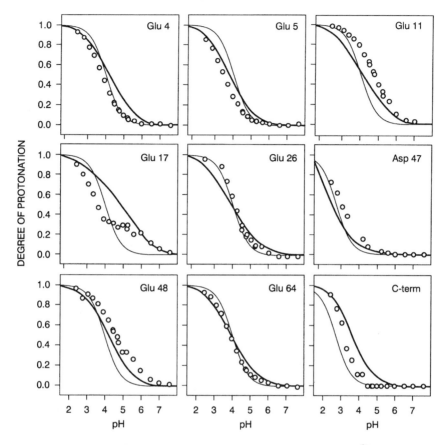

Figure 69. Site-specific titration curves of carboxylic residues in the apo form (Ca^{2+}-free) calbindin at a protein concentration of 0.5 mM and no salt added.[109] Points are from ^{13}C NMR data. Thick solid lines are results of Monte Carlo (MC) simulations and thin lines are the improved null model by Antosiewicz *et al.*[263] (see Table 17).

chains titrate as independent sites, which indicates that the effects of site–site interactions are weak. However, the apparent pK values are shifted up to 2 units. Table 25 compares the resulting p$K_{1/2}$ values with Monte Carlo simulation results, and Fig. 70a shows that the latter generally reproduces the pK values reasonably well, but typically overestimates somewhat their shift. In addition to the Ca^{2+} form of calbindin, Kesvatera *et al.*[68] have also studied the ionization of lysines in the apo form of this protein at two different concentrations. The experimental results were in reasonable agreement with simulations.

An interesting comparison of these results was recently provided by Juffer *et al.*[272] These authors have solved the DH equation for calbindin with a boundary

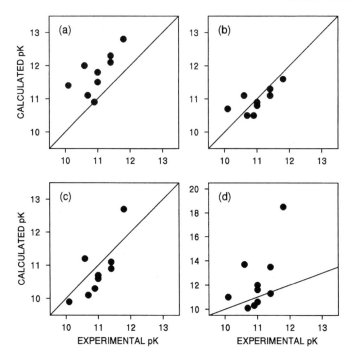

Figure 70.   Comparison of experimental and calculated apparent pK values of the lysines for the Ca^{2+} form of calbindin.[68] (a) Monte Carlo (MC) simulations from Kesvatera *et al.*[68] and Debye–Hückel (DH) calculations by Juffer *et al.*[272] at 0.1 M salt concentration at different relative permittivities of the interior of (b) $D_d \approx 80$, (c) $D_d = 20$, and (d) $D_d = 4$.

element method, and obtained the pK values of all basic residues. Their results for different values of the interior relative permittivity D_d are summarized in Table 25. The graphical comparison with experimental data in Fig. 70b shows the results for $D_d \approx 80$ and is therefore directly comparable to the Monte Carlo simulation shown in Fig. 70a. The DH model is in good agreement with experiment but underestimates the pK shifts slightly. While the Monte Carlo simulations incorporate the finite protein concentration with the cell model, the calculations of Juffer *et al.*[272] are carried out at an ionic strength of 0.1 M. Another difference between these approaches is that the DH calculation includes the shift due to the self-energy, while this contribution is neglected in the Monte Carlo approach. In spite of these differences, the mutual agreement between these fundamentally different approaches is gratifying. Figures 70c and 70d show the results of the analogous DH calculation for different values of the interior relative permittivity $D_d = 20$ and $D_d = 4$. While $D_d = 40$ still gives reasonable agreement with experiment, it is rather obvious that the same model for $D_d = 4$ yields unreasonably large pK shifts.

The analogous experimental and Monte Carlo simulation results for the carboxylic groups of the calbindin in the apo form (Ca^{2+}-free) show that the titration curves are substantially broader than predicted by the null model (see Fig. 69). The two-step titration curve for Glu 17 is most likely caused by a conformational transition and is probably not related to the effect of site–site interactions. While only the ^{13}C NMR chemical shift of the γ-carbon in the side chain was monitored, we suspect that due to their close proximity this chemical shift properly reflects the site-titration curve. The solid lines are results of Monte Carlo simulations, which satisfactorily reproduce the shifts as well as the broadening of these curves. In particular, for Glu 17 the simulation predicts a much broader titration curve than for all other residues. Table 26 summarizes the resulting $pK_{1/2}$ values and Hill exponents ν. The broadened titration curves can be reasonably well modeled by the Hill equation [see Eq. (207)] and therefore the Hill exponent represents an empirical measure of the widths of the titration curves. Similar Hill-exponent analysis was carried out for OMTKY3.[250]

Two-step titration curves were also observed for Asp 10 in ribonuclease HI isolated from *Escherichia Coli*[121] (see Fig. 17). Oda *et al.*[121] have suggested that this effect arises from site–site interactions and they proposed an interpretation based on a dielectric continuum model. However, this model uses a relatively low value for the interior relative permittivity $D_d = 10$ and is likely to overestimate the interaction strength. The majority of the ionizable residues in this protein do follow

Table 26. *Titration Data for Carboxylic Residues of Calbindin D_{9k} in the Apo Form (Ca^{2+}-Free) at Protein Concentration of 0.5 mM and No Added Salt[a]*

Residue	$pK_{1/2}$			ν	
	MC	Null	Exp.	MC	Exp.
Glu 4	4.2	4.0	3.8	0.52	0.80
Glu 5	3.7	4.0	3.4	0.60	0.77
Glu 11	4.1	4.0	4.7	0.44	0.70
Glu 17[b]	4.9	4.0	3.6	0.30	—
Glu 26	3.9	4.0	4.1	0.50	0.98
Asp 47	2.5	2.7	3.0	0.55	0.68
Glu 48	4.2	4.0	4.6	0.61	0.58
Glu 64	3.9	4.0	3.8	0.55	0.78
C-ter	3.6	2.7	3.2	0.72	1.17
rms	0.6	0.4			
max	1.3	0.7			

[a]Experimental and Monte Carlo (MC) simulation data on the primitive model level are from Kesvatera *et al.*[109] Hill exponents are denoted as ν. The corresponding site-titration curves are shown in Fig. 69. Root-mean-square (rms) and maximum (max) deviations for the pK values are given.
[b]This residue has a biphasic titration curve and no meaningfull Hill exponent can be extracted.

a simple one-step titration as expected from the null model, and it is also conceivable that a conformational transition might be responsible for the observed behavior of Asp 10. Effects of site–site interactions clearly represent an interesting issue, but in proteins they certainly are rather the exception than the rule. Ionizable sites in proteins usually titrate independently according to the null model.

There are several advantages of the Monte Carlo procedure as compared to the PB/DH approaches. One is that the mean-field approximation is avoided, the protein concentration is taken into account, and the method is surprisingly efficient. The main disadvantage is that a dielectric discontinuity is difficult to incorporate, except for simple geometries; consideration of negative polarizabilities might represent a tentative way.

Another interesting approach is the use of integral equations. Ullner et al.[288] have recently used the reference interaction site model together with a mean spherical approximation (RISM-MSA) to calculate pK shifts for calbindin. The RISM-MSA method includes the electrostatic interactions in the DH approximation, but treats the protein as an irregular body. The results were in extremely good agreement with Monte Carlo simulations. The advantage of the procedure is that the hard-core interactions are treated in a systematic fashion, and the molecular nature of the solvent can also be considered.

An important question is, of course, the accuracy of the dielectric continuum model. The obvious way to test the mean-field approximation inherent to the PB equation is to perform Monte Carlo on the same system. To our knowledge, there are rather few critical tests. The calcium binding to calbindin D_{9k} was used as a test case by Fushiki et al.,[281] who compared the results from the PB and DH equations with simulated free energies. The comparison was limited to the case with a uniform dielectric permittivity and the free energies from the three methods were indistinguishable in most cases. Significant differences were observed with divalent counterions. This phenomenon is well known in the electric double-layer theory[169] and was discussed in Section 3. It is interesting to note that, although the free energies are accurately reproduced in the DH approximation, the electrostatic potential and the concentration profile could be completely nonphysical; as a matter of fact negative concentrations might be predicted. In this respect, the PB equation is much better behaved. Da Silva et al.[289] have recently tested the Tanford–Kirkwood model with computer simulations for $D_d = 80$. The latter indicated that the binding of monovalent (and divalent) ions is well described by the Tanford–Kirkwood approach. Both salt effects and differences in charges on individual residues (e.g., by mutations) were handled reasonably well. The introduction of a dielectric discontinuity and a low dielectric interior of the protein may reduce the accuracy of the mean-field approximation, but only marginally. The reason for the limitation is the increased electrostatic coupling in the system. The dielectric model itself can in principle only be tested against a fully atomistic description, but such a test is not possible as yet.

There are two interesting ways of modulating electrostatic interactions in proteins, namely by site-directed mutants or in certain metal ion containing proteins by changing the redox status of the metal ion. In site-directed mutants, specific charged amino acids are replaced by unchanged ones (e.g., Asp with Ser, Asp with Asn, or Glu with Gln). Effects on pK shifts have been investigated for site-directed mutants in subtilisin.[290] In azurine, a copper-containing protein, the titration of two histidine residues has been studied both experimentally and theoretically with respect to the oxidation of copper (Cu^+ and Cu^{2+}). The pK shift found for His 32 and His 83 are fairly small, 0.1–0.3 pK units.[291] The experimental predictions using a low dielectric interior of the protein are in reasonable agreement with experiment, although being slightly too large.[292] Monte Carlo simulation with a high protein permittivity produces smaller but comparable numbers.[293]

5.6. Protein Folding

Proteins usually denaturate in strong acids or bases. The classical picture is that the strong Coulomb repulsion between numerous ionized residues of the same charge will counteract the attractive hydrophobic forces and unfold the protein.

If electrostatic interactions are indeed dominant, a close relationship between the stability of a folded protein and its titration curve must exist. The free energy of unfolding, ΔF, can be expressed as

$$\frac{a_{\mathrm{H}}}{N} \frac{\partial \Delta F}{\partial a_{\mathrm{H}}} = \theta_u - \theta_f \tag{244}$$

where θ_u and θ_f represent the degree of protonation in the unfolded and folded state, respectively. The relation follows from the derivative of the semi-grand partition function of the protein [cf. Eq. (67)] and taking the difference between these expressions for the unfolded and folded state. An interesting consequence of this relationship is that the free energy has an extremum for an equal degree of protonation (or charge) for both states. For the titration of the unfolded (denaturated) state one usually uses the null model, neglecting all site–site interactions. The model for the folded state includes the effect of site–site interactions, and possible shifts of pK values due to different molecular environments. The free energy is then obtained by integration of the differences between the computed titration curves.[260,271]

The result of such a calculation is shown for ribonuclease A in Fig. 71, in which the best site–site interaction model by Antosiewicz *et al.*[260] is compared with experimental data.[294] While the computations reflect the proper trend, quantitative agreement is lacking. The free energy of unfolding can be also obtained from experimental titration data on the folded and unfolded protein. Such data are available for ribonuclease and were discussed above (see Figs. 54 and 57). Integrating the difference of the corresponding titration curves yields a fairly good description of the experimental data.

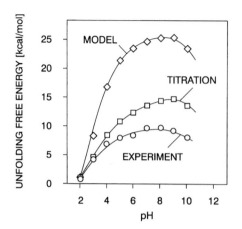

Figure 71. Unfolding free energies for ribonuclease A as a function of pH. Experimental data[294] are compared with the best high dielectric permittivity ($D_d = 20$) Poisson–Boltzmann (PB) model.[260] The titration result is obtained from the experimental data shown in Fig. 57.

Unfolding of proteins can be accomplished in many ways, such as by using high concentrations of urea or guanidinium chloride.[295,296] The major effect of these agents is to change the solvent hydrophobicity, but not of electrostatic interactions between charge residues in the protein. The addition of strong acid is also used as an unfolding agent[297]; that of a strong base is not common. Strong acids and bases will not only unfold the protein, but may also cause its hydrolysis and degradation. While the discussed electrostatic mechanism for protein unfolding is probably important for many proteins, it is essential to realize that some proteins retain their structure over a wide range of solution conditions. In such situations, other forces must be important as well.

6. Polyelectrolytes

Polyelectrolytes are charged macromolecules, which may have a linear or branched structure. Weak polyelectrolytes usually bear a large number of ionizable groups, typically $N > 100$. Their ionization properties are therefore not expected to depend on this number, and the consideration of an infinite chain usually suffices (large system limit, $N \to \infty$). The effects of site–site interactions for polyelectrolytes are interesting, since weakly charged systems can be treated within a mean-field approximation[150,298] while for highly charged polyelectrolytes the mean-field approximation breaks down due to the short-range nature of the interaction potential.[8,19,21] The situation simplifies substantially if the polyelectrolytes have equivalent ionizable groups, and in such cases rather detailed analysis of the ionization behavior can be put forward.[8,24,203] In other situations, the existence of nonequivalent groups (e.g., tacticity), conformational changes, and branching of the backbone complicates the problem substantially. All these issues will be addressed in the following.

6.1. Mean-Field Models

Polyelectrolytes may accumulate substantially higher charge densities than proteins, having often pronounced effects of site–site interactions. As a first step, we shall discuss the mean-field model, but due to the high charge densities it is essential to extend the model to the nonlinear regime of higher potentials. For the sake of simplicity, we shall restrict ourselves to the case of identical groups. In Section 5 we derived an isotherm in the mean-field approximation [cf. Eqs. (223) and (224)], which reads

$$\text{pH} = \text{p}K_{\text{eff}} + \log_{10} \frac{1 - \theta}{\theta} \tag{245}$$

with the effective pK value

$$\text{p}K_{\text{eff}} = \text{p}\hat{K}^{(\text{int})} - \beta e\psi_0 / \ln 10 \tag{246}$$

where ψ_0 is the uniform electrostatic potential on the surface. Within the constant capacitance model discussed in Section 5, Eq. (246) simplifies to [cf. Eq. (231)]

$$\text{p}K_{\text{eff}} = \text{p}\hat{K} - \bar{\varepsilon}\theta \tag{247}$$

These equations hold rather generally for any geometry. By expanding the free energy in a Taylor series Eq. (247) can be generalized as

$$\text{p}K_{\text{eff}} = \text{p}\hat{K} - \bar{\varepsilon}\theta - \frac{\bar{\lambda}}{2} \theta^2 + \cdots \tag{248}$$

where $\bar{\lambda}$ is the average triplet interaction parameter. This type of interpretation was initially suggested by Mandel.[299]

In the polyelectrolyte literature, it is quite common to express pK_{eff} as a function of the degree of protonation θ, instead of reporting the degree of protonation as a function of pH.[21,25,300] Both representations are equivalent. Nevertheless, in order to properly evaluate pK_{eff}, the protonation degree of the polyelectrolyte must be known accurately. This condition usually poses no restrictions as long as the polyelectrolyte can be fully titrated at both limits of full protonation and deprotonation. However, caution must be exercised, since small errors in the saturation values may result in large errors in pK_{eff}.

Two geometries are commonly used to discuss the charging characteristics of polyelectrolytes, namely cylindrical and spherical geometries. In the case of a uniformly charged cylinder, the interaction parameter can be written as [cf. Eq. (227)]

$$\bar{\varepsilon} = \frac{\beta e^2 N}{\ln 10\, C} = \frac{\beta e^2 \Lambda}{\ln 10\, \tilde{C}} \tag{249}$$

where $\Lambda = N/L$ is the site density per unit length and $\tilde{C} = C/L$ the specific capacitance per unit length (L being the contour length of the polyelectrolyte chain). The Debye–Hückel (DH) model for a charged cylinder with a smeared-out charge density leads to a diffuse layer capacitance[301]

$$\tilde{C} = \tilde{C}_d = \varepsilon_0 D_w \kappa \frac{K_1(\kappa a)}{K_0(\kappa a)} \quad \text{(cylinder)} \tag{250}$$

where D_w is the dielectric permittivity of water and $K_n(x)$ is a modified Bessel function of the second kind.[160] Alternatively, the electrostatic potential ψ_0 entering Eq. (246) can be directly evaluated from the numerical solution of the Poisson–Boltzmann (PB) equation for a cylinder. This approach was advocated by Nagasawa and coworkers[150,298] and is easily carried out on any modern personal computer.

The model for the spherical geometry, which is used to approximate a globular shape of a gel-like, possibly branched polyelectrolyte, was discussed within the DH approximation in Section 5 [see Eqs. (233) and (234)]. Within the PB approximation, the charge–potential relationship must be obtained numerically.

A closely related approach, which assumes a uniform distribution of charges within the polyelectrolyte, is yielding the *Donnan model*, which pictures the polyelectrolyte as a porous gel-like phase with a constant electrostatic potential ψ_0. For simplicity, we assume an excess of monovalent background electrolyte of bulk concentration c_s. The corresponding concentrations of cations and anions in the Donnan phase will be denoted by c_+ and c_-, respectively. Equating the electrochemical potentials of the cations and anions [cf. Eq. (17)] one obtains

$$c_\pm = c_s \exp(\mp \beta e \psi_0) \tag{251}$$

while the electroneutrality within the Donnan phase ensures that

$$e(c_+ - c_-) + Q/V_D = 0 \tag{252}$$

where Q is the total number of charges in the Donnan phase and V_D the corresponding Donnan volume. Eliminating c_+ and c_- from Eqs. (251) and (252) we obtain the charge–potential relationship for the Donnan model

$$Q = 2eV_D c_s \sinh(\beta e \psi_0) \tag{253}$$

Thus the charge increases exponentially with increasing potential. For small potentials, the equation can be linearized and the model reduces to the constant capacitance model with total capacitance

$$\tilde{C} = 2\beta e^2 V_D c_s \tag{254}$$

In contrast to all DH-type models, the Donnan capacitance is proportional to the electrolyte concentration. Strictly speaking, the Donnan model is only correct for a sufficiently large Donnan phase and this condition is not often met in practice.

6.2. Nearest-Neighbor Chain Interaction Models

Mean-field models are able to explain the titration of linear polyelectrolytes at low charge densities, and not for highly charged polyelectrolytes. Steiner[53] was probably the first to realize that the problem of titrating polyelectrolytes is closely related to the so-called Ising model on a linear chain with nearest-neighbor interactions.[55,59,60] Katchalsky,[21] Rice and Harris,[18] and Marcus[19] have further pursued this idea and shown that the typical two-step protonation behavior of a highly charged polyelectrolyte naturally emerges from such a model.

Let us number the sites along the chain consecutively ($i = 1, \ldots, N$) and assign to each ionizable site a state variable s_i ($s_i = 0$ if site i is deprotonated and $s_i = 1$ if it is protonated). Let us also assume that all sites along the chain are identical and have the same proton affinity. As discussed in Section 3, the free energy of such a linear chain can be derived from the cluster expansion method and written as [cf. Eq. (84)]

$$F(\{s_i\}) = -\mu \sum_i s_i + E \sum_i s_i s_{i+1} \qquad (255)$$

Here, all but the nearest-neighbor pair interactions are set to zero. As before, we have $\beta\mu / \ln 10 = p\hat{K} - pH$. Since we assume the sites to be identical, the average degree of protonation θ is given by the derivative of the partition function Ξ, namely

$$\theta = kT \frac{\partial \ln \Xi}{\partial \mu} = a_H \frac{\partial \ln \Xi}{\partial a_H} \qquad (256)$$

where the partition function is given by

$$\Xi = \sum_{\{s_i\}} e^{-\beta F(\{s_i\})} \qquad (257)$$

The evaluation of Eq. (257) is a classical problem in statistical mechanics. Let us briefly mention the *transfer matrix technique*, which can be generally used to solve such problems on a chain.[59,60] If we consider the limit $N \to \infty$, it is immaterial whether the chain has loose ends or whether the ends are tied together to a circular loop. For the latter case, the partition function can be written as a product of 2×2 transfer matrices with the elements[302]

$$T(s, s') = z^s u^{ss'} \qquad (258)$$

where $z = \hat{K} a_H$, and $u = e^{-\beta E}$ parameterizes the nearest-neighbor interaction energy E of one bond. The previously defined interaction parameter is given by $\varepsilon = \beta E/\ln 10 = -\log_{10} u$. Written out explicitly, the transfer matrix reads

$$T = \begin{bmatrix} 1 & 1 \\ z & zu \end{bmatrix} \qquad (259)$$

An inspection of Eq. (257) shows that the partition function for a closed loop can be written as

$$\Xi = \mathrm{Tr} T^N = \lambda_+^N + \lambda_-^N \tag{260}$$

where Tr denotes the trace of the matrix. The eigenvalues of the transfer matrix T are denoted as λ_+ and λ_- ($\lambda_+ \geq \lambda_-$). The $N \to \infty$ limit is solely determined by the larger eigenvalue and the partition function becomes

$$\Xi = \lambda_+^N \tag{261}$$

where the larger eigenvalue is given by

$$\lambda_+ = (1 + zu)/2 + [z + (1 - zu)^2/4]^{1/2} \tag{262}$$

The application of Eq. (256) gives the explicit expression for the titration curve of a linear polyelectrolyte chain with nearest-neighbor pair interactions

$$\theta = [2 + (\lambda_+/z)(1 - zu)/(1 - u + \lambda_+ u)]^{-1} \tag{263}$$

The N dependence disappears in the $N \to \infty$ limit. The same model was already introduced for finite linear chains in Section 4. On the basis of that analysis, we can also derive the corresponding result for an infinite chain.[21] The approach is similar to that presented in the derivation of the mean-field model in Section 5, except one deals with a double sum, which must be maximized with respect to two variables, namely the number of protonated sites n and the number of pairs m. By locating the maximum term in Eq. (188) one obtains[21]

$$\mathrm{pH} = \mathrm{p}\hat{K} + \log_{10}\left[\frac{u(\zeta + 1 - 2\theta)}{(\zeta - 1 + 2\theta)}\right] \tag{264}$$

where $\zeta = [1 + 4\theta(1 - \theta)(1 - u)]^{1/2}$ and u has the same meaning as above. While not immediately obvious, this result is identical to the one derived by the transfer matrix technique [cf. Eq. (263)].

The titration curves calculated using the expression for the nearest-neighbor Ising model are shown in Fig. 72, and they are compared with the results for finite chains, as discussed in Section 4. One observes that the limiting result for $N \to \infty$ is approached rather rapidly.

The characteristic double hump structure of the limiting curve can be understood as follows. During the first protonation step around $\mathrm{pH} \simeq \mathrm{p}\hat{K}$, the system circumvents all pair interactions by protonating every second site. At $\theta = 1/2$ that process is basically complete, and further protonation is only possible by overcoming two pair interactions; thus, the remaining sites protonate near $\mathrm{pH} \simeq \mathrm{p}\hat{K} - 2\varepsilon$. This protonation sequence is illustrated schematically in Fig. 73. Note that this picture is only correct when thermal fluctuations are negligible. When ε becomes

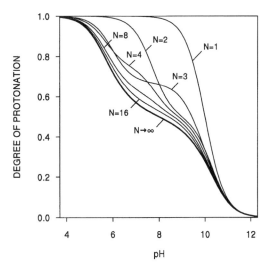

Figure 72. Titration curves for the linear Ising model on a chain with nearest-neighbor interactions for a different number of sites N. The parameters are $\varepsilon = 2$ and $p\hat{K} = 10$. For finite N the titration curves are calculated from formation constants given in Table 9 [cf. Eq. (188)], while for $N \rightarrow \infty$ Eq. (263) is used.

comparable to unity, the alternating minimum energy structure at $\theta = 1/2$ will be disordered through pairs of equally protonated sites.

This disorder can be quantified through the pair correlation function of the sites $\langle s_i s_{i+j} \rangle$. For the nearest-neighbor pair interaction model, there is a simple expression for the correlation function for $\theta = 1/2$. For $E > 0$ the result is[58,60]

$$\langle s_i s_{i+j} \rangle \propto (-1)^j e^{-j/\ell} + \text{const} \tag{265}$$

where the decay length ℓ is given by $\ell^{-1} = -\ln \tanh (\beta E/4)$. For strong site–site interactions the correlation function is purely oscillatory, which signals the alternating arrangement of the ionized sites. However, there is an additional decaying component, and the decay length ℓ can be interpreted as the average length of an

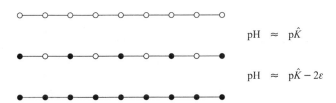

$$\text{pH} \approx p\hat{K}$$

$$\text{pH} \approx p\hat{K} - 2\varepsilon$$

Figure 73. Schematic protonation sequence of a linear chain with strong nearest-neighbor pair interactions. Protonated sites are shown as filled circles and deprotonated sites as open circles. At high pH, all sites are deprotonated. At intermediate pH every second site is protonated, while at low pH all sites protonate. The indicated transition regions between these states are related to the microscopic ionization constant $p\hat{K}$ and the nearest-neighbor pair interaction parameter ε.

alternating sequence without defects. Typical interaction parameters ϵ for polyelectrolytes are in the range 1–3, which corresponds to ℓ in the range of 2–15. Thus, weakly interacting polyelectrclytes will be disordered, while for strong interactions, substantial ordering along the chain is expected. For $\theta \neq 1/2$ the results are more complicated.

6.3. Discrete Charge Model

Let us now discuss the analog of the Tanford–Kirkwood discrete charge model introduced for proteins,[70] as sketched in Fig. 74. Ionizable sites are assumed as point charges, embedded in a dielectric medium. While the Tanford–Kirkwood model uses a sphere to model the shape of a protein,[61] the polyelectrolyte is mimicked by a dielectric cylinder of radius a and relative permittivity D_d. This cylinder is immersed in an aqueous electrolyte solution of relative permittivity D_w and Debye screening length κ^{-1}. For simplicity, the ionizable sites are assumed to be identical and arranged along the cylinder axis at equal spacing. For a given configuration, this system has the free energy [cf. Eq. (92)]

$$F(\{s_i\}) = -\mu \sum_i s_i + \frac{1}{2} \sum_{i,j} E_{ij} s_i s_j \qquad (266)$$

where $\beta\mu/\ln 10 = \text{pH} - p\hat{K}$ with $p\hat{K}$ being the pK value of the sites given all sites are deprotonated, and the pair interaction parameters $E_{ij} = W(|\mathbf{r}_i - \mathbf{r}_j|)$ are related to the electrostatic interaction potential $W(r)$ of two sites located at $\mathbf{r}_i$ and $\mathbf{r}_j$ at a distance $r = |\mathbf{r}_i - \mathbf{r}_j|$.

The site–site interaction potential can be found from the solution of the DH equation in the cylindrical geometry; for details see Appendix A. The result reads[70,303]

$$W(r) = \frac{e^2}{4\pi\varepsilon_0 D_d} \left[\frac{1}{r} - \frac{2}{\pi} \int_0^\infty \frac{pD_w K_1(pa)K_0(ka) - kD_d K_1(ka)K_0(pa)}{pD_w K_1(pa)I_0(ka) + kD_d I_1(ka)K_0(pa)} \cos kr \, dk \right] \qquad (267)$$

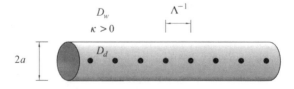

Figure 74. Schematic representation of the discrete charge cylinder model. The polyelectrolyte is modeled as a dielectric cylinder of radius a and relative permittivity D_d immersed into an electrolyte solution with relative permittivity D_w and Debye screening length κ^{-1}. The charges are uniformly spaced at a distance Λ^{-1}.

where $p = (k^2 + \kappa^2)^{1/2}$; modified Bessel functions are denoted as $I_n(z)$ and $K_n(z)$.[160] This interaction potential has several characteristic features. For small distances ($r \rightarrow 0$), the presence of the electrolyte is not felt and the charges interact by a simple Coulomb law

$$W(r) = \frac{e^2}{4\pi\varepsilon_0 D_d} \frac{1}{r} . \qquad (268)$$

At large distances ($r \rightarrow \infty$), screening sets in and leads to a rapid decay of the potential. In the physically relevant case where $D_d \ll D_w$, the interaction potential behaves for large distances as

$$W(r) = \frac{e^2}{4\pi\varepsilon_0 D_w} \frac{\exp(-\kappa r)}{r} \qquad (269)$$

which is the screened Coulomb potential between two point charges in an electrolyte solution [cf. Eq. (38)]. This limiting behavior follows from the observation that the presence of the cylinder can be neglected at separation distances much larger than the cylinder radius.

The interaction potential for a cylinder can be readily calculated numerically and is plotted as a function of distance in Fig. 75, taking a cylinder radius $a = 0.25$ nm and relative permittivity $D_d = 3$. The cylinder is immersed in a 0.5 M monovalent

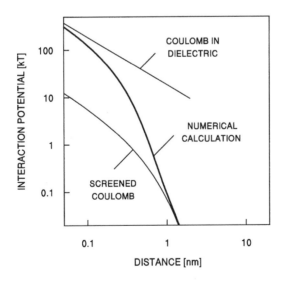

Figure 75. Site–site interaction potential for a cylinder of radius $a = 0.25$ nm and relative permittivity $D_d = 3$ in a 0.5 M monovalent electrolyte solution ($\kappa^{-1} = 0.43$ nm and $D_w = 80$). The result is obtained numerically and compared with the Coulomb potential in the dielectric [cf. Eq. (268)] and screened Coulomb potential [cf. Eq. (269)].

electrolyte solution ($\kappa^{-1} \simeq 0.43$ nm and $D_w = 80$). The result is compared with the limiting laws, Eqs. (268) and (269). Since the potential crosses from strong interactions to much weaker interactions, the potential decays very rapidly at distances comparable with the cylinder radius. Such a rapidly decaying interaction potential is short-range, and can be effectively approximated by nearest-neighbor interactions.

The ionization behavior of the discrete charge cylinder model is summarized in Fig. 76 at an ionic strength 0.5 M. Figure 76a shows the degree of protonation as a function of pH, while Fig. 76b shows pK_{eff} as commonly done in polyelectrolyte literature. The distance between the charges is taken to be 0.35 nm. The squares are the Monte Carlo results, which represent accurate solutions of the discrete charge model. One observes that this model predicts a two-step titration curve, which originates from the short-range nature of the interaction potential. Taking only the nearest-neighbor interactions into account should therefore lead to a good approximation of the Monte Carlo data. This approximation was discussed for the Ising model on a linear chain above, and the resulting isotherm [cf. Eq. (263)] is also shown in Fig. 76. Indeed, this model leads to a fairly accurate description of the Monte Carlo data.

Figure 77 shows the protonation behavior in the same setting, but the charges are placed at a larger separation distance of 0.55 nm. Clearly, the effect of interactions is much weaker and the two-step behavior is no longer apparent.

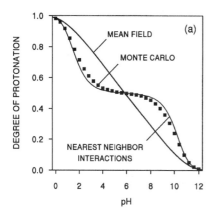 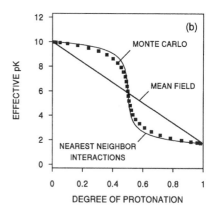

Figure 76.   Titration curves for the discrete charge model at high charge.[70] The cylinder has a radius $a = 0.25$ nm, relative permittivity $D_d = 3$, and is immersed in a 0.5 M monovalent electrolyte solution ($\kappa^{-1} \simeq 0.43$ nm and $D_w = 80$). The basic ionizable groups with $p\hat{K} = 10$ are arranged equidistantly along the cylinder axis at a distance of $\Lambda^{-1} = 0.35$ nm. The Monte Carlo results are compared with the nearest-neighbor interaction model [cf. Eq. (263)] and with the mean-field model [cf. Eqs. (245) and (247)]. (a) Degree of protonation as a function of pH, and (b) effective pK defined by Eq. (245) as a function of the degree of protonation.

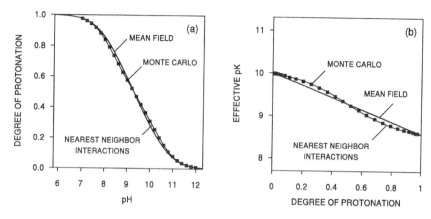

Figure 77. Titration curves for the discrete charge model at low charge. The cylinder has a radius $a = 0.25$ nm, relative permittivity $D_d = 3$, and is immersed in 0.5 M monovalent electrolyte solution ($\kappa^{-1} \approx 0.43$ nm and $D_w = 80$). The basic ionizable groups with $p\hat{K} = 10$ are arranged equidistantly along the cylinder axis at a distance of $\Lambda^{-1} = 0.55$ nm. Monte Carlo results are compared with the nearest-neighbor interaction model [cf. Eq. (263)] and with the mean-field model [cf. Eqs. (245) and (247)]. (a) Degree of protonation as a function of pH, and (b) effective pK defined by Eq. (245) as a function of the degree of protonation.

6.3.1. Mean-Field and Smearing-Out Approximations

The mean-field model predicts a broad, featureless protonation curve. For a linear geometry, the strength of the mean field is given by

$$\overline{E} = \frac{1}{N} \sum_{i \neq j} W(|\mathbf{r}_i - \mathbf{r}_j|) = \Lambda \int g(z)W(z)dz \tag{270}$$

where Λ is the interfacial site density (per unit length) and $g(z)$ is the pair distribution function of sites along the cylinder axis (z-axis). In the above equation the actual site–site interaction potential has to be used. For the case of equidistant ionizable sites along a line, Eq. (270) can be written as

$$\overline{E} = 2 \sum_{n=1}^{\infty} W(n/\Lambda) \tag{271}$$

The mean-field titration curve is given by Eqs. (245) and (247) with the mean-field interaction parameter $\overline{\varepsilon} = \beta E/\ln 10$, and pK_{eff} becomes a linear function of θ (see Fig. 76b). Within the mean-field model, pK_{eff} has a precise physical interpretation; it is the microscopic pK of any group at given degree of ionization. For the nearest-neighbor interaction model (or any other type of interaction), pK_{eff} is a convenient representation of the data with no relation to microscopic pK's. For example, in the nearest-neighbor interaction model there are only three possible microscopic pK's, namely p$\hat{K}$, p$\hat{K} - \varepsilon$, and p$\hat{K} - 2\varepsilon$.

For the strongly interacting scenario (see Fig. 76a) the mean-field model fails to generate the two-step titration curve, while in the case of the weakly interacting system (see Fig. 76b) it does an excellent job in predicting the Monte Carlo data. This situation is generic; for weak or moderate interactions, the mean-field model represents an excellent approximation for polyelectrolytes, while it fails for stronger interactions. The exact strength of the interaction at which the mean-field description fails depends on the range of the site–site potential. For a short-range potential, as used in the present example, already moderate interactions lead to substantial deviations from the mean-field behavior. On the other hand, if the interaction potential is longer-range (i.e., at lower ionic strengths or for higher permittivity of the cylinder), the mean-field approximation remains valid within the regime of stronger interactions.

A frequently used site–site interaction potential in polyelectrolyte literature is the screened Coulomb potential [cf. Eq. (269)]. As evident from Fig. 75, this potential represents an excellent approximation for interactions of point charges within a dielectric cylinder for distances larger than the cylinder radius ($a \ll r$), provided the cylinder is thin with respect to the Debye length ($a \ll \kappa^{-1}$). This potential is therefore expected to represent interactions in polyelectrolytes with lower charge densities quite well. Inserting Eq. (269) into Eq. (271) and introducing a few approximations, one can show that for the screened Coulomb potential the mean-field interaction parameter becomes[203,304,305]

$$\bar{E} = \frac{\beta e^2}{2\pi\varepsilon_0 D_w a} \ln[1 - \exp(-\kappa a)] \tag{272}$$

As can be verified by Monte Carlo simulations, the mean-field approximation for the screened Coulomb potential is excellent for the entire parameter range relevant for polyelectrolytes in water.[203] The screened Coulomb potential is sufficiently long-range to induce mean-field behavior in this regime. (Deviations from mean-field behavior do occur, but only for unrealistic parameters.)

The mean-field interaction energy can be also evaluated for more complicated geometries. For example, the all-*trans* conformation with fixed bond angles is very similar to the model of the linear chain.[24] Other geometries, such as a series of *trans–gauche* configurations, or a flexible chain can also be considered. While this kind of analysis was carried out in the literature,[148] it is somewhat problematic, since the chain is initially coiled and tends to expand as it ionizes. This aspect will be discussed below.

It is instructive to analyze the relationship between the discrete site model and the smeared-out charge model discussed above. Smeared-out charge corresponds to no spatial correlations of the ionizable sites along the chain. Inserting $g(z) = 1$ into Eq. (270) we recover the smeared-out charge result for the cylinder given in

Eq. (249). The capacitance per unit length can be interpreted as originating from two parallel plate capacitors in series

$$\tilde{C}^{-1} = \tilde{C}_{dl}^{-1} + \tilde{C}_{in}^{-1} \tag{273}$$

The first term $\tilde{C}_{dl}$ is given exactly by the diffuse layer capacitance for the cylinder [cf. Eq. (250)], while the second term in Eq. (273) actually diverges. This divergence originates from smearing-out the charges along the cylinder axis; if the charges are smeared-out along a second internal cylinder of radius $b < a$, the corresponding term gives $\tilde{C}_{in}^{-1} = a \ln(a/b)/(\varepsilon_0 D_w)$ and does not diverge. The diffuse layer model neglects this contribution entirely.

For the strong-coupling situation shown in Fig. 76, the mean-field interaction energy is characterized by a pair interaction parameter $\bar{\varepsilon} \simeq 8.27$, while the double layer model gives $\varepsilon_{dl} \simeq 1.76$. The latter model thus seriously underestimates the strength of the electrostatic interactions. For the weak-coupling situation shown in Fig. 77, the corresponding parameters are $\bar{\varepsilon} \simeq 1.42$ and $\varepsilon_{dl} \simeq 1.13$. In this case, the diffuse layer model already represents a reasonable approximation.

We could also incorporate a discussion of pK shifts, and address whether an ionizable group attached to a polyelectrolyte chain has the same ionization constant as the same group on a small molecule in solution. The discrete charge cylinder model can be used to address this question in an entirely analogous fashion as for the protein. If the permittivity of the interior is smaller than that of water ($D_d < D_w$), the analysis again shows that an acidic group would be somewhat more basic on the polyelectrolyte than in solution, while a basic group would be more acidic. The predicted shifts are too large, but smaller than in the protein geometry (or, as we shall see in Section 7, for the planar interface). The analysis of experimental titration curves of polyelectrolytes shows unambiguously that the ionization constants of groups on polyelectrolytes are quite similar to the corresponding ionization constants in solution.

6.3.2. Chain Flexibility

As the charge density on a flexible chain increases, the chain will expand due to increasingly strong electrostatic repulsion. This effect is particularly relevant for polyelectrolytes with poorly soluble backbone. As a result, the effects of site–site interactions weaken and thereby influence the titration behavior. Therefore, the coupling between ionization and conformation must be treated in a self-consistent manner.[24,203,306] Such effects can play an important role in conformational transitions of polyelectrolytes, leading to structural transitions or collapse upon loss of charge. While the relationship between the chain conformational and titration behavior was discussed extensively, it became possible to treat this problem with computer simulations in a consistent fashion only recently. The technique involves a simulation in the (semi-) grand canonical ensemble, where all possible conformations of a chain are sampled from a thermal distribution, but at the same time the chemical potential of the protons is kept fixed, and leads to protonation and deprotonation reactions of individual ionization sites.[24,203,306] Such simulations are rather time-consuming, as it is difficult to properly equilibrate long polymer chains.

Let us discuss simulation results of a fully flexible titratable chain, where the site–site interaction potential was modeled by screened Coulomb interaction.[203] These Monte Carlo simulations used a pivot algorithm, where the chain is rotated as a rigid body around a randomly chosen bond within the chain.[203,307] The procedure greatly improved the sampling efficiency compared to just moving one monomer at a time, and allowed simulations of long chains. The technique was used extensively for flexible chains in both low-salt[152,309] and high-salt[203] conditions. Similar studies were carried out for models with discrete conformation states,[24] while Sassi *et al.*[306] studied a lattice model.

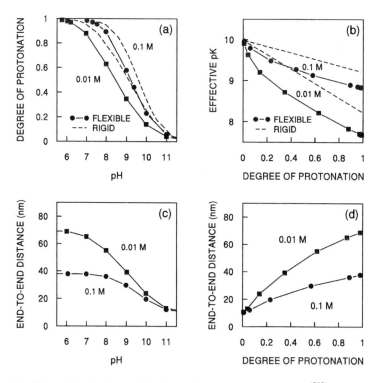

Figure 78. Monte Carlo simulation of the titration behavior of flexible chains.[203] The chain of 320 subunits is fully flexible with rigid bonds of 0.6 nm length. The ionized sites interact according to a screened Coulomb potential. Quantities are plotted as a function of pH (left column) and degree of protonation (right column). (a) Degree of protonation, (b) pK_{eff}, (c) and (d) end-to-end distance. The points represent the Monte Carlo simulation results and the solid lines serve to guide the eye. Dashed line shows the corresponding theoretical result for the rigid chain.

Figure 78 shows the calculated titration curves and average end-to-end distance for a fully flexible chain with 320 sites and rigid bonds of 0.6 nm in length.[203] The sites mimicked amine sites of $p\hat{K} = 10$, which are neutral in the deprotonated state and positively charged in the protonated state. Figures 78a and 78b show the average degree of protonation and the end-to-end distance as a function of pH, while Figs. 78c and 78d display the pK_{eff} and the end-to-end distance as a function of the degree of protonation. As the chain ionizes, the end-to-end distances increase from about 10 nm to a final size, which depends strongly on the ionic strength. Note that the chain remains substantially coiled, since the end-to-end distance for a fully stretched chain would be about 190 nm.

The titration data obtained by Monte Carlo are compared with the mean-field result for the linear rigid chain. The differences are most obvious in Fig. 78c. As we have already discussed for the screened Coulomb potential of a rigid chain, the mean-field approximation is excellent. Figure 78 demonstrates that chain flexibility leads to a nonlinear dependence of pK_{eff} as a function of θ already on the DH level.

The increase of the end-to-end distance upon ionization can be understood from the following simple argument due to Flory.[308] Let us present this argument for a cationic, weak polyelectrolyte chain of N monomers with average radius R. The free energy of the chain can be written as the sum of chain entropy and electrostatic energy[309,310]

$$F = \frac{3kT}{2}\left(\frac{R}{Na}\right)^2 + \bar{E}N\frac{\theta^2}{2} \tag{274}$$

where θ is the degree of protonation and a the length of a single bead. The mean-field energy per site $\bar{E}$ is given by Eq. (270) and is approximated by a constant concentration of sites in a sphere of radius R. Replacing the sum by an integral and extending the upper limit to infinity, we obtain an expression for $\bar{E}$ which, inserted into Eq. (274), leads to

$$F = \frac{3kT}{2}\left(\frac{R}{Na}\right)^2 + \frac{3e^2N}{8\pi\varepsilon_0 D_w R^2 \kappa^2}\frac{\theta^2}{2}. \tag{275}$$

By taking the derivative with respect to R, this expression can be minimized to give

$$R \propto \theta^{2/5} c_s^{-1/5} N^{3/5} a^{2/5} \tag{276}$$

where c_s is the salt concentration. Equation (276) represents a scaling relationship between the end-to-end distance and various chain parameters, such as the degree of protonation or the salt concentration c_s. The dependence on the number of monomers reflects the classical Flory exponent.[308,311] The validity of this relationship has been tested for a variety of situations by Monte Carlo simulation[203] and the results are shown in Fig. 79. While the true functional form is not a simple proportionality relationship, the data obtained for a variety of chain lengths, salt

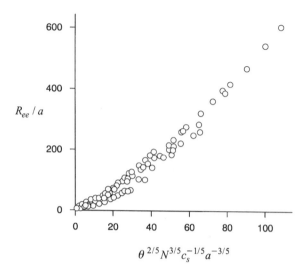

Figure 79. Scaling relationship for the end-to-end separation from Monte Carlo simulations for different system conditions.[203]

concentrations, and degrees of protonation indeed follow a common scaling relationship. The actual dependencies on various variables were analyzed in detail, and it turns out that the simple Flory argument yields the proper values of the exponents with the exception of the bead length a, the dependence on which is somewhat weaker than predicted. A similar analysis can be carried out for the salt-free case, and one can even obtain an expression for the titration curve from the Flory approach.[309]

6.4. Polyamines

Figure 4 summarizes the structures of various polyamines, namely linear poly(ethylene imine) (LPEI), linear poly(vinyl amine) (PVA), and dendritic poly(propylene imine) (DPPI). Another industrially important polyamine is branched poly(ethylene imine) (BPEI), which has a randomly branched structure (see Fig. 80). All these molecules are weak bases, which are nearly uncharged at pH > 11. At lower pH, their charge increases to rather high values and leads to a breakdown of the mean-field approximation. Nevertheless, the charging behavior can be understood on the basis of short-range interactions and will be discussed in the following.

6.4.1. Linear Polyamines

Let us first focus on linear polyamines, and particularly LPEI.[25,222,312] The titration behavior of this polyelectrolyte is known accurately. As shown in Fig. 81,

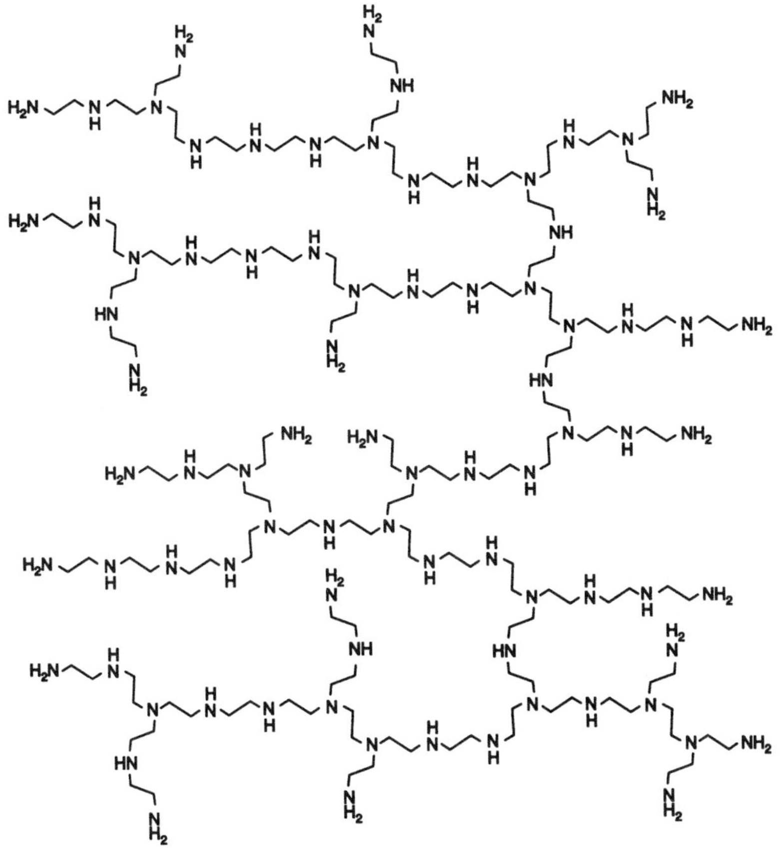

Figure 80.   Hypothetical structure of randomly branched poly(ethylene imine) (BPEI). Reprinted from Ref. 312 with permission from Steinkopff Verlag.

the titration curves display two characteristic features. We observe the characteristic two-step behavior with an intermediate plateau at $\theta = 1/2$, while with decreasing ionic strengths the titration curve is shifted to the left.

In Section 4 we have given a quantitative interpretation of the ionization of small oligomers of LPEI (see Table 11). Based on the cluster expansion, a consistent description of the macroscopic pK values for an entire homologous series of molecules could be obtained, but such a description required consideration of nearest-neighbor pair and triplet interactions. We can try to use parameters derived from the oligomers and attempt to predict the titration behavior of LPEI. Since the investigated polyelectrolyte consists of about 500 ionizable groups, it is sufficient

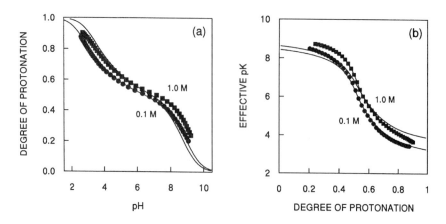

Figure 81.   Titration of linear poly(ethylene imine) (LPEI) at ionic strengths of 0.1 M and 1.0 M. Experimental data points are taken from Smits *et al.*[25] The solid line is the Ising model prediction with independently derived parameters from small molecules (see Table 7). For 0.1 M we have $p\hat{K}^{(II)} \simeq 8.44$, $\varepsilon \simeq 1.97$, and $\lambda \simeq 0.42$, while for 1.0 M, $p\hat{K}^{(II)} \simeq 8.63$, $\varepsilon \simeq 1.75$, and $\lambda \simeq 0.42$. (a) Degree of protonation as a function of pH and (b) pK_{eff} as a function of degree of protonation.

to consider the infinite chain limit ($N \to \infty$). In this limit primary amine terminal groups become immaterial, and we have to consider only a chain of identical secondary amine groups with $p\hat{K}^{(II)}$, including nearest-neighbor pair and triplet interactions. The partition function for the infinite chain can be evaluated either in analogy to combinatorial treatment of the linear chain or with transfer matrix techniques.[25,223,302] With the parameters obtained from the oligomers, we predict the titration curves of LPEI without adjustable parameters. For 0.1 M we use $p\hat{K}^{(II)} \simeq 8.44$, $\varepsilon \simeq 1.97$, and $\lambda \simeq 0.42$, while for 1.0 M, $p\hat{K}^{(II)} \simeq 8.63$, $\varepsilon \simeq 1.75$, and $\lambda \simeq 0.42$. As evident from Fig. 81, the result compares favorably with experimental data for 0.1 M, while there is some disagreement for 1.0 M.[222] Nevertheless, given the fact that the parameters were derived from small molecules, the predictions are reasonable. Moreover, the model also predicts broadening of the second protonation step in the pH range 3–4 due to triplet interactions. A pair interaction model will always generate a symmetrical titration curve around $\theta = 1/2$.

Titration behavior of poly(vinyl amine) (PVA) is not too different from LPEI, but the effect of interactions is somewhat weaker. This polyelectrolyte was studied in the classical work of Katchalsky *et al.*[21] and their experiments did actually provide the first experimental evidence of the two-step titration behavior of polyelectrolytes. While Katchalsky *et al.*[21] interpreted the PVA titration curve with the nearest-neighbor pair Ising model, they have pointed out that the data cannot be consistently accounted for within this framework. As in the case of LPEI, the titration curves of PVA are asymmetrical, but the reason for this behavior for PVA is most likely different than for LPEI. While in the latter case the triplet interactions are responsible for the asymmetry, in PVA the asymmetry probably has to do with a random arrangement of the amine groups along the chain. However, such *tacticity* effects are poorly understood for PVA. Poly(lysine) is an example of a polyamine with even weaker interactions. While racemic poly(DL-lysine) shows very weak effects of interactions, poly(L-lysine) displays a sharp helix-coil transition.[313] Similar effects can be pronounced for polycarboxylates and will be discussed later in more detail.

6.4.2. Branched Polyamines

Polyamines are commonly branched, where tertiary amines represent the branch points. The standard polymerization procedure of ethylene imine produces a randomly branched structure, as sketched in Fig. 80. Almost perfectly regular branched dendritic structures became available recently[39,83]; see Fig. 4c. The titration behavior of these molecules is known, and we shall see that it can be predicted from the ionization behavior of oligomers to a reasonable degree of accuracy.

Before we discuss the actual systems in more detail, let us employ the nearest-neighbor pair interaction model to illustrate some characteristic features of the titration behavior of branched polyelectrolytes.[302,312] Within this model, titration curves can be readily calculated for a wide variety of structures with a variant of the transfer matrix technique. We assume that all ionizable sites are equivalent and have the same microscopic pK value 10. We further assume only nearest-neighbor interactions and set the pair interaction parameter $\varepsilon = 2$. Recall that this model roughly mimics LPEI, and the corresponding results for the linear chain were discussed above. Figure 82 compares the titration curves for the linear chain with two structures, namely a dendrimer and a comb. The linear chain shows the classical two-step behavior, with an intermediate plateau at $\theta = 1/2$. The dendrimer shows an intermediate plateau at $\theta = 2/3$ and a wider splitting between both protonation steps than the linear chain. The comb shows an intermediate behavior between these two extremes.

In order to understand these differences, recall the origin of the intermediate plateau at $\theta = 1/2$ for the linear chain (cf. Fig. 73). The first half of the sites protonate at pH 10 (pH $\simeq$ p$\hat{K}$) since no interactions are involved, while the second half protonate at pH 6 (pH $\simeq$ p$\hat{K} - 2\varepsilon$) because two pair interactions must be overcome.

Consider now the dendrimer as shown in Fig. 83a. The intermediate stable configuration corresponds to every second shell being protonated and leads to a plateau in the titration curve at $\theta \simeq 2/3$ (for every deprotonated site there are two protonated ones). In analogy to the linear case, 2/3 of the sites protonate at pH 10 (pH $\simeq$ p$\hat{K}$) since no interactions are involved, while the remaining 1/3 protonate at pH 4 (pH $\simeq$ p$\hat{K} - 3\varepsilon$) because three pair interactions must be overcome. For this reason, the splitting between the protonation steps for the dendrimer is wider than in the linear arrangement.

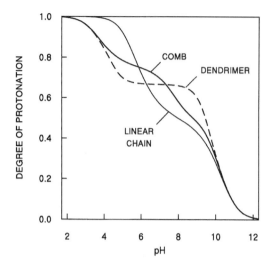

Figure 82. Model calculations of titration curves of various branched polyelectrolytes are compared with the linear chain. We use the nearest-neighbor pair interaction model with $p\hat{K} = 10$ and $\varepsilon = 2$. The structures of the dendrimer and comb are shown in Fig. 83.

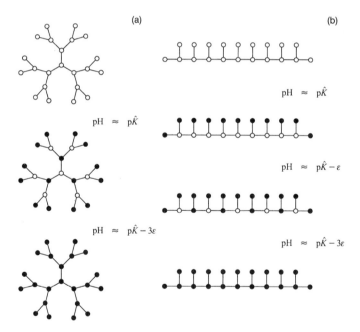

Figure 83. Schematic protonation sequences for branched structures with strong nearest-neighbor pair interactions. Protonated sites are shown as filled circles and deprotonated sites as open circles. At high pH, all sites are deprotonated. At intermediate pH, various low energy structures are preferred as shown. At low pH, all sites protonate. The transition regions are also indicated, and they are related to the microscopic ionization constant $p\hat{K}$ and the nearest-neighbor pair interaction parameter ε. (a) Dendrimer and (b) comb-like structure.

The situation for the comb is more complicated, as illustrated in Fig. 83b. The titration curve displays two intermediate plateaus, namely at $\theta = 1/2$ and $\theta = 3/4$. This pattern can be understood as follows. At pH around 10, all primary groups protonate since no nearest-neighbor pair interactions have to be invoked. The stability of this protonation state gives rise to the first plateau at 1/2. Further binding of protons invokes the protonation of every second tertiary site, which leads to the intermediate plateau at 3/4. During this process one has to overcome one pair interaction, and that is why this protonation step happens at pH around 8 ($\mathrm{pH} \simeq \mathrm{p}\hat{K} - \varepsilon$). In the final protonation step, three nearest-neighbor interactions must be overcome causing the last step to be located at pH 4 ($\mathrm{pH} \simeq \mathrm{p}\hat{K} - 3\varepsilon$).

Let us now discuss experimental data for branched polyamines and focus on dentritic poly(propylene imine) (DPPI, see Fig. 4c). The potentiometric titration curves of such dendrimers were studied in detail by van Duijvenbode *et al.*[83] As an example, in Fig. 84 the titration results are given for the largest available dendrimer with 64 primary amine groups (128 amine groups in total). The titration curves indeed show the characteristic intermediate plateau at $\theta = 2/3$ as suggested by the simple nearest-neighbor pair model for the dendrimer. Since in a poly(propylene imine) the amine groups are further separated than in poly(ethylene imine), the pair interaction parameter is about a factor of two smaller, and consequently the splitting of the two protonation steps for DPPI and LPEI becomes comparable. The solid line shown in Fig. 84 represents a result of the calculation based on the nearest-neighbor interaction model, with parameters $\mathrm{p}\hat{K}^{(\mathrm{I})} \simeq 10.33$ (primary amine), $\mathrm{p}\hat{K}^{(\mathrm{III})} \simeq 9.1$ (tertiary amine), and $\varepsilon \simeq 0.97$. With the exception of $\mathrm{p}\hat{K}^{(\mathrm{III})}$ all parameters are derived from small molecules.

While the picture put forward for the description of the titration curve seems consistent, it does not provide an unambiguous proof that the protonation follows

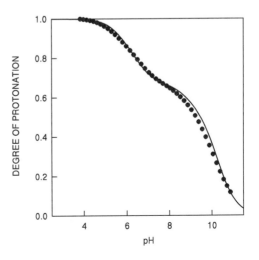

Figure 84. Potentiometric titration of dendritic poly(propylene imine) (DPPI) with 126 ionizable sites at an ionic strength of 1.0 M. The solid line is the result of a model calculation with parameters for $\mathrm{p}\hat{K}^{(\mathrm{I})} \simeq 10.33$ (primary amine), $\mathrm{p}\hat{K}^{(\mathrm{III})} \simeq 9.1$ (tertiary amine), and $\varepsilon \simeq 0.97$. With the exception of $\mathrm{p}\hat{K}^{(\mathrm{III})}$ all parameters are obtained from small molecules. The structure of the molecule is shown in Fig. 4c. Reprinted from Ref. 83 with permission from Elsevier Science.

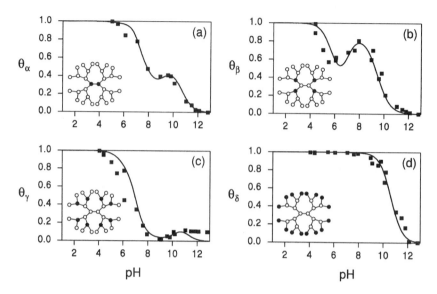

Figure 85.   Site-specific titration of the dendritic poly(propylene imine) (DPPI) with 30 ionizable sites as monitored with ^{15}N NMR. The solid lines are calculations with the nearest-neighbor pair interaction model.[108]

the suggested odd–even shell pattern. This picture has been confirmed, however, by means of ^{15}N NMR,[108] which was used to measure the titration curves of the individual protons, and the result is shown in Fig. 85. The shells are labeled from the inside out by α, β, γ, and δ. One observes very clearly that the outermost δ-shell protonates around first, while the second γ-shell protonates only at much lower pH values. Due to the smallness of the dendrimer employed, the β-shell (consisting of 4 nitrogens) and α-shell (consisting of 2 nitrogens) no longer follow this trend precisely, but one can still recognize the same features. The solid lines in Fig. 85 are derived from a nearest-neighbor pair interaction model and one observes that the measured site-titration curves can be understood quantitatively. However, it is not possible to use the same parameters to model the ^{15}N NMR data and potentiometric titration data. While the potentiometric titration data were obtained at sufficiently high dilutions of the dendrimer, the NMR experiments had to be carried out at rather high concentrations, since the natural abundance of the ^{15}N isotope is rather low. While dendrimer–dendrimer interactions probably do not influence substantially the ionization behavior, the solvent is no longer a pure aqueous salt solution. For this reason, the microscopic ionization constants are modified by solvent shifts.

Let us now turn to the titration of branched poly(ethylene imine) (BPEI). As illustrated in Fig. 80, BPEI is a random, branched polymer with primary, secondary, and tertiary amine groups at a proportion of roughly 1:2:1. The titration data from

Baliff *et al.*[314] at an ionic strength 0.5 M indicate that even at pH 3 the protonation is not yet complete. For a quantitative interpretation we need a consistent set of parameters at this ionic strength. The microscopic pK values of the primary and secondary amine groups turn out to be p$\hat{K}^{(I)} \simeq 9.71$ and p$\hat{K}^{(II)} \simeq 8.59$, respectively. We must also estimate the microscopic pK value of the tertiary amine group, which shall be denoted as p$\hat{K}^{(III)}$. From group additivity relationships[6] we obtain p$\hat{K}^{(III)} \simeq 7.50$. The nearest-neighbor interaction parameter is for pairs $\varepsilon \simeq 1.87$ and for triplets $\lambda \simeq 0.42$. We also have to consider a next-nearest-neighbor pair interaction parameter for groups adjacent to a tertiary amine group, with a value $\varepsilon' \simeq 0.27$ (see Table 7). This model cannot be solved analytically for a randomly branched polymer, but we can use the Monte Carlo technique to calculate the titration curves and the result is shown in Fig. 86. Once the polymer is sufficiently large, such results are essentially identical for different realizations of the random polymer. Indeed, the model explains quantitatively that BPEI is much more difficult to protonate at low pH than LPEI. At higher degrees of protonation of the tertiary groups, one must overcome three pair and three triplet interactions for full protonation.

The protonation pattern of BPEI is somewhat similar to the comb-like structure mentioned earlier. At high pH no interactions are involved and 1/2 of all sites are protonated around pH $\simeq$ p$\hat{K}$. Further protonation proceeds similarly as for the comb, recognizing that secondary amine groups are also present in BPEI. The protonation thus proceeds in three "steps," namely around pH of p$\hat{K} - \varepsilon$, p$\hat{K} - 2\varepsilon$, and

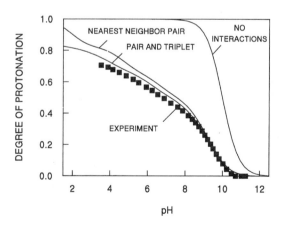

Figure 86. Titration of branched poly(ethylene imine) (BPEI) at an ionic strength of 0.5 M. The experimental data are taken from Baliff *et al.*[314] The solid line is the Ising model prediction with independently derived microscopic pK values p$\hat{K}^{(I)} \simeq 9.71$ (primary amine), p$\hat{K}^{(II)} \simeq 8.59$ (secondary amine), and p$\hat{K}^{(III)} \simeq 7.50$ (tertiary amine). Interaction parameters are for pairs $\varepsilon \simeq 1.87$ (nearest), $\varepsilon' \simeq 0.25$ (next-nearest groups to tertiary amine groups), and for triplets $\lambda \simeq 0.42$. Results of calculations with only nearest-neighbor pair interactions and without interactions are also shown.[302]

$p\hat{K} - 3\varepsilon$. In reality these "steps" overlap, and are further smeared-out due to the inherent randomness of the polyelectrolyte backbone. Therefore, one obtains a steady rise of the titration curve for $\theta > 1/2$ as observed experimentally.

6.5. Polycarboxylates

Most polycarboxylates have a rather low charge density. For such molecules the effect of site–site interactions must be weak, and the titration curves should be well described by mean-field and smeared-out charge models. Some polycarboxylates do have substantial charge densities, and for these systems deviations from the mean-field behavior are expected.

6.5.1. Weakly Charged Linear Polycarboxylates

Probably the best studied molecule with a low charge density is hyaluronic acid, a polysaccharide with carboxylic side groups on every second saccharide ring (see Fig. 7a). The experimental titration data by Cleland et al.[315] are shown in Fig. 87, and compared with the mean-field PB model for a cylinder with a smeared-out charge on its surface. The intrinsic ionization constant is $p\hat{K}^{(int)} = 2.75$, the cylinder radius $a = 1$ nm, and the spacing between the charges $\Lambda^{-1} = 1$ nm. The latter distance compares reasonably well with the geometrical separation of the ionizable residues in the molecule. The PB model properly captures the overall shape of the titration curve, as well as the ionic strength dependence. Cleland[148] pointed out that the data can be equally well interpreted in terms of discrete charges, either on a line or in a worm-like chain. The fact that all predictions are consistent with the data signals that the effect of site–site interactions is so weak, that all models treat these interactions in an adequate fashion. One also expects the data to be consistent with the flexible chain model.

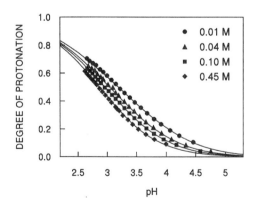

Figure 87. Comparison of experimental titration curves of hyaluronic acid[315] with the smeared-out charge model involving the PB equation for a cylinder.[315] The parameters are $p\hat{K}^{(int)} = 2.75$, radius $a = 1$ nm, and spacing between the charges $\Lambda^{-1} = 1$ nm.

A polycarboxylate with a somewhat higher charge density is poly(glutamic acid) (see Fig. 53). The stereoregular poly(D-glutamic acid), a polypeptite, undergoes a sharp conformational transition upon protonation, and forms a helix at low charge density, the content of which decreases when salt is added.[316] Poly(DL-glutamic acid) polymerized from a racemic mixture still has the tendency to contract, but no longer forms a helix.[317] The experimental titration data are shown in Fig. 88, plotted as $\Delta pK_{eff} = pK_{eff} - p\hat{K}'$ as a function of degree of protonation θ. This representation is more suited to appreciate the differences between models.[203,317] Carboxymethyl cellulose behaves very similarly to poly(DL-glutamic acid).[318]

The data are compared with the flexible chain model, which assumes a spacing of 0.7 nm between the charges. The model reproduces the curvature of the experimental data rather well, and moreover the spacing is in good agreement with the geometric distance between the charged residues. For comparison, the result of the PB model for a uniformly charged cylinder of radius 0.5 nm and spacing between charges of 0.5 nm is also shown. This model reproduces the data at higher ionic strength more satisfactorily, but underestimates the curvature at lower salt levels. Moreover, the parameters employed are less realistic. The high curvature of the plot near $\theta \simeq 1$ can indeed be attributed to flexibility. Note that the PB solutions typically require shorter length parameters than the flexible model to reproduce the correct shift in the apparent dissociation constant. The flexible model introduces stronger interactions than the rigid rod due to shorter average distances.

While there is only one structural formula of poly(acrylic acid) (PAA) (see Fig. 3), various different geometric isomers are possible, which arise since the carboxylic group can be attached to the carbon backbone from two different sides. This isomerism is usually referred to as *tacticity*, and two of these isomers are shown in Fig. 89. When all carboxylic side groups point to the same side of the chain, we refer to an *isotactic* structure, while an alternating arrangement is called a *syndiotactic* structure. Various other structures are possible, of course. For example, the

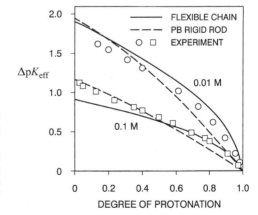

Figure 88. Comparison of experimental titration curves for poly(DL-glutamic acid) with PB model plotted as $\Delta pK_{eff} = pK_{eff} - p\hat{K}'$. The parameters for the model are cylinder radius $a = 0.5$ nm and spacing between the charges $\Lambda^{-1} = 0.5$ nm. The flexible chain model uses a spacing of 0.7 nm.[203,317]

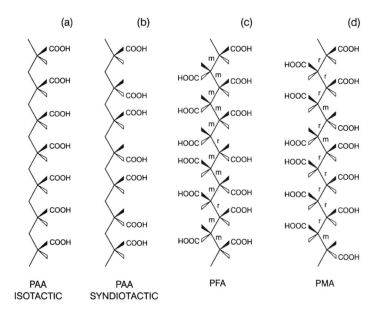

Figure 89.   Structures of various linear polycarboxylates. (a) Isotactic poly(acrylic acid) (PAA), (b) syndiotactic PAA, (c) poly(fumaric acid) (PFA), and (d) poly(maleic acid) (PMA). In PFA every second bond is mesomeric (m) and about 1/2 from the remaining ones are racemic (r). For PMA every second bond is racemic and about 1/3 from the remaining ones are mesomeric.

atactic structure corresponds to random arrangement of the side groups. As mentioned above, the titration behavior of PVA is also likely to be influenced by tacticity effects.

There are numerous literature data sets on the titration behavior of PAA,[299,319,320] but they are not always extrapolated to low polymer concentrations, resulting in substantial differences between the different data sets. The extrapolated data for atactic PAA by Nagasawa *et al.*[320] is shown in Fig. 90 and resembles previous polycarboxylates. However, the effect of site–site interactions is stronger. The data can be reasonably modeled by the PB model with a uniformly charged cylinder of radius $a = 0.37$ nm and spacing between the charges of $\Lambda^{-1} = 0.55$ nm. The model does overestimate somewhat the ionic strength dependence, and the curvature is not properly captured.

These differences can be illustrated more clearly by plotting $\Delta pK_{eff} = pK_{eff} - p\hat{K}'$ as a function of the degree of protonation, as done in Fig. 91 with the data from Kawaguchi and Nagasawa.[319,321] These data also indicate that tacticity has some effect on the titration behavior of PAA, the isotactic structure being slightly more basic than the syndiotactic one. Ullner *et al.*[203] have compared the experimental data with the flexible chain model and PB smeared-out charge model. The flexible chain model uses a bond length of 0.6 nm, while the PB model

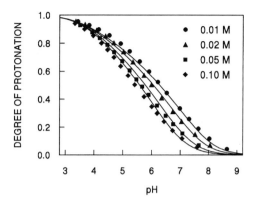

Figure 90. Comparison of experimental titration curves for atactic poly(acrylic acid) (PAA)[320] with the PB model for a cylinder of radius $a = 0.37$ nm and spacing between the charges $\Lambda^{-1} = 0.55$ nm.

uses a cylinder with a radius of 0.66 nm and a spacing of the charges of 0.2 nm. The fit of the flexible model is quite good at 0.01 M, but it shows a too strong ionic strength dependence. The assumption of a linear response implied by the use of a screened Coulomb potential might not be valid for such a highly charged polyelectrolyte. A detailed analysis of the data shows that the results can be best fitted to the cylinder PB as well as to a rigid rod at high salt concentrations.[203] The success of the latter is perhaps not all that surprising considering that PAA is very soluble in water. Water acts as a good solvent and the chain is extended even at low degrees of dissociation. This picture is somewhat different from the flexible chain simulations, where the uncharged chain has a more compact conformation. We suspect that the chain is already somewhat pre-expanded at low charge, and at high salt concentrations the electrostatic interactions may not be sufficient to drive the expansion any further. The polyelectrolyte then acts as an effective rigid rod; the data at 0.1 M in fact are in perfect agreement with a rigid rod that has a bond length

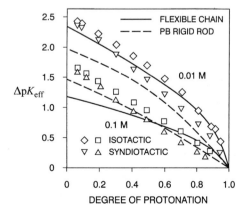

Figure 91. Comparison of experimental titration curves for poly(acrylic acid)[320] with the PB model and Monte Carlo simulations.[203] The data are plotted as $\Delta pK_{eff} = pK_{eff} - p\hat{K}'$. The flexible chain has a spacing of 0.6 nm between the charges (solid). The PB model involves a cylinder with radius a = 0.66 nm and a smeared-out charge density equivalent to a spacing between charges of $\Lambda^{-1} = 0.2$ nm (dashed).

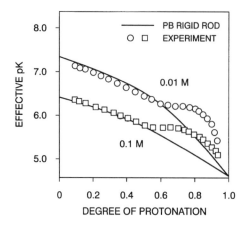

Figure 92.   Experimental data for pK_{eff} for isotactic poly(methacrylic acid) (PMAA). The solid line is the PB model with a cylinder of radius 0.55 nm and the same spacing between the charges.[320]

of 0.4 nm. At higher salt levels, the electrostatic interactions might take over and promote the expansion of the chain, making the flexible model applicable with a larger charge separation. The fact that the data can be fitted very well with a bond length of 0.6 nm supports this picture. The expansion of PAA with increasing pH can be directly confirmed by specific viscosity measurements.[322]

We have seen that chain flexibility can affect the titration behavior, but in several cases actual conformational changes can be observed during ionization. We have already mentioned the helix-coil transition of poly(D-glutamic acid).[317] Poly(methacrylic acid) (PMAA) provides another example of such an effect.[319] Its structure is very similar to PAA, with only a difference of additional methyl groups at the backbone. The titration curve of isotactic PMAA shown in Fig. 92 is reasonably well reproduced by the PB model with a cylinder of radius 0.55 nm and the same spacing between the charges at higher degree of ionization.[319] However, the pronounced hump in the titration curve at low degree of ionization suggests the existence of a conformational transition within the chain. As discussed earlier, such a transition is not driven by a simple swelling of the chain, but the chain must rather rearrange itself into an entirely different conformation.

6.5.2. Highly Charged Linear Polycarboxylates

The protonation behavior of polycarboxylates could be explained, so far, with smeared-out charge models, or with chain flexibility, as long as site–site interactions were weak to moderate and the titration curves mean-field like. The main reason for this behavior is that carboxylic groups have large separations. Carboxylic groups occur as side chains, and tacticity effects enhance long-range interaction potentials, which promotes the mean-field behavior.

As soon as the carboxylic groups are more closely spaced, the typical two-step titration curves and strong deviations from mean-field behavior are observed. The most prominent examples for this behavior are poly(maleic acid) (PMA) and poly(fumaric acid) (PFA).[153,300] This behavior is similar to our discussion of the polyamines, which were approached with a discrete site model with short-range interactions, and where pronounced deviations from the mean-field behavior were reported. (The charge density must be high for polyamines in order to remain soluble in water, and therefore the mean-field regime cannot be attained.)

Ionization reactions of linear polycarboxylates, however, are complicated by tacticity effects. While PMA is predominantly isotactic, PFA is mostly syndiotactic due to the structure of the monomers. In PMA, a pair of carboxylic groups always points in the same direction (racemic), while the configuration of the second pair is random, but has a probability of about 2/3 to point to the same side as the previous pair, as deduced from ^{13}C NMR.[323] In PFA, a pair of carboxylic groups point in opposite directions (mesomeric), and the next pair is about equally likely to point to either side[324] (see Fig. 89). Since these groups are so closely spaced, the difference in tacticity has a profound effect on the titration behavior, as indicated by the titration curves of PMA and PFA at ionic strength 0.1 M in Fig. 93. Their functional form is rather different, and the effect of interactions in PMA is larger than in PFA. Qualitatively, this trend is immediately understood, since PMA is predominantly isotactic while PFA is mostly syndiotactic. A quantitative interpretation can be given in terms of the nearest-neighbor interaction model.[153] However, one has to consider the randomness in the tacticity of every other bond. A simple approach is to consider a chain with nearest-neighbor pair interactions, but with two different interaction parameters. We assign a pair interaction parameter ε_r to the racemic bond and ε_m to the mesomeric bond ($\varepsilon_r > \varepsilon_r$). With the measured tacticity discussed above as input, a fit of the titration data yields $\varepsilon_r \simeq 2.9$, $\varepsilon_m \simeq 1.4$, and $p\hat{K}' \simeq 3.7$ for PMA, while we obtain $\varepsilon_r \simeq 4.2$, $\varepsilon_m \simeq 1.4$, and $p\hat{K}' \simeq 3.6$ for PFA. These parameters can be compared to the corresponding diprotic acids; one finds

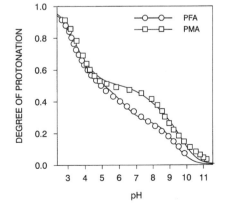

Figure 93. Titration curves of poly(maleic acid) (PMA) and poly(fumaric acid) (PFA) in 0.1 M NaCl. The solid lines are best fits with the nearest-neighbor pair interaction model with disorder.[153]

from the pK values of maleic and fumaric acids the following estimates for $\varepsilon_r \approx 3.5$ and $\varepsilon_m \approx 0.7$ at 0.1 M (see Table 5 for values at other ionic strengths). Due to the presence of the double bond these values are expected to differ from the polymeric analogs, but they are certainly comparable. While the interaction parameter for the mesomeric bond is virtually the same for PMA and PFA, there is a substantial difference for the racemic bonds. In reality we should probably incorporate additional higher order interactions, which were not considered here. It is also not entirely clear why the maximum swelling of the PMA chain occurs at $\theta = 1/2$ as inferred from viscosity measurements.[153]

Another aspect, which has yet to be properly addressed, is the strong effect of counterions. A few examples at an ionic strength of 0.05 M are shown in Fig. 94. While the first protonation step depends strongly on the type of the counterion, the second one is almost independent of it. As shown in Fig. 94, these effects can be captured with the discrete site model, but the reasons for the substantial changes of the interaction parameters with the type of counterion have yet to be understood.

6.6. Humic Acids

As an example of a branched anionic weak polyelectrolyte, let us briefly focus on humic acids, a possible structure of which is shown in Fig. 7b. These compounds occur naturally and represent the major part of the organic carbon in humus, soils, and surface waters.[42] Humic acids are operationally defined as the fraction of organic compounds which are soluble in strong base. These polyelectrolytes are very heterogeneous, highly branched, and have molecular weights around $10^4 - 10^5$. Their brown-yellow color originates from extended aromatic polycyclic structures. The ionizable groups on these compounds are mostly carboxylic and phenolic, while amines contribute to a lesser extent.

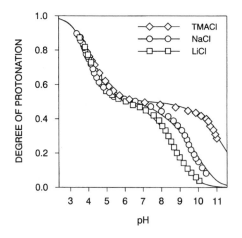

Figure 94. Titration curves of poly(maleic acid) (PMA) in different background electrolytes of 0.05 M, namely LiCl, NaCl, and tetramethylammonium chloride (TMACl). The solid lines are best fits based on the nearest-neighbor pair interaction model with disorder.[153]

The protonation behavior of these compounds was studied in detail recently.[228,325,326] As an example consider the titration behavior of a humic acid, which was isolated from peat, and is rather typical for this class of compounds. Figure 95 shows the experimental data and model calculations as discussed by Kinniburgh *et al.*[325] One observes two broad protonation steps, one around pH 4 and the other around pH 9. In analogy to findings on other anionic polyelectrolytes, changes in the ionic strength lead mainly to parallel shifts of the titration curves.

The data have been rationalized using a Donnan model discussed before, where one pictures the ionizable sites to be uniformly distributed within a charged gel-like phase (Donnan phase). The charge–potential relationship was given earlier [cf. Eq. (253)]. The protonation behavior of the humic acid is assumed to follow a superposition of two Langmuir–Freundlich isotherms

$$\theta = \theta_1 \frac{(a'_H \overline{K}_1)^{\nu_1}}{1 + (a'_H \overline{K}_1)^{\nu_1}} + \theta_2 \frac{(a'_H \overline{K}_2)^{\nu_2}}{1 + (a'_H \overline{K}_2)^{\nu_2}} \qquad (277)$$

as discussed in Section 4 [cf. Eq. (206)]. However, this equation now incorporates the Donnan potential ψ_0 through $a'_H = a_H \exp(-\beta e \psi_0)$. The set of equations is solved self-consistently by invoking the charge–potential relationship.

The present Donnan model is able to rationalize the experimental data shown in Fig. 95 rather well, using the binding constants $pK_1 \simeq 2.9$ and $pK_2 \simeq 8.8$, parameters $\nu_1 \simeq 0.54$ and $\nu_2 \simeq 0.25$, fractional site concentrations $\theta_1 \simeq 0.35$ and $\theta_2 \simeq 0.65$, and a total site concentration of 6.12 mmol/g. The specific volume of the Donnan phase must be assumed to increase with increasing ionic strengths.[228]

Recently, Avena *et al.*[326] showed that such titration data can be also rationalized by measuring the specific volume of the Donnan phase by viscosity measurements, which lead to gyration radii on the order of a few nm. With such independent

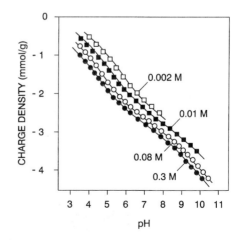

Figure 95. Titration behavior of humic acid isolated from peat in KNO_3 as background electrolyte.[228] Points are experimental data and the solid lines are best fits according to a Donnan model.

measurements of the specific volumes one finds that the Donnan model fails to describe the ionic strength dependence of the titration data. The viscosity data are, however, consistent with a rigid sphere PB model, as discussed in Section 5. This picture suggests that humic acid molecules are rather compact, with the majority of the ionizable groups residing on the surface—a picture rather analogous to a protein molecule. Due to the compactness of the molecule, the Donnan model appears to somewhat overestimate the interpenetration of the small ions. While the proper interpretation of the ionization behavior and its ionic strength dependence may still need further analysis, the Donnan model represents a simplified frame-work to discuss the ionic strength effects in such systems.

7. Ionizable Interfaces

Ionizable interfaces occur mostly at the solid–water boundary in colloidal suspensions, or as macroscopic surfactant monolayers. Such an interface usually consists of a large number of ionizable sites (say $N \gg 1000$), and to consider the large system limit is usually adequate ($N \to \infty$). While a large number of interacting sites can be only approached with computer simulation rigorously, an accurate description of their ionization behavior is possible by means of a mean-field approximation since the site–site interactions at interfaces are typically long-range. We have seen that a mean-field model breaks down for highly charged polyelectro-lytes, but this breakdown has not been observed for interfaces as yet.

Most approaches that treat planar ionizable interfaces rely on the diffuse layer model and its generalizations, and invoke the mean-field approximation for the site–site interactions as well as a smearing-out approximation of the surface charge. The validity of these approximations will be addressed for a few special cases. Application of the models to real systems, such as ionizable surfactant monolayers, latex, and oxide particles, will be discussed later. The subject of ionizable interfaces was previously reviewed by Healy and White[50] and, in this series, by James and Parks.[95]

7.1. Diffuse Layer Model and Its Generalization

Let us briefly derive the diffuse layer model based on a popular argument invoking the constancy of the chemical potential. This approach is commonplace for interfaces, and it yields the same mean-field results as discussed previously (cf. Sections 5 and 6). The derivation starts with the ionization reaction[30,50,95,327]

$$A + H \rightleftharpoons AH \tag{278}$$

where AH and A are the protonated and deprotonated surface groups, respectively. This reaction may represent the protonation of a single amine or carboxyl group on the surface. Amphoteric oxide–water interfaces can be often modeled with the simple one-step protonation reaction of a single oxygen coordinating a metal center[40]

$$Me - OH^{-1/2} + H^+ \rightleftharpoons Me - OH_2^{+1/2} \tag{279}$$

This reaction follows from valence bond considerations for a trivalent metal center; such a surface reverses the sign of its charge upon protonation. When the description of the ionizable interface is based on the one-step reaction [cf. Eq. (278)] one refers to a 1-pK model.[40,69] The mass action law for this ionization reaction is now written as

$$\hat{K}^{(int)} = \frac{\Gamma_{AH}}{\Gamma_A c_H(0)} \tag{280}$$

where Γ_{AH} and Γ_A are the surface concentrations of AH and A, $c_H(0)$ is the concentration of protons at the surface, and $\hat{K}^{(int)}$ is the intrinsic association constant. The concentration of protons at the surface is obtained by equating the chemical potential at the surface and in the bulk

$$kT \ln c_H(0) + e\psi_0 = kT \ln a_H \tag{281}$$

[Some authors refer to Eq. (281) as the electrochemical potential.] The mass action law given in Eq. (280) now becomes

$$\hat{K}^{(int)} \exp(-\beta e\psi_0) = \frac{\Gamma_{AH}}{\Gamma_A a_H} = \frac{\theta}{(1-\theta)a_H} \tag{282}$$

where we have introduced the degree of protonation θ and $\beta = 1/(kT)$ is the inverse thermal energy. This relation can be rewritten, and we arrive at

$$pH = p\hat{K}^{(int)} - \beta e\psi_0/\ln 10 + \log_{10}\frac{1-\theta}{\theta} \tag{283}$$

as already obtained previously [cf. Eq. (222)], or

$$pH = pK_{eff} + \log_{10}\frac{1-\theta}{\theta} \tag{284}$$

with the effective pK

$$pK_{eff} = p\hat{K}^{(int)} - \beta e\psi_0/\ln 10 \tag{285}$$

which depends on the electrostatic potential ψ_0. This quantity is simply the microscopic pK of an ionizable group at a given surface potential. The present analysis evidently leads to the same mean-field result, but the inherent assumptions are less obvious (see Section 5).

Michal Borkovec et al.

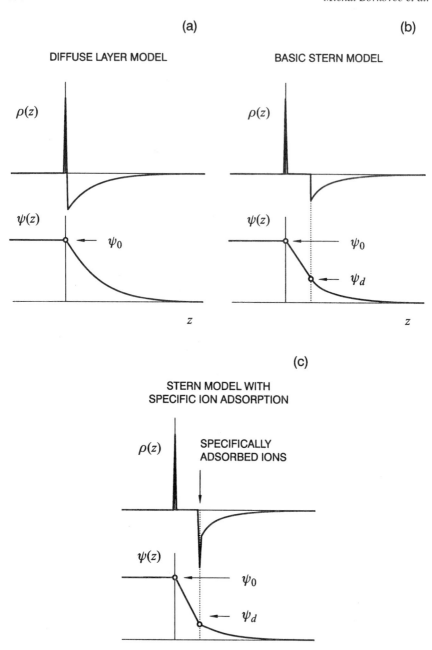

Figure 96. Potential and charge distributions for various models for planar interfaces. (a) Diffuse layer model, (b) basic Stern model, and (c) Stern model with specific ion binding.

The surface potential can be calculated from the surface charge and a known charge–potential relationship. The surface charge density σ_0 is linearly related to the degree of protonation

$$\sigma_0 = e\Gamma_0(\theta + Z) \tag{286}$$

where Γ_0 is the total number of sites per unit area and Z the charge of the group in its deprotonated state (in units of the elementary charge e). For $Z = 0$ one deals with a basic group (e.g., amine), for $Z = -1$ with an acidic group (e.g., carboxyl). The amphoteric oxide–water interface is obtained for $Z = -1/2$ [cf. Eq. (279)].

Given the surface charge, the surface potential can be evaluated from the charge–potential relationship. For interfaces, the surface potentials could be large, and the Debye–Hückel (DH) approximation may fail. For a planar interface, the charge–potential relationship can be evaluated analytically within the Poisson–Boltzmann (PB) approximation which leads to the Gouy–Chapman equation [cf. Eq. (41), see Section 3]. Identifying the surface potential ψ_0 with the diffuse layer potential ψ_d we obtain

$$\psi_0 = \frac{2}{e\beta} \operatorname{arcsinh} [\beta e\sigma_0/(2\varepsilon_0 D_w \kappa)] \tag{287}$$

where σ_0 is the surface charge density given by Eq. (286). When this relation is inserted into Eq. (283), one obtains the *diffuse layer model*. The resulting profiles for the charge density and electrostatic potentials are shown in Fig. 96a. For submicron particles, it might be necessary to correct Eq. (287) for the effects of curvature.[328,329]

For small surface potential, a simpler linear charge–potential relationship will hold. In this DH regime we obtain

$$\sigma_0 = \tilde{C}\psi_0 \tag{288}$$

where $\tilde{C}$ is the specific capacitance per unit area. In this case, the titration curve can be written as

$$\mathrm{pH} = \mathrm{p}\hat{K} - \bar{\varepsilon}\theta + \log_{10}\frac{1-\theta}{\theta} \tag{289}$$

where $\mathrm{p}\hat{K} = \mathrm{p}\hat{K}^{(\mathrm{int})} - z\bar{\varepsilon}$ and the interaction parameter

$$\bar{\varepsilon} = \frac{\beta e^2}{\ln 10}\frac{\Gamma}{\tilde{C}} \tag{290}$$

is inversely proportional to the surface capacitance. When Eq. (289) is applied to planar interfaces, one usually refers to a *constant capacitance model*, which for small surface potential is equivalent to the diffuse layer model. The corresponding capacitance is that for the diffuse layer capacitance in the DH regime

$$\tilde{C} = \tilde{C}_d = \varepsilon_0 D_w \kappa. \tag{291}$$

Figure 97 illustrates the charging behavior within the diffuse layer model, with Fig. 97a showing the protonation of a hypothetical amine surface ($Z = 0$) with $p\hat{K}^{(\text{int})} = 10$ and a site density $\Gamma_0 = 1 \text{ nm}^{-2}$ for three ionic strengths. With decreasing ionic strength the electrostatic interactions become stronger, which broadens the titration curve and shifts its midpoint to the left. The constant capacitance model (DH theory) predicts much stronger interactions, while the diffuse layer PB model

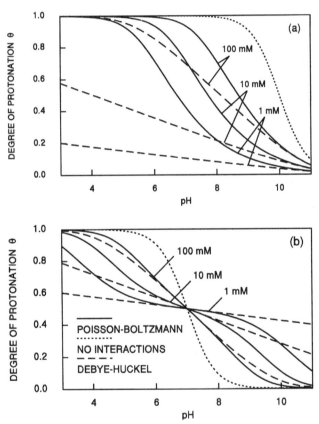

Figure 97. Typical dependence of the degree of protonation for interfaces as a function of pH for different ionic strengths. Solid lines are calculated from the Poisson–Boltzmann approximation (diffuse layer model) and dashed lines from the corresponding Debye–Hückel approximation (constant capacitance model). The site density is 1 nm^{-2}. The dotted lines are the titration curves without site-site interactions. (a) Amine surface ($z = 0$) with $p\hat{K}^{(\text{int})} = 10$, and (b) amphoteric surface ($z = -1/2$) with $p\hat{K}^{(\text{int})} = 7$.

leads to narrower and asymmetrical titration curves. As is customary in protein literature (see Section 5), some authors[87] have referred to the midpoint of such curves as $pK_{1/2}$ (i.e., pK_{eff} for $\theta = 1/2$). For a basic surface ($Z = 0$) $pK_{1/2}$ decreases with decreasing ionic strength, while the reverse trend is observed for an acidic surface ($Z = -1$).

Figure 97b displays the analogous situation for an amphoteric surface ($Z = -1/2$) with $p\hat{K}^{(int)} = 5$ and the same site density, which shows the characteristic common crossing point corresponding to the point of zero charge (PZC). The DH approximation (constant capacitance model) leads to titration curves closely resembling straight lines, while the PB approximation (diffuse layer model) leads to sigmoidal curves with an intermediate plateau at the PZC.

7.1.1. Nernstian Surface

The effects of site–site interactions are usually very large at many types of interfaces, leading to a weak dependence of the degree of protonation on pH, particularly at low ionic strengths. To a first approximation, one can assume that the surface concentrations Γ_{AH} and Γ_A remain constant—the basis of the Nernst approximation. From Eq. (282) it then follows that the surface potential is a linear function of the logarithm of the proton activity[99,327]

$$\psi_0 = \frac{1}{e\beta} \ln a_H + \text{const} \tag{292}$$

or, more conveniently,

$$\psi_0 = \frac{\ln 10}{e\beta} (\text{pH}_{PZC} - \text{pH}) \tag{293}$$

where $\ln 10/(e\beta) \simeq 59$ mV at 25 °C and pH_{PZC} is the PZC. For $Z = -1/2$ we have $\text{pH}_{PZC} = p\hat{K}^{(int)}$. This approximation is good as long as the interactions between the ionizable groups are strong and the degree of protonation changes little over the pH range considered. For the constant capacitance model one can show, using a Taylor expansion of Eq. (282), that the Nernst approximation is applicable to large interaction parameters $\bar{\varepsilon} \gg 1$. When $\bar{\varepsilon}$ is comparable to unity, corrections must be introduced.

The titration curve of a *Nernstian surface* is solely determined by the charge–potential relationship. In the case of the diffuse layer model, it is given by the Gouy–Chapman relation [cf. Eq. (287)], while for the constant capacitance model it is a simple linear function [cf. Eq. (288)]. In the latter case, the titration curve becomes a straight line.

7.1.2. Basic Stern Model

In spite of its simplicity, the diffuse layer model often provides an excellent description of the ionization of interfaces, particularly for not too large site densities, as exemplified later. However, for higher charge densities modifications often become necessary.

Let us first discuss the *basic Stern model*, which introduces an additional potential drop within the Stern layer. As in the diffuse layer model,

$$\sigma_d + \sigma_0 = 0 \tag{294}$$

where σ_0 is the surface charge and σ_d the diffuse layer charge. However, the model allows the potentials ψ_0 and ψ_d to be different. This condition requires an additional charge–potential relationship, which is simply assumed to be linear

$$\sigma_0 = \tilde{C}_s(\psi_0 - \psi_d) \tag{295}$$

where $\tilde{C}_s$ is the Stern capacitance per unit area. With Eqs. (283) and (295) the surface potential can now be expressed as

$$\psi_0 = \frac{\sigma_0}{\tilde{C}_s} + \frac{2}{e\beta} \text{ arcsinh } [\beta e\sigma_0/(2\varepsilon_0 D_w\kappa)]. \tag{296}$$

Inserting this charge–potential relation into Eq. (283) one obtains the titration curve for the basic Stern model. Corresponding charge density and electrostatic potential profiles are shown in Fig. 96b.

The Stern capacitance is often interpreted as originating from a plate capacitor, which is given by $\tilde{C}_s = \varepsilon_0 D_s/L$, where L is the thickness of the Stern layer and D_s its relative permittivity. Because of the preferred orientation of water molecules on the surface, this relative permittivity has been suggested to be lower than the relative permittivity of pure water.[82,330,331] However, this interpretation of the basic Stern model might be misleading. In analogy to the interpretation of the constant capacitance model, the Stern capacitance should be rather viewed as a lumping parameter, which incorporates the shortcomings of the smearing-out approximation.

For low surface charge densities, Eq. (296) can be again linearized and we arrive at the constant capacitance model. The overall capacitance $\tilde{C}$ is now given by

$$\tilde{C}^{-1} = \tilde{C}_d^{-1} + \tilde{C}_s^{-1} \tag{297}$$

where $\tilde{C}_d$ and $\tilde{C}_s$ are the diffuse layer and Stern layer capacitances, respectively. The interface acts as two capacitances in series. When the Stern capacitance becomes very large, the basic Stern model reduces to the diffuse layer model.

7.1.3. Specific Counterion Binding

An extension of the basic Stern model is to include specific counterion adsorption. These ions are assumed to reside directly at the plane of origin of the diffuse layer and contribute an additional surface charge σ_c, which satisfies

$$\sigma_0 + \sigma_c + \sigma_d = 0. \tag{298}$$

Introducing specific ion binding calls for the consideration of an additional chemical equilibrium. As known from solution chemistry, ligands do bind salt ions, and this concept is carried over to surface reactions. One assumes that a surface site may also bind salt anions or cations according to the scheme

$$A + C \rightleftharpoons AC \tag{299}$$

where C is the counterion or co-ion. For a carboxylic surface group, this reaction reads

$$-COO^- + Na^+ \rightleftharpoons -COO^- \cdot Na^+ \tag{300}$$

while for an oxide–water interface such specific binding reactions are often written as[82,332,333]

$$-Me - OH^{-1/2} + Na^+ \rightleftharpoons -Me - OH^{-1/2} \cdot Na^+ \tag{301}$$

$$-Me - OH_2^{+1/2} + Cl^- \rightleftharpoons -Me - OH_2^{+1/2} \cdot Cl^- \tag{302}$$

In analogy to Eq. (280) the corresponding mass action law is

$$K_c^{(ion)} \exp(-eZ_c\beta\psi_c) = \frac{\Gamma_{AC}}{\Gamma_A a_c} \tag{303}$$

where a_c is the solution activity of the counterion or co-ion of charge Z_c (in units of elementary charge e) and ψ_c is the potential at the adsorption plane of the ions. For an amphoteric surface, where adsorption of anions and cations might take place, two mass action relations have to be incorporated into the model. The corresponding charge density and electrostatic potential profiles with specific ion binding are shown in Fig. 96c.

The basic Stern model was extended to the *triple layer model*[334] by displacing the adsorption plane for the specifically adsorbed ions closer to the surface. Another modification was introduced in the *three-layer model*[82] or a *three-plane model*,[335] where in the basic Stern model an additional equipotential plane was introduced. Further extensions of this kind have been discussed; for example, four different equipotential planes were introduced in the literature.[336,337] A proper interpretation of all the parameters can be difficult, however.

7.2. Discrete Charge Model

Within the DH theory, pair interactions between individual point charges close to the interface can be evaluated explicitly. As for proteins, the interactions between the individual ionizable sites at the solid–water interface will depend on their mutual separation.

Let us illustrate these concepts with a simple model of a planar ionizable interface where discrete ionizable groups are treated explicitly.[69,70] This treatment was pioneered by Tanford and Kirkwood[61] for proteins (see Section 5) and was also applied to polyelectrolytes (see Section 6). Discreteness of charge for ionizable interfaces was already discussed to some extent previously.[35,338]

The discrete site model for a planar interface is sketched in Fig. 98. The solid is represented by a semi-infinite dielectric of relative permittivity D_d, in contact with an aqueous electrolyte solution. The ionizable sites are assumed to be identical and are modeled as point charges located within the dielectric at a distance a from the dielectric–water boundary. For simplicity, we shall assume that the sites are all equivalent and arranged on a square lattice. The same techniques could be used to study situations where the point charges are placed within the high permittivity medium, as appropriate for proteins (see Section 5).

For a given configuration of ionized sites, this system has the free energy [cf. Eq. (92)]

$$F(\{s_i\}) = -\mu \sum_i s_i + \frac{1}{2} \sum_{i,j} E_{ij} s_i s_j \tag{304}$$

where s_i are the state variables (zero for a deprotonated site and unity for a protonated site), $\beta\mu/\ln 10 = p\hat{K} - pH$ with $p\hat{K}$ being the pK value of the sites assuming all sites to be deprotonated, and E_{ij} are the pair interaction energies. Due to the linearity of the DH approximation, no higher-order interaction parameters enter and the pair interaction energies are given by the interaction potential $W(r)$, which depends on the mutual distance r as $E_{ij} = W(|\mathbf{r}_i - \mathbf{r}_j|)$. Thereby, sites i and j

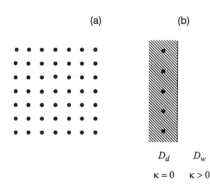

Figure 98. Schematic representation of the discrete site model of the interface with a dielectric discontinuity. The sites are arranged in a square lattice. (a) Top view and (b) side view.

are located at $\mathbf{r}_i$ and $\mathbf{r}_j$ with a distance $|\mathbf{r}_i - \mathbf{r}_j|$ in between. As discussed in Appendix A, the interaction potential reads[181–183]

$$W(r) = \frac{e^2}{4\pi\varepsilon_0 D_d}\left[\frac{1}{r} - \int_0^\infty \frac{D_w p - D_d k}{D_w p + D_d k} J_0(kr)e^{-2ak}\, dk\right] \qquad (305)$$

where r is the lateral distance between charges, $p = (k^2 + \kappa^2)^{1/2}$, and $J_0(x)$ is the Bessel function of the first kind.[160]

Two interesting limiting forms can be deduced from Eq. (305). For small distances $(r \to 0)$, the presence of the electrolyte is not felt and the charges interact by a simple Coulomb law

$$W(r) = \frac{e^2}{4\pi\varepsilon_0 D_d}\frac{1}{r}. \qquad (306)$$

At large distances $(r \to \infty)$, the interaction decays as

$$W(r) = \frac{e^2(1 + D_w a\kappa/D_d)^2}{2\pi\varepsilon_0 D_d \kappa^2}\frac{1}{r^3}. \qquad (307)$$

The interaction potential for interfaces and the corresponding limiting laws are shown in Fig. 99. We use the parameters $a = 0.25$ nm, $D_w = 80$, $D_d = 3$, and an ionic strength of 0.5 M $(\kappa^{-1} \approx 0.43$ nm$)$.

The $1/r^3$ dipolar decay of the interactions can be understood as follows. Each charge near the surface will generate an image charge. Far away, the charge and its image will act like a dipole normal to the surface, and the interaction of a point charge with such a dipole leads to the $1/r^3$ dependence. Due to this dependence,

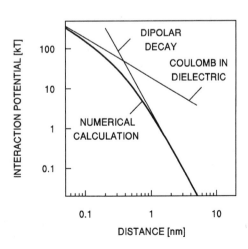

Figure 99. Interaction potential for a plane. The solid line is the result of a numerical evaluation of Eq. (305). The Coulomb interaction in the dielectric [cf. Eq. (306)] and the dipolar decay [cf. Eq. (307)] are also shown. The ionizable groups are modeled as point charges located $a = 0.25$ nm within a dielectric with relative permittivity $D_d = 3$, and in contact with water of $D_w = 80$ and an ionic strength of 0.5 M $(\kappa^{-1} \approx 0.43$ nm$)$.

the potential decays very slowly with distance (long-range potential). This feature is characteristic for an interface, and contrasts with polyelectrolytes, where the interaction potential may decay quickly (short-range potential).

The titration behavior of a square lattice, where the sites interact with this particular potential, is shown in Fig. 100. The square lattice is assumed to have a lattice spacing of 1.1 nm ($\Gamma \approx 0.8$ nm^{-2}). The squares shown in Fig. 100 are Monte Carlo results, which represent numerically accurate results for the discrete site model. The solid line (labeled as 1-pK model) is the prediction of the constant capacitance model [cf. Eq. (289)] with the mean-field interaction parameter calculated from the mean-field interaction energy

$$\overline{E} = \frac{1}{N} \sum_{i \neq j} W(|\mathbf{r}_i - \mathbf{r}_j|) = \Gamma \int g(\mathbf{r})W(r) \, d^2r \qquad (308)$$

where $g(\mathbf{r})$ is the pair distribution function of the sites in the interface and $\bar{\varepsilon} = \beta\overline{E}/\ln 10$.

As one can see from Fig. 100, a simple mean-field approximation is very good even at a high ionic strength of 0.5 M. This agreement may be surprising, since the screening is very effective ($\kappa^{-1} \approx 0.43$ nm) and one might expect deviations from mean-field behavior to be more pronounced. The reason for the good agreement is that the site–site interaction potential given by Eq. (305) decays slowly with distance and even distant pairs of sites contribute to the interaction. Due to this long-range interaction, the mutual arrangement of the protonated sites is essentially random and no spatial correlations exist. Figure 101 illustrates a typical configuration at $\theta = 1/2$.

Simulations for different parameter values reveal that the mean-field model remains an excellent approximation, if we change the lattice spacing or use

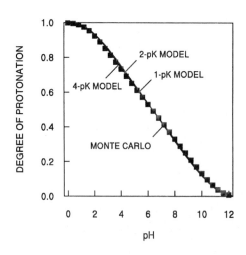

Figure 100. Discrete site model of the titration of a planar lattice. Comparison of Monte Carlo results (points) with the constant capacitance model (1-pK model), where the capacitance was obtained by the exact evaluation of the lattice sum. On the scale of the graph, the various model predictions for the 1-pK model, 2-pK model, and 4-pK model coincide. The ionizable groups are modeled as point charges located $a = 0.25$ nm within a dielectric with relative permittivity $D_d = 3$ arranged on a square lattice with a site density of 0.8 nm^{-2}, p$\hat{K} = 10$, and $z = 0$. This interface is in contact with a salt solution with an ionic strength of 0.5 M ($D_w = 80$ and $\kappa^{-1} \approx 0.43$ nm).

Figure 101. Configuration of ionized sites on an interface at $\theta = 0.5$. Protonated and deprotonated sites are denoted by filled and empty circles, respectively. The parameters are the same as in Fig. 100.

different, physically realistic, interaction potential parameters. The same conclusion has been verified for other types of lattices (e.g., honeycomb, triangular) and is expected to hold for a random (disordered) arrangement of sites. Within the DH approximation it is basically impossible to find a situation for the planar interface where the mean-field approximation breaks down.[70] At higher potentials the interactions appear to be even more mean-field like.

As discussed in Appendix B, the long-range interaction potential also eliminates any order–disorder phase transitions on the surface, which are well known on surfaces dominated by short-range interactions.[339–342] An interesting candidate for the breakdown of the mean-field approximation for an ionizing interface might be a bilayer, where the site–site interactions are probably not as long-range as for a monolayer.

7.2.1. Smearing-Out Approximation

In contrast to the mean-field approximation, the smearing-out approximation of the charges along the interface is not necessarily appropriate. This approximation replaces the actual point charges with a laterally uniform charge distribution of the same overall charge density, and is commonly invoked in descriptions of ionizable planar interfaces. Equation (308) reveals that the mean-field model discussed here reduces to the smeared charge situation only if spatial correlations between the ionizable sites are neglected. Indeed, on setting $g(r) = 1$ in Eq. (308) one recovers the smeared-charge result for the plane [cf. Eq. (290)]. The capacitance per unit area is now

$$\tilde{C}^{-1} = \tilde{C}_d^{-1} + \tilde{C}_i^{-1} \tag{309}$$

where $\tilde{C}_d$ is the diffuse layer capacitance given by Eq. (291), while $\tilde{C}_i = \varepsilon_0 D_d / a$ is the inner capacitance which arises from the dielectric medium between the smeared-out charge and the planar interface. The total capacitance can be again interpreted as originating from two parallel plate capacitors in series. In the diffuse layer model, the second contribution in Eq. (308) is neglected leaving only the double layer capacitance $\tilde{C}_d$.

In a lattice, however, substantial spatial correlations exist between the positions of the sites and $g(\mathbf{r}) \neq 1$. Therefore, the smearing-out approximation may fail, as was also discussed in the electrochemical literature.[343] The strength of the mean-field $\bar{E}$ depends on the geometrical arrangement through the pair correlation function $g(\mathbf{r})$ and thus this parameter will depend on the lattice geometry. The main effect of the correlation function is to introduce an excluded volume around the origin (correlation hole) and to weaken the strength of the mean field.

For the square lattice (as shown in Fig. 100) we obtain $\bar{\varepsilon} \simeq 7.37$. This parameter might vary by about 10% from lattice to lattice for the same site density (e.g., triangular or honeycomb lattice). The strength of the mean field is poorly estimated by the smeared-out charge model, which predicts $\bar{\varepsilon}_s \simeq 22.4$ (overestimate), as well as the classical double layer model, which yields $\bar{\varepsilon}_d \simeq 1.37$ (underestimate). The value of the interaction parameter could, of course, be interpreted by introducing an appropriate value of the Stern capacitance, but the physical origin is different; the additional interactions originate from site–site interactions. The interesting aspect of such a model is that the relative permittivity of the solid enters as well, an aspect also stressed by Sverjensky.[344] The smearing-out approximation improves for low charge densities.

7.2.2. 1-pK versus 2-pK Models

While agreement between the mean-field models shown in Fig. 100 is already very good, but it is still possible to invent better approximation schemes. A popular approach is the so-called mean-field cluster expansion, which treats the chemical equilibria within an entire cluster of sites explicitly and includes the effect of the remaining sites with the mean-field approximation. Such a mean-field cluster expansion was used for proteins[265] and here we discuss the application to our problem of an ionizable interface.[69]

The procedure is illustrated in Fig. 102. The lattice is subdivided into identical imaginary clusters, which may consist of one or more sites. The mean-field cluster expansion treats all sites within this cluster explicitly as if they were part of a

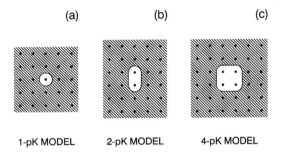

Figure 102. Schematic representation of the 1-pK model, 2-pK model, and 4-pK model. Within the mean-field cluster expansion the encircled sites are considered explicitly, while the shaded ones are treated within a mean-field approximation. Reprinted with permission from Ref. 69. Copyright (1997) American Chemical Society.

polyprotic molecule. One further assumes that all sites, which do not belong to this cluster, influence the sites inside the cluster only through their average degree of protonation. Here we introduce an approximation, however, since each site is influenced by the actual protonation state of the individual sites and not just through their average.

The technical details of such a mean-field cluster expansion are given in Appendix B. The main result is that for clusters consisting of just a single site, we recover the simple mean-field model discussed above [cf. Eq. (289)]. This approximation can be modeled by a one-step ionization process [cf. Eq. (278)] and is also referred to as the 1-pK model. Considering a cluster of two sites, one recovers precisely the classical surface ionization model with two consecutive protonation steps according to the scheme

$$X + H \rightleftharpoons XH \tag{310}$$

$$XH + H \rightleftharpoons XH_2 \tag{311}$$

where X now represents a pair of surface sites. This description introduces two equilibrium constants and the effects of the remaining sites are treated in a mean-field approximation as discussed as above. A description based on a two-step reaction [cf. Eqs. (310) and (311)] is referred to as a 2-pK model. Depending on the size of the cluster treated in the mean-field cluster expansion, a 3-pK model, 4-pK model, etc. can be devised as well. For a sufficiently large cluster, the approach becomes equivalent to the exact solution of the full many-body problem.

The results of the mean-field cluster expansion on a square lattice are given in Fig. 100, which shows that the 1-pK model works just as well as the more complicated 2-pK model. Any higher approximation schemes, such as the 4-pK model, produce similar results. For this reason, the additional complexity introduced by the 2-pK model is not warranted.

One might be tempted to suspect that the 2-pK model could properly represent the two-step protonation behavior of strongly interacting polyelectrolytes, which is not the case; the 2-pK model performs as poorly as the simple 1-pK model in this situation. This fact is known in the statistical mechanics literature, where it was remarked that the convergence of a mean-field cluster expansion is very slow.[264]

The 2-pK model is still the preferred description for oxide–water interfaces in the literature. The proper interpretation of this model is that both ionization steps occur on two *different* oxygens, for example, as the protonation of two oxygens coordinating a single metal center[327,345–347]

$$[-Me(OH)_2]^- + H^+ \rightleftharpoons [-Me(OH_2)(OH)] \tag{312}$$

$$[-Me(OH_2)(OH)] + H^+ \rightleftharpoons [-Me(OH_2)_2]^+ \tag{313}$$

or, similarly, as the protonation of two oxygens coordinating two neighboring metal centers. The site density which enters the problem is just 1/2 of the total site density on the surface. However, the 2-pK model introduces more parameters and does not offer any advantages over the simpler 1-pK model.[40,348]

Within the 2-pK model of metal–oxide interfaces, some researchers refer to the scheme

$$Me - O^- + H^+ \rightleftharpoons Me - OH \tag{314}$$

$$Me - OH + H^+ \rightleftharpoons Me - OH_2^+ \tag{315}$$

This scheme assumes that both protonation steps occur on a single oxygen. Two consecutive protonation reactions of an oxygen atom have usually *very* different pK values and thus only a single protonation step will be operational in the experimentally accessible window (see also Table 28 and the discussion in Section 4). For this reason, we suspect that the scheme in Eqs. (314) and (315) does not represent the proper interpretation of the 2-pK-model.[40]

7.2.3. pK Shifts

From the present discrete charge model one can also estimate the differences between the pK values of a surface group and the same group attached to a small solute. The physical reason for this disparity lies in the different dielectric environments of the ionizable group. In solution the group can be thought to be in a small dielectric cavity, while at the surface it resides in a semi-infinite dielectric. The predictions of such pK shifts will be semiquantitative at best, since we are using a continuum approach down to molecular distances.

The calculation of the change of the intrinsic pK in two different media has been illustrated for proteins and polyelectrolytes. However, in the present case of a planar interface, analytical expressions can be given. The ionization constant of a group i on the surface $p\hat{K}_i^{(int,surf)}$ and of the same group in water $p\hat{K}_i^{(int,aq)}$ are related by [cf. Eq. (58)]

$$p\hat{K}_i^{(int,surf)} = p\hat{K}_i^{(int,aq)} - (1 + 2Z_i)\frac{\beta}{\ln 10}(W_i^{(s,surf)} - W_i^{(s,aq)}) \qquad (316)$$

where $W_i^{(s,surf)}$ and $W_i^{(s,aq)}$ refer to the self-energy contributions at the surface and in water, and Z_i is the charge in the deprotonated state (see Section 3). Again, we split off the ionic strength dependent part and compare the contribution at zero ionic strength first. Dropping the subscript i, at zero ionic strength we have [cf. Eq. (58)]

$$p\hat{K}^{(0,surf)} = p\hat{K}^{(0,aq)} - (1 + 2Z)\frac{\beta(W_0^{(s,surf)} - W_0^{(s,aq)})}{\ln 10} \qquad (317)$$

where the self-energy contributions are evaluated without the electrolyte ($\kappa = 0$). For the planar interface, this contribution simply corresponds to the potential generated by an image charge[160]

$$W_0^{(s,surf)} = \frac{e^2}{8\pi\varepsilon_0 D_d a}\frac{D_d - D_w}{D_d + D_w} \qquad (318)$$

where a is the distance of the charge from the interface. The contribution from the isolated molecule $W_0^{(s,aq)}$ can be approximated by the Born self-energy of a point charge in a dielectric sphere [cf. Eq. (36)]. For simplicity, we may set the radius of this sphere to be equal to the distance of the charge from the surface.

Detailed analysis of the ionic strength dependent contribution shows that the DH theory predicts a very similar dependence for the sphere and the plane, and essentially cancels for different geometries. Thus the ionic strength dependence of the intrinsic pK of surface groups should be very similar to its counterpart on a small molecule.

The same pattern was already discussed for proteins and polyelectrolytes. In the common situation of lower relative permittivity for the dielectric than for water ($D_d < D_w$), a basic group on the surface will be more acidic than the same group in solution, while an acidic group will be more basic. No shift is expected for an ionizable group with an effective charge of $Z_i = -1/2$.

7.3. Latex Particles

Spherical, monodisperse poly(styrene) latex particles can be synthesized with various functional groups on the surface,[31] the most common being sulfate groups, originating from the initiator of the polymerization process. Carboxylated latex particles have all (or most) sulfate groups replaced by a carboxyl group. Mixed functionality latex particles with carboxyl, amine, or aldehyde groups are also available. While there is a reasonable degree of certainty about the nature of the

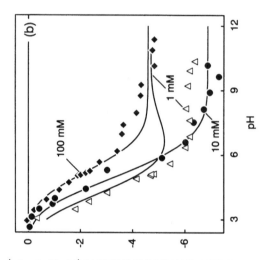

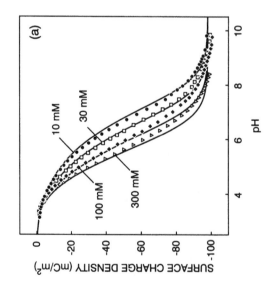

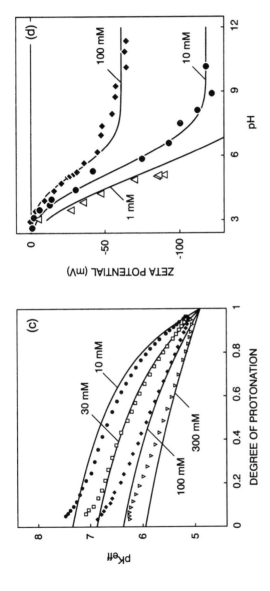

Figure 103. Charging behavior of carboxylated latex particles 309 nm in diameter in KCl as background electrolyte. [351] Experimental data (points) are compared with a simple diffuse layer model (lines). A site density of 0.6 nm^{-2} and p$K^{(int)}$ = 4.9 has been used. Primary experimental data from (a) potentiometric titration and (b) electrophoretic mobility measurements. Derived quantities (c) from the titration data is the pK_{eff} representation, and (d) from mobility data the electrostatic potential at the plane of shear (ζ-potential). The plane of shear is assumed at a constant distance of 0.25 nm from the surface.

functional groups on the latex particle surface, their detailed geometrical arrangement is unknown. The functional groups are likely to be distributed randomly on the surface and they could be either located within the poly(styrene) body close to the surface or rather reside in solution being attached to short flexible chains (hairs). Ionization of such particles can be easily monitored by potentiometric titrations and a substantial body of data exists.[50,71,94,349–351]

Moreover, such particles are usually close to perfect spheres for which theories for electrophoretic mobilities[101,102] and electrical conductivities[95–97] have been developed. Particularly, from electrical conductivity measurements the total number of dissociable groups can be determined.[93,94,97] The quantitative interpretation of the mobility and conductivity data is not always straightforward, since these quantities relate to time-dependent processes whose description calls for additional assumptions.[352–354]

Consider first the charging behavior of a carboxyl latex.[351] Potentiometric titration and electrophoretic mobility data on monodisperse latex particles of 309 nm in diameter are shown in Fig. 103a and 103b, respectively. The data are compared with a simple diffuse layer model with a site density of 0.6 nm^{-2} and p$\hat{K}^{(\text{int})} = 4.9$. The electrophoretic mobility data were calculated by assuming the plane of shear at a distance of 0.25 nm from the surface and employing the procedure of O'Brien and White[101] to convert the surface potentials to electrophoretic mobilities.

For small charge densities, the titration data are very well described with the diffuse layer model, although deviations occur at higher charge densities, which are more apparent in the plot of pK_{eff} shown in Fig. 103c. The model predicts a somewhat stronger ionic strength dependence than observed experimentally. These deviations may be due to specific ion binding, discrete charge effects, or rearrangement of the polymer tails on the surface. The pK value of the surface carboxyl group is slightly higher than the corresponding value one would expect for a similar group in solution; the electrostatic model predicts a shift in the same direction. The electrostatic potential at the plane of shear (ζ-potential) is shown in Fig. 103d. Note that not all electrophoretic data points could be converted to the zeta potential, since the experimentally observed mobilities sometimes exceed the maximum mobilities in the elektrokinetic model.[101]

As a further example, consider the titration of polyfunctional latex with carboxylic and amine groups on the surface. In Fig. 104 we show the titration data from Harding and Healy.[349] The titration curves have a common crossing point around pH 8, corresponding to the PZC, as confirmed by electrophoretic mobility measurements.[349] For pH values above the PZC the ionic strength dependence is typical for a negatively charged system, while below the PZC the reverse ionic strength dependence is observed.

The fits of the diffuse layer model are compared with experimental data in Fig. 104. Two surface reactions must be considered, namely involving carboxyl and amine groups

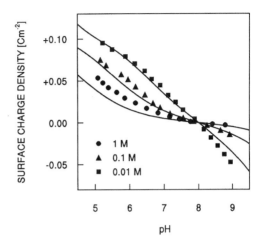

Figure 104. Charging behavior of carboxyl–amine latex of diameter 207 nm.[349] Experimental data are compared with the diffuse layer model. The groups are assumed to have pK values of 5.4 and 8.2, and site densities of 0.95 and 1.55 nm^{-2}, respectively.

$$-COO^- + H^+ \rightleftharpoons -COOH \quad p\hat{K}^{(int)} \simeq 5.4 \qquad (319)$$

$$-NH_2 + H^+ \rightleftharpoons -NH_3^+ \quad p\hat{K}^{(int)} \simeq 8.2 \qquad (320)$$

The ionization constants were proposed by Harding *et al.*[349] to explain their titration data, but these constants are quite different from the low molecular weight counterparts. These pK shifts could well be due to the different dielectric environment on the latex surface. Indeed, on the surface the acidic groups appears more basic while the basic group is more acidic. The site densities for carboxyl and amine groups of 0.95 and 1.55 nm^{-2}, respectively, give a good representation of the data. The ionic strength dependence is again somewhat overestimated by the model. It is conceivable that the present data set could be equally well represented with somewhat different parameters, and thus the proposed pK values must be viewed as tentative.

7.4. Ionizable Monolayers

The excellent control of the molecular architecture of an ionizable interface can be achieved with an ionizable surfactant monolayer. The latter may reside on the water–air, water–oil interfaces or be cast onto solid–water interfaces.[33–37,87,88] While details of the geometrical arrangement of the ionizable groups are sometimes not fully established, their chemical identity is known exactly. However, the determination of the average degree of protonation of the interface is nontrivial, and is mostly achieved by means of spectroscopic or contact angle techniques (see Section 2). Various studies of this kind have been performed with contact angle and capillary rise measurements,[87–91] chromophoric or fluorescent indicators,[36,37,128–131] second harmonic generation,[132] neutron reflection,[92] and infrared (IR) spectroscopy.[33,34,88,134] Surfaces of micelles, surfactant monolayers in microemulsion droplets, or bilayers in vesicles or lamellar liquid crystals may show similar behavior as ionizable monolayers provided the surfactants contain weakly acidic or basic groups.[266,355]

As a representative example of the ionization of an anionic monolayer, consider Fig. 105. Lovelock *et al.*[33] have studied the degree of dissociation as a function of pH for different ionic strengths for 4-heptadecyl-7-hydroxycoumarin in a poly(vinyloctadecyl ether) monolayer cast onto quartz, by means of Fourier transform IR spectroscopy[33,34] and compared with predictions of the diffuse layer model [cf. Eqs. (282) and (287)]. The parameters entering the model are an intrinsic ionization constant $p\hat{K}^{(int)} = 8.2$ and the density of ionizable groups $\Gamma_0 = 1.0$ nm^{-2}.

The diffuse layer model is able to describe the general trends of the pH dependence as well as the ionic strength effect on the ionization behavior rather well. Deviations between experimental data and the diffuse layer model are apparent at low charge densities and high ionic strengths, to a lesser extent in the titration data (see Fig. 105a), but more clearly as a minimum in the pK_{eff} representation (see Fig. 105b). Based on results of different deposition techniques, Lovelock *et al.*[33] conclude that the distribution of the ionizable sites remains random within the monolayer, and that a lateral rearrangement represents an unlikely possibility for the observed discrepancy. This hypothesis is also supported by the arguments put forward above, where we have stated that for a planar interface the mean-field approximation should remain valid for any lateral arrangements of the sites. The most likely explanation for the deviation is related to a conformational transition within the monolayer. The ionization behavior of poly(methacrylic acid), which

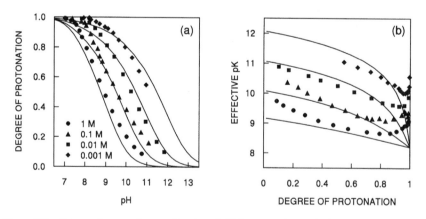

Figure 105. Degree of protonation as a function of pH for a monolayer of 4-heptadecyl-7-hydroxy-coumarin in poly(vinyl octadecyl ether) in aqueous NaCl solutions. Symbols are experimental data[33] and solid lines are predictions with the diffuse layer model. Parameters used are $p\hat{K}^{(int)} = 8.2$ and a site density of 1.0 nm^{-2}. Representation as (a) degree of protonation and (b) pK_{eff}.

undergoes a conformational transition, shows similar ionization behavior (see Section 6 and Fig. 92).

For acidic amphiphilic indicators $p\hat{K}^{(int)}$ can be up to one pK unit higher than the corresponding value in solution.[33,131] The reverse trend was observed for the weak base hexadecyl aniline; the $p\hat{K}^{(int)}$ on the surface was lower than in solution.[132] Analogous shifts of about one pK unit was confirmed with nonionic micelles.[266] These shifts of the pK values are in qualitative agreement with the predictions from electrostatics. An acidic group residing on an interface becomes more basic than the corresponding solution species (pK larger), while a basic group becomes more acidic on the surface (pK smaller).

An interesting confirmation of the mean-field relation [cf. Eq. (283)] was achieved by studying doped charged monomolecular surfactant films.[36,128,129,266] Fluorescent amphiphilic coumarin indicator was used in mixed monolayers containing methylstearate and long-chain sulfates (negatively charged films) and quaternary ammonium ions (positively charged films). The degree of dissociation of the indicator as a function of pH and ionic strength was determined by fluorescence measurements. Since the surface concentration of the indicator was very low, the classical ionization behavior of monoprotic acid was observed, but in this setup one can tune ψ_0 of the surface independently of the degree of protonation by adjusting the amount of the ionic surfactant. In other words, the apparent pK of the indicator is nothing but pK_{eff} introduced in Eq. (285). The experimental results were in excellent agreement with the Gouy–Chapman relation [cf. Eq. (287)] for positively charged films, while only semiquantitative agreement was observed for negatively charged films.[129] The remaining deviations were possibly due to specific adsorption of cations, either as simple or polyvalent cations from trace impurities, but could also originate from the simplified treatment of the dielectric discontinuity within the PB approach.

7.5. Metal Oxide and Metal Hydroxide Particles

Titration results for a variety of metal oxide and hydroxide particles are known, and much of this work has been compiled by Lyklema.[99] Unfortunately, there are not that many reliable data sets as it may seem. In many cases, samples were contaminated by carbon dioxide from air, dissolved silica, or the powders were poorly characterized.[41,356] A large number of models were proposed to explain these observations. These models were addressed in much detail by different authors.[40,51,95,357,358] Here we shall only review the most important aspects relating to the molecular mechanisms of the proton adsorption to oxides.

7.5.1. Experimental Aspects

A characteristic feature of titration curves of most metal oxides and hydroxides is that the titration curves for different ionic strengths cross in a common intersection point, which is identified as the PZC, as discussed above. Experimentally determined PZC values for some common oxides are summarized in Table 27. The only notable exception is silica, whose ionization behavior resembles an acidic surface. In simple electrolytes, the common crossing point of the titration curves usually coincides with a common crossing point in electrophoretic mobility, and thus represents the true PZC. A few exceptions to this rule were described and interpreted,[103,336,337] but it is difficult to ascertain that impurities were properly excluded in these studies. Recently, electrophoretic mobility measurements at high electrolyte concentrations have indicated that the zero mobility point for particles may not coincide with the PZC. These results suggest that the electrolyte anions are somewhat more strongly bound than cations.[103]

Another issue concerns the determination of the total number of ionizable sites. Due to the strong effects of site–site interactions, one would expect that oxides do not titrate fully and, in spite of claims to the contrary, no clear plateaus are apparent in the titration curves. Moreover, at extreme pH values most oxides and hydroxides

Table 27. Point of Zero Charge (PZC) for Various Oxides and Hydroxides[a]

Substance	Formula	PZC
Silica	SiO_2	2.0
Cassiterite	SnO_2	5.5
Rutile	TiO_2	5.8
Ruthenium oxide	RuO_2	5.8
Anatase	TiO_2	6.0
Baddelyte	ZrO_2	6.2
Magnetite	Fe_3O_4	6.5
Akaganeite	$\beta\text{-FeOOH}$	7.1
Lepidocrocite	$\gamma\text{-FeOOH}$	8.0
Cerium oxide	CeO_2	8.1
Alumina	$\gamma\text{-Al}_2O_3$	8.5
Zinc oxide	ZnO	9.0
Lead oxide	PbO_2	9.2
Goethite	$\alpha\text{-FeOOH}$	9.3
Hematite	$\alpha\text{-Fe}_2O_3$	9.3
Lanthanum oxide	La_2O_3	9.6
Gibbsite	$\alpha\text{-Al(OH)}_3$	10.0
Nickel hydroxide	$Ni(OH)_2$	11.2
Nickel oxide	NiO	11.3
Cobalt oxide	Co_3O_4	11.3
Cobalt hydroxide	$Co(OH)_2$	11.4

[a]The values refer to temperatures of 20–30 °C and simple monovalent background electrolytes. The error bar is about ±0.4. From Refs. 99, 333, and 358.

will tend to dissolve. In our opinion, reliable estimates of site densities of oxides cannot be obtained from titration data. From crystallographic considerations to be discussed below, one usually infers site densities, which are much higher than experimentally accessible. We suspect that one only titrates a fraction of the available sites, typically 20–40%.

Substantial effort was also put into investigations of the temperature dependence of titration curves and the related calorimetric heats of adsorption.[82,359–363] Such experiments can provide additional information about ionization processes at oxide–water interfaces. The temperature dependence of intrinsic ionization constants of different surface groups is of particular interest. This quantity, related to the reaction enthalpy, is more directly accessible from molecular and quantum-mechanical models than the equilibrium constant.[235–237,239,364,365] However, the interpretation of temperature-dependent and calorimetric data has yet to be fully established.[360,363,366,367] For example, while titration curves measured at different temperatures superpose for some oxides[362] (congruence), no such congruence was observed for rutile over a wider temperature range.[82]

From the various spectroscopic techniques discussed previously, IR spectroscopy has the potential to yield useful information on the ionization properties of oxide–water interfaces. It would be of course highly desirable to extract similarly detailed information about the protonation of individual sites, as was done, for example, with proteins. Unfortunately, such results are not available for oxides so far.

7.5.2. Data Interpretation

Proper interpretation of the titration behavior of oxides and hydroxides from the molecular point of view turns out to be difficult. The main problems are as follows: (i) The crystal habitus nature of dominant crystal faces and defect structure on the surfaces is mostly unknown. (ii) The types of ionizable groups and their geometrical arrangement on the surface are not well established. The same uncertainty applies to intrinsic pK values of these groups. (iii) The approximations entering the commonly used surface-complexation models (such as the mean-field and smearing-out approximations) are poorly tested on a statistical-mechanical basis. (iv) Since these systems behave close to Nernstian, the titration curves are utterly insensitive to molecular details incorporated into the models.

Most oxide particles investigated in the literature are crystalline, but a few amorphous solids, such as silica or ferrihydrite, have also been investigated. Quite often, the oxide powders are mixtures of different minerals; for example, TiO_2 often consists of a mixture of rutile and anatase. Even pure minerals are often polycrystalline and consist of aggregates of nanometer-sized crystallites (e.g., hematite). For gibbsite, and also for goethite, it is possible to obtain some information about the dominant crystal faces from electron micrographs, particularly from the crystal shape. However, this kind of information is always incomplete, and it is very difficult to infer anything about the presence of point defects, kink sites, steps, or terraces.[368–370] Very promising are oriented single crystals,[331] but the determination of the degree of ionization for the individual faces has not been attempted as yet.

The structure of various crystal faces of different oxides and hydroxides has been investigated in great detail. The simplest working hypothesis is that the dividing plane is located at the surface terminated by oxygens, which are parts of underlying fully coordinated metal centers.[41,331,333,364,368–372] Each surface oxygen atom may be coordinated with a different number of metal atoms, possibly resulting in singly, doubly, and triply coordinated oxygens on the oxide–water interface. These oxygen atoms represent the proton binding sites, and each of them will have a different proton affinity. A few examples are given in Fig. 106.

While this working hypothesis seems reasonable, little independent experimental verification is available and the real situation might be more complicated. Various nonequivalent types of oxygen atoms with the same coordination number can be present on the surface; again, the proton affinities of these groups are expected to differ.[219,333,358,359] Furthermore, it is uncertain whether the structure of the solid persists all the way to the surface. While the symmetry of the crystal lattice determines the symmetry of a crystal surface, it is conceivable that restructuring occurs, and that the actual surface groups might differ from those one would infer from crystallographic considerations. In some cases, the surface might be amorphous or recrystallized—bearing no relation to the bulk crystal whatsoever. Unfortunately, little is known about these points.

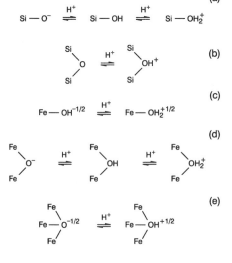

Figure 106. Possible surface groups and reactions anticipated for water–oxide interfaces for silica, iron oxides, and iron hydroxides. Reactions of a singly (a) and doubly (b) coordinated oxygen on a silica surface. Reactions of a singly (c), doubly (d), and triply (e) coordinated oxygen on an iron(III) oxide or hydroxide.

The valence bond method is essential to establish the contributions of the individual groups to the overall surface charge, which turn out to be fractional.[40,41,230,233,234,331,371] For common reactions, these charges are summarized in Table 28 and they are essential in calculating the correct charge of the mineral–water interface. The valence bond method was used to rationalize the PZC values of oxides[373] and their protonation behavior,[40,41,331,344,357] making possible predictions for the likely protonation state of surface groups.[331] However, reliable estimates of the intrinsic proton affinities of different surface groups are difficult to obtain, due to the lack of experimental methods for direct determination of these parameters. Even though the valence bond method appears to be a very promising approach, the resulting proton affinities can be tested only indirectly through the experimental titration curves. This test is not quite satisfactory, since the Nernstian behavior makes the resulting titration curves sometimes hardly dependent on the individual proton affinities (see also below). In fact, it is only the position of the PZC that contains clear molecular information.

The valence bond approach to estimate pK values of solution species was briefly reviewed in Section 4. Hiemstra *et al.*[40,41] have used a very similar method to estimate proton affinities of different surface oxygens at the oxide–water interfaces. This multisite complexation (MUSIC) model assumes that all surface oxygens with one particular coordination will lead to the same proton affinity. The resulting pK values from the original MUSIC model are summarized in Table 28, which can be considered as reasonable estimates of the range of the pK values for

Table 28. *Estimated Intrinsic pK Values for Various Types of Surface Groups with the Original MUSIC Model*[40]

Surface reaction	p$\hat{K}^{(int)}$
$AlOH^{-1/2} + H^+ \rightleftharpoons AlOH_2^{+1/2}$	10
$Al_2O^{-1} + H^+ \rightleftharpoons Al_2OH$	12
$Al_2OH + H^+ \rightleftharpoons Al_2OH_2^{+1}$	−2
$Al_3O^{-1/2} + H^+ \rightleftharpoons Al_3OH^{+1/2}$	2
$FeOH^{-1/2} + H^+ \rightleftharpoons FeOH_2^{+1/2}$	11
$Fe_2O^{-1} + H^+ \rightleftharpoons Fe_2OH$	14
$Fe_2OH + H^+ \rightleftharpoons Fe_2OH_2^{+1}$	0
$Fe_3O^{-1/2} + H^+ \rightleftharpoons Fe_3OH^{+1/2}$	4
$TiOH^{-1/3} + H^+ \rightleftharpoons TiOH_2^{+2/3}$	6
$Ti_2O^{-2/3} + H^+ \rightleftharpoons Ti_2OH^{+1/3}$	5
$Ti_3O + H^+ \rightleftharpoons Ti_3OH^{+1}$	−8
$SiO^{-1} + H^+ \rightleftharpoons SiOH$	12
$SiOH + H^+ \rightleftharpoons SiOH_2^+$	−2
$Si_2O + H^+ \rightleftharpoons Si_2OH^{+1}$	−17

various surface oxygens. Contescu *et al.*[219] have pointed out the difference in proton affinity of AlOH groups in tetrahedral and octahedral environments, and calculated the expected differences in the pK values according to the MUSIC model. This model is also in qualitative agreement with the results of Bargar *et al.*[331] and confirms that two successive protonation steps of one particular oxygen atom are unlikely.

After the publication of the original MUSIC model, evidence started to accumulate that nonequivalent surface oxygens with the same coordination number might have different proton affinities. Hiemstra *et al.*[333,335,358] have refined the original MUSIC approach by considering nonequivalent oxygens with the same number of coordinating metal atoms and by including hydrogen bonds and more specific consideration of the bond length and bond valence relationship. In most cases, the predicted pK values with the revised MUSIC model did turn out to be in the same range. Nevertheless, substantial variations of pK values were predicted for oxygens with the same coordination number, but residing in different local environments and on different faces of various kinds of minerals. The most notable difference between the prediction of the original MUSIC model and the improved valence bond approach concerns the very different proton affinities of two kinds of triply coordinated oxygens on the 110 goethite surface.

7.5.3. Goethite

As far as its charging behavior is concerned, goethite (α-FeOOH) is probably the best investigated mineral.[41,333,345,356,374–376] Crystallographic considerations lead to the conclusion that a perfect single goethite (α-FeOOH) crystal can be made out of 110 and 021 faces only.[372] Electron microscopy and atomic force microscopy indicate that the 110 face is dominant, but other faces such as 021, 100, or 010 might be present as well.[335,372,377] The surfaces are also to some degree imperfect; the 110 face is known to have a large number of steps.

Let us accept the hypothesis that the protonation behavior of goethite is mainly determined by the ionization of a perfect 110 face. The most likely structure of such a face, shown in Fig. 107, is tentative, as it is based on the assumption that the crystal structure pertains all the way to the surface, and that the cut occurs at oxygens, which are part of complete metal ion coordination shells underneath.[372] One further assumes that no surface relaxation occurs. All these assumptions concerning the structure of the 110 face are reasonable, but untested.

The unit cell of the 110 face has dimensions of about 0.3 nm $\times$ 1.1 nm, which leads to a site density of 3.0 nm^{-2} for any individual surface oxygen. Within each unit cell six different kinds of oxygens are shown: one is singly coordinated, one doubly and four are triply coordinated. One oxygen out of the four triply coordinated oxygens is inert, since it has the same coordination as within the bulk crystal. The potentially reactive ones are labeled as FeO, Fe$_2$O, Fe$_3$O$_{I'}$, Fe$_3$O$_{I''}$, and Fe$_3$O$_{II}$. The expected protonation states and charges below

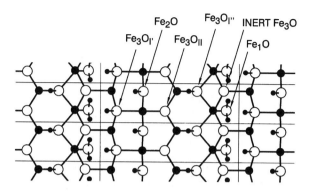

Figure 107. Possible surface structure for the 110 face of goethite. All six nonequivalent oxygen atoms are labeled. The expected protonation states and charges below the PZC are $FeOH_2^{+1/2}$, Fe_2OH, $Fe_3O_{I'}H^{+1/2}$, $Fe_3O_{I''}H^{+1/2}$, and $Fe_3O_{II}^{-1/2}$.

the PZC are $FeOH_2^{+1/2}$, Fe_2OH, $Fe_3O_{I'}H^{+1/2}$, $Fe_3O_{I''}H^{+1/2}$, and $Fe_3O_{II}^{-1/2}$, as also indicated in Fig. 107. In the bulk crystal the oxygen from the FeOH group is bound to the oxygen of the $Fe_3O_{I'}$ group by an internal hydrogen bond. At the surface this hydrogen bond is probably broken, but enhances the protonation of the $Fe_3O_{I'}$ site. The doubly coordinated oxygen in Fe_2O will be protonated at high pH and remain so over the entire pH window of interest. Two types of triply coordinated groups are present. The first two, $Fe_3O_{I'}$ and $Fe_3O_{I''}$, are nonequivalent. Both groups are expected to protonate at high pH, since they bind a proton through an internal hydrogen bridge within the bulk crystal. The Fe_3O_{II} group is expected to remain deprotonated over the entire pH window, since this group is not protonated in the crystal. Table 29 summarizes the proton affinities derived from valence bond considerations.[333,335,358] This approach suggests

Table 29. *Estimated Intrinsic pK Values for the Different Surface Groups in Each Unit Cell on the 110 Face of Goethite by Hiemstra et al.*[333,358]a

Surface reaction	$p\hat{K}^{(int)}$
$FeOH^{-1/2} + H^+ \rightleftharpoons FeOH_2^{+1/2}$	7.7
$Fe_2O^{-1} + H^+ \rightleftharpoons Fe_2OH$	12.3
$Fe_2OH + H^+ \rightleftharpoons Fe_2OH_2^{+1}$	0.4
$Fe_3O_{I'}^{-1/2} + H^+ \rightleftharpoons Fe_3O_{I'}H^{+1/2}$	11.7
$Fe_3O_{I''}^{-1/2} + H^+ \rightleftharpoons Fe_3O_{I''}H^{+1/2}$	11.7
$Fe_3O_{II}^{-1/2} + H^+ \rightleftharpoons Fe_3O_{II}H^{+1/2}$	0.2

aSimple valence bond charges are given. The site density for each surface group is 3.0 nm^{-2}.

that the $Fe_3O_{I'}$ and $Fe_3O_{I''}$ groups behave precisely in the same fashion, an aspect which awaits independent verification.

Rustad *et al.*[364] have proposed that all these different surface oxygens have very similar proton affinities. Their results are based on correlations between proton binding energies and proton affinities of aqueous complexes. This proposition is in marked contrast to the results based on the valence bond principle, from which one concludes that surface oxygens have very different protons affinities.[40,331,333] One possible source for the discrepancy is that Rustad *et al.*[364] have not taken solvation processes into account. We suspect that values of Hiemstra *et al.*[40,333] are more reliable, since the revised MUSIC model rationalizes the PZC for many different minerals. Nevertheless, the final resolution of this discrepancy must be postponed until independent measurements of the proton affinities become possible.

Based on these pK values the titration curves of goethite can be predicted. The result of such a calculation is shown Fig. 108a, based on the basic Stern model with specific ion binding. The five different surface groups with the proton affinities given in Table 29 are used in the calculation.[333,358] As explained above, the site density is 3.0 nm^{-2} for each group. With a Stern capacitance of 1.35 F/m^2 and a binding constant for the salt anions and cations of 0.1 M^{-1}, one obtains a good description of the data. The model also predicts site specific titration curves. The groups Fe_2OH and Fe_3O_{II} are predicted to be essentially inert over the entire pH window. The charging behavior is dominated by the very gradual protonation of FeOH and Fe_3O_I groups. The protonation of

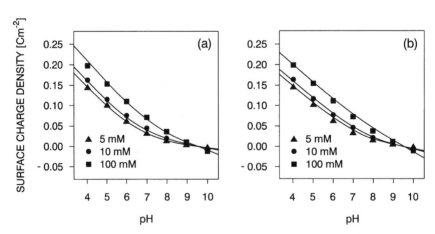

Figure 108.   Titration data of goethite.[333] Comparison of experimental data (points) with calculations using the basic Stern model including symmetric specific binding of salt ions. (a) Consideration of all reactive surface groups and site densities, as shown in Table 29. Further model parameters are $\tilde{C}_s = 1.35$ F/m^{-2} and an ion binding constant of 0.1 M. (b) Simple 1-pK model with $p\hat{K}^{(int)} = 9.5$, site density 6 nm^{-2}, and $\tilde{C}_s = 0.9$ F/m^{-2}.

the FeOH group sets in around pH $\simeq$ 10, while the protonation of Fe_3O_I is close to complete at this pH.

The result of a simplified model is shown in Fig. 108b. In the latter model one considers only a single reaction [cf. Eq. (279)] with a site density of 6.0 nm^{-2}, an intrinsic proton affinity with $p\hat{K}^{(int)} = 9.5$, a Stern capacitance of $\tilde{C}_s = 0.9\ F/m^2$, and the same ion binding constant. The basic Stern model with specific ion binding provides a good description of the titration data.

In comparing Figs. 108a and 108b one immediately realizes the fundamental difficulty in the verification of such models on the basis of titration curves alone. While the molecular basis and the number of surface species in both models is rather different, the fit of the experimental data is quite similar and does not really allow one to discriminate between the models. The reason for the results to be insensitive to the details of the model has to do with the fact that the oxide–water interface behaves like a Nernstian surface to a good approximation. Recall that the titration curve for the ideal Nernstian surface is solely given by the charge–potential relationship and is independent of the surface protonation mechanisms. The ultimate test of a model would be a comparison of site specific titration curves with experimental data. However, such data are not available at present.

7.5.4. Hematite

Another frequently studied iron oxide is hematite (α-Fe_2O_3).[332,362,374,378–381] The advantage of this system is that the particles can be prepared in spheroidal

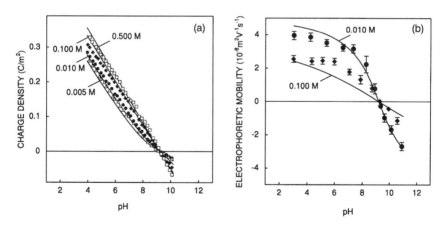

Figure 109. Titration and mobility data for hematite in KNO_3 electrolyte.[332] The data were modeled in terms of the 1-pK basic Stern model with specific adsorption of salt ions. The model assumes a site density of 8 nm^{-2}, an intrinsic constant $p\hat{K}^{(int)} = 9.2$, a Stern capacitance $\tilde{C}_s = 1.1\ Fm^{-2}$, and symmetric binding of salt ions with a binding constant with $K^{(ion)} \simeq 2\ M^{-1}$ for both cations. (a) Potentiometric titration data and (b) electrophoretic mobility data. The plane of shear was assumed to coincide with the plane of origin of the diffuse layer.

shape and electrophoretic mobilities can be interpreted in a reasonably straightforward manner. On the other hand, the particles have been shown to be aggregates of smaller crystallites and the surface structure is largely unknown. Probably, the 001, 110, and 210 are the most common faces and contribute to the overall charging behavior.[358] Recently, it was suggested that the 001 face of this mineral could display a different charging behavior than a typical oxide surface, rather resembling silica.[389]

Figure 109 shows recent titration and electrophoretic mobility data for hematite particles in KNO_3 as background electrolyte.[332] Very similar titration data were reported elsewhere.[380] The rather uniform, spheroidal particles have a diameter of about 90 nm. The mobilities were interpreted in terms of the theory of O'Brien and White for spherical particles.[101] These charging data were modeled in terms of the 1-pK basic Stern model with specific adsorption of salt ions. The model assumes a site density of 8 nm^{-2}, an intrinsic constant $p\hat{K}^{(int)} = 9.2$, a Stern capacitance $\tilde{C}_s = 1.1$ F/m^{-2}, and symmetric binding of salt ions with a binding constant $K^{(ion)} \simeq 2$ M^{-1} for both ions. The employed site density is consistent with crystallographic considerations. The electrophoretic mobility was calculated by assuming that the potential at the slipping plane (ζ-potential) equals the diffuse layer potential ψ_d. While not perfect, this model is able to provide a reasonable interpretation of the titration and mobility data.

Titration data of hematite typically show an anomalously weak salt dependence, which cannot be explained with the basic Stern model. This peculiar salt dependence has been rationalized here by assuming rather strong specific binding of salt ions.[332] Another possible interpretation can be based on weaker specific binding of salt ions, but considers different crystal faces with different PZC, both contributing to the charging behavior.[358]

7.5.5. Rutile and Anatase

The titration behavior of rutile and anatase, which are modifications of titatinum oxide (TiO_2), has also been thoroughly investigated.[49,82,134,334,381,383–386] An interesting data set was recently published by Macheski *et al.*[82] These authors have investigated the titration behavior of rutile in NaCl solutions over a wide range of temperatures. At 25 °C the PZC is at pH 5.4, while at 250 °C it has dropped to 4.2. The shape of the titration curves also depends on temperature.

The experimental data at 25 °C are shown in Fig. 110 and are compared with the 1-pK model using an intrinsic constant $p\hat{K}^{(int)} = 5.4$, a site density 12.5 nm^{-2}, and a binding constant of salt ions $K^{(ion)} \simeq 0.046$ M^{-1} for both cations. The interesting point is that the data are not symmetrical around PZC. In order to account for this asymmetry, the authors introduced a three-plane with another equipotential plane, and assumed salt anions to adsorb at the origin of the diffuse layer and salt cations closer to the surface. This model then requires two capacitances, which were chosen as 1.92 F m^{-2} for the outer and 3.06 F m^{-2} for

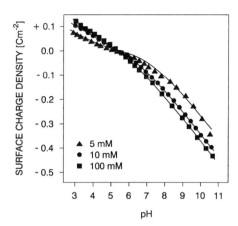

Figure 110. Titration data of rutile at 25 °C in NaCl electrolyte.[82] The three-plane 1-pK model uses an intrinsic constant p$\hat{K}^{(int)} = 5.4$, a site density 12.5 nm^{-2}, and a binding constant of salt ions $K^{(ion)} \approx 0.046$ M^{-1} for both cations.

the inner layer. This *three-layer model* was first introduced by Bousse *et al.*[387] and a similar model was used for the modeling of anion adsorption by Hiemstra *et al.*[335] The basic Stern model as well as the triple layer model[96,334] assume that both types of salt ions adsorb at a common plane, and leads to titration curves symmetric around the PZC.

The three-layer model also predicts an asymmetry in the specific adsorption of salt ions, as observed for anatase by measurements by means of isotopes.[386] Simultaneous titration and mobility data were interpreted,[383,384] but the difficulty lies in the irregular morphology of such particles; the assumption of spherical shape, which is needed for a proper analysis of the electrophoretic data, cannot be made with confidence. For the same reason, the molecular picture is probably more complex. The charging behavior is mainly determined by the protonation of singly and doubly protonated oxygens[40,41]

$$Ti - OH^{-1/3} + H^+ \rightleftharpoons Ti - OH_2^{+2/3} \qquad (321)$$

$$Ti_2 - O^{-2/3} + H^+ \rightleftharpoons Ti_2 - OH^{+1/3} \qquad (322)$$

These singly and doubly coordinated oxygens typically occur with equal site densities on different faces, as shown, for example, for the 110 face of rutile in Fig. 6. Triply coordinated oxygens are present, but are located within the bulk crystal and are therefore expected to be unreactive.[333]

7.5.6. Gibbsite

Gibbsite [γ-Al(OH)$_3$] is another interesting model system. The titration behavior has been studied in considerable detail and substantial information is available about the crystallographic faces from electron micrographs.[382,388] The crystallites are plate-like and the dominant planar face is 001, while the edges are 010, 100, or 110.

For gibbsite it was clearly shown that different crystal faces have different charging characteristics. Hiemstra *et al.*[382] have prepared gibbsite platelets of different thickness. These samples show different titration behavior when normalized to the total surface area. However, when the titration data were normalized to the surface area of the edges, the curves coincided, confirming that the charging of gibbsite occurred on the edges while the planes did not seem to contribute.

This behavior is nicely explained by the MUSIC model. The planar 001 face consists of doubly coordinated oxygens, which are expected to be uncharged over the entire pH window (see Table 28). On the other hand, the edge faces 010, 100, or 110 are dominated by singly coordinated oxygens and these groups are expected to contribute to the charging.

7.5.7. Silica

The titration curves of silica differ markedly from most other common oxides or hydroxides.[41,100,390] While these curves for most oxides at high ionic strength follow almost a straight line through the PZC, those for silica are curved and approach the PZC in a nonlinear fashion (see Fig. 111).

This difference is readily understood from the MUSIC model. The ionizable surface groups of most oxides are protonated at the PZC; all groups are deprotonated for silica. While the detailed surface structure of silica is largely unknown, the following aspects can be established. In the bulk crystal, every oxygen is coordinated with two silica atoms. Doubly coordinated oxygens will be present on the surface, but they are essentially equivalent to those inside the bulk crystal and can be assumed to be inert. Singly coordinated oxygens, being also present on the surface, are thought to be responsible for the observed charging behavior. From

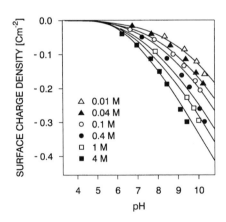

Figure 111. Charging behavior of silica.[41,100] The solid lines are calculations based on the basic Stern model with a site density of 8 nm^{-2}, a binding constant p$\hat{K}^{(\text{int})} \approx 7.5$, and a Stern capacitance $\tilde{C}_s \approx 2.9$ F m^{-2}.

crystallographic considerations of various silica modifications, site densities of the singly coordinated groups should be around $5-8 \, nm^{-2}$. Valence bond considerations reveal that Si–OH at the surface is neutral and thus the dominant surface reactions have to be[41]

$$-SiO^- + H^+ \rightleftharpoons -SiOH \qquad (323)$$

$$-SiOH + H^+ \rightleftharpoons -SiOH_2^+. \qquad (324)$$

For completeness, both protonation steps on the same oxygen atom are considered. However, as evident from Table 28, the $p\hat{K}^{(int)}$ values of these two reactions are very different. Taking only the surface deprotonation reaction [cf. Eq. (323)] into account, the basic Stern model is able to provide a good description of the experimental titration data. In Fig. 111, Bolt's results[100] are compared with predictions of the basic Stern model[41] using a site density of $8 \, nm^{-2}$, a binding constant $p\hat{K}^{(int)} \simeq 7.5$, and a Stern capacitance $\tilde{C}_s \simeq 2.9 \, Fm^{-2}$. While $p\hat{K}^{(int)}$ for Eq. (323) is lower than the value estimated with the original MUSIC model, the refined MUSIC model predicts $p\hat{K}^{(int)} \simeq 7.9$, which is in reasonable agreement with the fitted value. The refined MUSIC model also predicts $p\hat{K}^{(int)} \simeq -4$ for Eq. (324). Combination of both predicted ionization constants leads to a PZC around pH 1.9, which is in excellent agreement with the experimental value of 2.0 ± 0.3.[333]

An additional point of interest is that other types of singly coordinated oxygens could be contributing to the charging of silica. While Eqs. (323) and (324) assume that only one of the four coordinating oxygens is exposed to the solution, it is also conceivable that a pair of oxygens are solvated and ionizing. Such pairs of singly coordinated oxygens are found on the 100 face of cristobalite.[369] In this case, one would have to consider a two-step dissociation reaction

$$SiO_2^{2-} + H^+ \rightleftharpoons Si(OH)O^- \qquad (325)$$

$$Si(OH)O^- + H^+ \rightleftharpoons Si(OH)_2. \qquad (326)$$

Since two different oxygens are being protonated, the splitting between the pK values of these reactions would be only a few pK units. This picture suggests the possibility of additional reactive groups protruding from the surface, and it is conceivable that an entire Si–O–Si(OH)$_3$ surface group could be solvated and participate in the ionization mechanism. This view would suggest a more porous, gel-like silica surface.[391,392]

8. Discussion

In this review we have covered the protonation behavior of various systems, ranging from small molecules with a small number of ionizable sites all the way to large systems, such as polyelectrolytes and interfaces. The essential part of the analysis is the treatment of interactions between the ionizable sites, since the proton affinity of a given site usually changes when another site is being protonated in its neighborhood.

With the assumption that the proton binding is localized, the ionization problem can be mapped on a well-known problem in statistical mechanics. A simple example is the so-called Ising model on a chain, but the approach is more general and treats the ionization problem as a discrete site model. This framework was adopted throughout the review. The interactions between the sites can be introduced in a systematic fashion with the cluster expansion approach. The first-order model neglects all interactions and leads to the *null model*, which is very important for proteins. The second-order model introduces site–site interactions, the distance dependence of which has a profound influence on the ionization pattern of the system under consideration; most characteristic differences occur in the ionization behavior between interfaces and polyelectrolytes at sufficiently high charge densities. For interfaces, the site–site interaction potential is long-range and leads to a featureless titration curve. The ionizable sites protonate in an *uncorrelated* fashion (i.e., random) and the situation can be well described by a mean-field approximation. For highly charged polyelectrolytes, on the other hand, the site–site interaction potential has a dominant short-range component, with the consequence that polyelectrolytes often show a two-step titration curve. In this case, the sites protonate in a strongly *correlated* fashion (i.e., alternating) and the mean-field approximation breaks down.

The onset of correlations between ionizable sites and the corresponding breakdown of the mean-field description is poorly understood. For sufficiently distant groups, the site–site interactions are mainly of electrostatic origin and they can be estimated in a straightforward manner by solving the Poisson–Boltzmann (PB) or Debye–Hückel (DH) equation for the geometries of interest. Such a continuum approach is probably sufficient for systems with moderate interactions, but is expected to fail for systems with small separations between ionizable groups with strong site–site interactions. In such situations, a purely continuum model may reproduce the behavior qualitatively, but will prove inadequate quantitatively. Currently, it is not clear whether one should use some hybrid mean-field pair interaction models or systematically invoke higher-order interaction parameters.[25,162] It also remains to be seen how to proceed beyond the PB approximation in order to answer this type of question on the primitive model level,[68] or even including the molecular structure of water.[8,10]

In proteins, the effect of site–site interactions is moderate; the ionization of most sites within a protein typically displays a one-step titration curve. Two-step or more complicated titration curves, which are characteristic for strongly interacting systems, are seldomly observed. The main effect of the interactions in proteins are shifts in the ionization constants. In proteins with known geometry, sufficiently large shifts can be explained with electrostatic models semiquantitatively. Nevertheless, several assumptions within these models are still inadequately tested, such as, for example, the value of the dielectric constant of the protein interior or the role of side chain flexibility.

The important issue of molecular flexibility is poorly understood, not only for proteins, but also for small molecules or polyelectrolytes. Recall that the cluster expansion approach involves configurational averages and the conformational degrees of freedom are introduced implicitly. The interaction parameters represent free-energy differences and also include entropic contributions. While such effects are absent for rigid systems, in the case of flexible molecules the estimates of interaction parameters may be inaccurate if based on rigid structures. Polyelectrolytes, for example, may swell appreciably as they ionize, causing the interactions to weaken. Similar conformational degrees of freedom have also been addressed for gas adsorption and can sometimes be interpreted as higher-order interaction parameters.[393]

In the near future, much progress can be expected in the *ab initio* quantum mechanical predictions of pK values of ionizable groups.[10,237] While empirical approaches still represent the most powerful methods, the situation might change rapidly. *Ab initio* techniques for the calculation of ionization constants of small molecules have become increasingly reliable. This progress is not only due to more accurate computational schemes but, more importantly, due to the improved methods in treating the solvent in a self-consistent fashion. The *ab initio* tools may soon surpass continuum approaches to estimate pK shifts of ionizable groups in different environments, allowing for more reliable predictions of interaction parameters. For example, the application of *ab initio* techniques is very interesting for the oxide–water interface, since the pK values of the surface oxygen atoms in various coordination geometries are rather uncertain at present.

Our understanding of proton interactions rests on the assumption of *localized* binding, which means that a given proton binds to a single ionizable site. This assumption appears to be an excellent approximation in most cases, even though the existence of hydrogen bonds points toward the fact that delocalization may occur to some extent. Even if hydrogen bonding becomes important, it is usually still possible to assign a given proton to a single ionizable site and hydrogen bonds can be considered separately.[23] A possible candidate for proton delocalization could be oxide–water interfaces, since the binding sites are situated close to one another. However, protons bound to these interfaces exhibit infrared absorption bands, a fact that supports the idea of localized binding.

Whether phase transitions occur in ionizing systems is another fundamentally interesting question. In two-dimensional systems, such as ionizable interfaces, second-order phase transitions are conceivable from the theoretical point of view.[60,340] Due to the repulsive nature of these interactions, one expects phase transitions of the order–disorder type and to manifest itself as a discontinuity in the slope of the titration curve, or as a λ-type singularity in the differential capacitance. Various ordered superlattices on the surface may occur.[340–342] While such order–

disorder transitions have been observed in gas adsorption and electrochemical systems, they have not yet been reported in ionizable systems. The reason might be related to the long-range character of site–site interactions, since for a sufficiently long interaction range no phase transitions are to be expected. The latter are more likely to occur in systems with short-range potentials. Ionizable bilayers could represent a two-dimensional system with sufficiently short-range interactions, leading to phase transitions. A related question deals with spin glass phases in random systems,[394] and whether, for example, the surface latex particle may represent a realization of such a system represents an interesting issue.

The understanding of intermolecular interactions in such systems also represents an open challenge. We have mostly focussed on dilute systems, whose protonation behavior is independent of the concentration of the ionizable species. For higher concentrations, however, interactions between the ionizable species may become important and influence the protonation behavior. These effects have already been discussed by Hill[54] in detail. Low-salt conditions in a polyelectrolyte (or macromolecular) solution is probably the best known manifestation, and under such conditions the ionization behavior depends on the polyelectrolyte concentration, mainly through the screening introduced by the presence of counterions. This situation can be approached within the cell model.[68] Macroion–macroion interaction will be also substantially influenced by dissociation reactions of ionizable surface groups,[162] and spontaneous fluctuations of the macroion dipole moment may lead to additional attractive forces. At short distances, the interaction potential may be diminished as the repulsive charge–charge interaction between individual ionizable groups favors the uncharged state.

ACKNOWLEDGMENTS

We would like to thank D. Bedeaux, S. Behrens, C. Chassagne, M. Cohen-Stuart, J. Daicic, C. Erkelens, T. Hiemstra, D. Horn, T. Kesvatera, D. Kinniburgh, J. Kleimann, E. Matijević, D. Stam, U. Steuerle, M. Ullner, R. van Duijvenbode, M. van Genderen, W. van Riemsdijk, and J. Westall for fruitful collaboration, interesting discussions, and helpful comments on the manuscript.

Appendix A.   Electrostatics of Point Charges

Here we summarize the derivation of the charging energy of an assembly of point charges in a dielectric medium, which is immersed into an indifferent electrolyte. This problem can also be handled numerically for arbitrary geometries. We focus on linear systems and simple geometries where analytical solutions are possible.

We first derive the potential of a single charge. The potentials are written as eigenfunction expansions of the Poisson equation and the linearized Poisson–Boltzmann equation. The Coulomb potentials originating from the point charges are expanded in the same eigenfunctions, and from the necessary boundary conditions one can determine the expansion coefficients. Since the system is linear, one can obtain the potential of an arbitrary assembly of charges by applying the superposition principle and summing up the individual contributions of all charges. The energy is then evaluated from the sum of the potentials evaluated at the location of the individual charges.

A.1. Planar Interface

Let us find the potential of a single charge near a planar interface. Consider a planar semi-infinite dielectric medium with relative permittivity D_d (Region I) which is in contact with an electrolyte solution. The point charge of magnitude q' is located in this medium at $\mathbf{r}'$ at a distance d from the interface. Adjacent to the dielectric medium we also assume a Stern layer of thickness ℓ and relative permittivity D_w (Region II). The actual electrolyte solution has the same relative permittivity D_w and a screening length κ^{-1}. The electrostatic potential $\psi(\mathbf{r})$ in the dielectric and in the Stern layer satisfies

$$\nabla^2\psi(\mathbf{r}) = 0 \qquad \text{Region I and II} \tag{A.1}$$

except at $\mathbf{r}' = \mathbf{r}$ where the charge is located. In the electrolyte solution we have

$$\nabla^2\psi(\mathbf{r}) = \kappa^2\psi(\mathbf{r}) \qquad \text{Region III} \tag{A.2}$$

These equations can be solved by expanding the solution in plane waves. Let us define the z-coordinate to be normal to the interface, and write the spatial position vector $\mathbf{r}$ as $(\mathbf{t}, z)$ where $\mathbf{t}$ is the projection of the vector $\mathbf{r}$ onto the interface. We now expand the potentials into plane waves

$$\psi(\mathbf{t}, z) = \frac{1}{(2\pi)^2} \int e^{i\mathbf{k}\cdot\mathbf{t}} \hat{\psi}(\mathbf{k}, z)d^2\mathbf{k} \tag{A.3}$$

with the inverse relation

$$\hat{\psi}(\mathbf{k}, z) = \int e^{-i\mathbf{k}\cdot\mathbf{t}} \psi(\mathbf{t}, z)d^2\mathbf{t}. \tag{A.4}$$

Substitution of Eq. (A.3) into Eqs. (A.1) and (A.2) leads to an expression, which applies in the dielectric and in the Stern layer, namely

$$\frac{\partial^2\hat{\psi}}{\partial z^2} = k^2\hat{\psi} \qquad \text{Region I and II} \tag{A.5}$$

where $k = |\mathbf{k}|$. The solution of this equation can be written as $\hat{\psi} \propto e^{\pm kz}$. In the electrolyte we have

$$\frac{\partial^2\hat{\psi}}{\partial z^2} = \alpha^2\hat{\psi} \qquad \text{Region III} \tag{A.6}$$

where $\alpha^2 = k^2 + \kappa^2$. Similarly, the solution of this equation can be written as $\hat{\psi} \propto e^{\pm \alpha z}$.

The solution is obtained as a sum of a particular and a general solution of the homogeneous system. The particular solution is given by the Coulomb potential in an infinite dielectric

$$\psi^{(c)}(\mathbf{r}) = \frac{q'}{4\pi\varepsilon_0 D_d} \frac{1}{|\mathbf{r} - \mathbf{r}'|} \tag{A.7}$$

Inserting this expression into Eq. (A.3) we obtain

$$\hat{\psi}^{(c)}(\mathbf{k}, z) = \frac{q'}{2\varepsilon_0 D_d} \frac{1}{k} e^{-k|z+d|} \tag{A.8}$$

Now we add this particular solution to the general solution of the homogeneous Eq. (A.5). The potential can now be expressed as

$$\hat{\psi}_I = A_1 e^{kz} + \frac{q'}{2\varepsilon_0 D_d} \frac{1}{k} e^{-k|z+d|} \qquad \text{Region I, } z < 0 \tag{A.9}$$

where A_1 is an unknown constant. Since this potential vanishes as $z \to -\infty$, the term e^{-kz} must be omitted. Note that we have indicated by a subscript that this potential $\hat{\psi}_I$ applies only in Region I. In the Stern layer we have

$$\hat{\psi}_{II} = A_2 e^{kz} + A_3 e^{-kz} \qquad \text{Region II, } 0 < z < \ell \tag{A.10}$$

where A_2 and A_3 are unknown constants. Finally, in the electrolyte we have

$$\hat{\psi}_{III} = A_4 e^{-\alpha z} \qquad \text{Region III, } \ell < z \tag{A.11}$$

where A_4 is an unknown constant. The term $e^{\alpha z}$ does not appear, since the potential vanishes for $z \to \infty$.

The coefficients $A_1, A_2, A_3,$ and A_4 can be determined by applying the boundary conditions. Between Regions I and II we have

$$\hat{\psi}_I(0, \mathbf{k}) = \hat{\psi}_{II}(0, \mathbf{k}) \tag{A.12}$$

and

$$D_d \frac{\partial \hat{\psi}_I}{\partial z}\bigg|_{z=0} = D_w \frac{\partial \hat{\psi}_{II}}{\partial z}\bigg|_{z=0} \tag{A.13}$$

while between Regions II and III the boundary conditions read

$$\hat{\psi}_{II}(\ell, \mathbf{k}) = \hat{\psi}_{III}(\ell, \mathbf{k}) \tag{A.14}$$

and

$$\left.\frac{\partial \hat{\psi}_{\text{II}}}{\partial z}\right|_{z=\ell} = \left.\frac{\partial \hat{\psi}_{\text{III}}}{\partial z}\right|_{z=\ell} \tag{A.15}$$

Inserting Eqs. (A.9), (A.10), and (A.11) into the boundary conditions [cf. Eqs. (A.12), (A.13), (A.14), and (A.15)] we obtain a system of linear equations, which can be solved in a straightforward fashion. The result for coefficient A_1 reads

$$A_1 = \frac{q'}{2\varepsilon_0 D_d} \frac{1}{k} e^{-kd} R(k) \tag{A.16}$$

with the definition

$$R(k) = \frac{k_+ D_- + k_- D_+ e^{-2k\ell}}{k_+ D_+ + k_- D_- e^{-2k\ell}} \tag{A.17}$$

where $k_\pm = k \pm \alpha$ and $D_\pm = D_d \pm D_w$. The electrostatic energy of a pair of elementary charges ($q' = e$) located in Region I at $\mathbf{r}''$ and $\mathbf{r}'$ ($\mathbf{r}'' \neq \mathbf{r}'$) is given by

$$W(\mathbf{r}'', \mathbf{r}') = e\psi_1(\mathbf{r}'') \tag{A.18}$$

Using Eqs. (A.3), (A.9), and (A.16) we obtain the final expression

$$W(\mathbf{r}'', \mathbf{r}') = \frac{e^2}{4\pi\varepsilon_0 D_d} \int_0^\infty J_0(kr)[1 + e^{-2kd}R(k)]dk \tag{A.19}$$

where $r = |\mathbf{r}'' - \mathbf{r}'|$. Taking the limit $\ell \to 0$ we obtain the result discussed in Section 7. The self-energy is given by a similar expression to Eq. (A.19), but taking $r \to 0$ and omitting the first term in the integrand.

A.2. Sphere

The above discussion of the planar interface is closely analogous to the original calculation by Kirkwood,[11] where he evaluated the electrostatic energy of an assembly of point charges in a dielectric sphere. Here we summarize the results; details are given elsewhere.[11]

Consider a single point charge q' at $\mathbf{r}'$ located in a sphere of radius $|\mathbf{r}'| \leq b$ and relative permittivity D_d (Region I). The sphere is surrounded by a Stern layer of thickness ℓ with relative permittivity D_w and $\kappa = 0$ (Region II). This sphere of radius $a = b + \ell$ is suspended in a salt solution of relative permittivity D_w and screening length κ^{-1} (Region III). As above, Eqs. (A.1) and (A.2) apply also in this situation. Here, we shall expand the solutions into spherical harmonics[160]

$$\psi(\mathbf{r}) = \sum_{l,m} \tilde{\psi}(r; l, m) Y_{lm}(\vartheta, \varphi) \tag{A.20}$$

where $r = |\mathbf{r}|$ and $Y_{lm}(\vartheta, \varphi)$ are spherical harmonics evaluated at polar angle ϑ and azimuthal angle φ. The inversion relation reads

$$\tilde{\psi}(r; l, m) = \int_0^{2\pi} \int_0^\pi \psi(\mathbf{r}) Y^*_{lm}(\vartheta, \varphi) \sin \vartheta \, d\vartheta \, d\varphi \qquad (A.21)$$

where the asterisk denotes the complex conjugate. In the dielectric and the Stern layer

$$\frac{1}{r} \frac{\partial}{\partial r}\left(r \frac{\partial \tilde{\psi}}{\partial r}\right) = l(l+1)\tilde{\psi} \qquad \text{Region I and II} \qquad (A.22)$$

applies, which has general solutions $\tilde{\psi} \propto r^l$ and $\tilde{\psi} \propto r^{-l-1}$. In the electrolyte

$$\frac{1}{r} \frac{\partial}{\partial r}\left(r \frac{\partial \tilde{\psi}}{\partial r}\right) = [l(l+1) + \kappa^2]\tilde{\psi} \qquad \text{Region III} \qquad (A.23)$$

holds, with the relevant solution $\tilde{\psi} \propto C_l(\kappa r) e^{-\kappa r}/r$. This solution vanishes as $r \to \infty$. Also, there is a second solution of this equation, which is of no interest here. The polynomials $C_l(x)$ were introduced by Kirkwood[11] by the recursion relation

$$(2l+1)(2l-1)[C_{l+1}(x) - C_l(x)] = x^2 C_{l-1}(x) \qquad (A.24)$$

with $C_0(x) = 1$ and $C_1(x) = 1 + x$. Their derivative $C'_l(x)$ satisfies

$$(2l+1)[C_l(x) - C'_l(x)] = x^2 C_{l-1}(x). \qquad (A.25)$$

The $C_l(x)$ are given explicitly by

$$C_l(x) = \sum_{s=0}^{l} \frac{2^s \Gamma(l+1)\Gamma(2l-s+1)}{\Gamma(s+1)\Gamma(2l+1)\Gamma(l-s+1)} x^s. \qquad (A.26)$$

These polynomials are closely related to Bessel functions. The use of the doubling formula for the Gamma function yields

$$C_l(x) = \frac{x^{l+1/2} e^x}{2^{l-1/2}\Gamma(l+1/2)} K_{l+1/2}(x) \qquad (A.27)$$

where $K_n(x)$ are modified Bessel functions of the second kind.[160] To incorporate the point charge, we use the same approach as above. The Coulomb potential [cf. Eq. (A.7)] is expanded in spherical harmonics[160]

$$\tilde{\psi}^{(c)}(r; l, m) = \frac{4\pi}{2l+1} \frac{(r')^l}{r^{l+1}} Y^*_{lm}(\vartheta', \varphi') \qquad (A.28)$$

for $r > r'$. Near the boundary the potential can be written as

$$\tilde{\psi}_I(r; l, m) = A_1 r^l + \frac{4\pi}{2l+1} \frac{(r')^l}{r^{l+1}} Y_{lm}^*(\vartheta', \varphi') \qquad \text{Region I, } r' < r \le b. \qquad (A.29)$$

In the Stern layer we have

$$\tilde{\psi}_{II}(r; l, m) = A_2 r^l + A_3 \frac{1}{r^{l+1}} \qquad \text{Region II, } b \le r \le a \qquad (A.30)$$

while in the electrolyte we can write

$$\tilde{\psi}_{III}(r; l, m) = A_4 C_l(\kappa r) \frac{e^{-\kappa r}}{r^{l+1}} \qquad \text{Region III, } a \le r. \qquad (A.31)$$

The unknown coefficients A_1, A_2, A_3, and A_4 are determined by applying the boundary conditions. Between Regions I and II we have

$$\psi_I(b) = \psi_{II}(b) \qquad (A.32)$$

and

$$D_d \frac{\partial \psi_I}{\partial r}\bigg|_{r=b} = D_w \frac{\partial \psi_{II}}{\partial r}\bigg|_{r=b} \qquad (A.33)$$

while analogous conditions hold between Regions II and III. The linear system of equations can be solved for the unknown coefficients. Following the same arguments as for the plane, we arrive at the expression

$$W(\mathbf{r}'', \mathbf{r}') = \frac{e^2}{4\pi\varepsilon_0 D_d} \frac{1}{|\mathbf{r}'' - \mathbf{r}'|} + \sum_{l=0}^{\infty} (r'' r')^l S_l P_l(\cos\gamma) \qquad (A.34)$$

where γ is the angle between the vectors $\mathbf{r}''$ and $\mathbf{r}'$, $P_l(x)$ is a Legendre polynomial,[160] and the coefficient S_l is given by

$$S_l = \frac{e^2}{D_d b^{2l+1}} \frac{D_d(l+1)U_l - D_w V_l}{D_d l U_l + D_w V_l} \qquad (A.35)$$

where

$$U_l = (l+1)a^l C_{l+1}(\kappa a) + l b^{2l+1}[C_{l+1}(\kappa a) - C_l(\kappa a)] \qquad (A.36)$$

and

$$V_l = a^l C_{l+1}(\kappa a) - b^{2l+1}[C_{l+1}(\kappa a) - C_l(\kappa a)] \qquad (A.37)$$

The self-energy is obtained by omitting the first term in Eq. (A.34) and letting $\mathbf{r}'' \to \mathbf{r}'$.

We note that for $\kappa = 0$ we have $C_l(0) = 1$ and the above expression for the coefficient S_l can be simplified to

$$S_l = \frac{1}{D_d b^{2l+1}} \frac{(l+1)(D_d - D_w)}{lD_d + (l+1)D_w} \tag{A.38}$$

For $l = 0$, Eq. (A.38) gives the Born solvation energy of a charge in a spherical cavity [cf. Eq. (30)] while for $l = 1$, Eq. (A.38) gives the reaction field contribution of a dipole.[8,186]

A.3. Cylinder

We now turn to cylindrical geometry. The approach is similar to the development by Soumpasis.[303] For simplicity, we neglect the Stern layer and consider a dielectric cylinder of radius a and relative permittivity D_d embedded in an electrolyte solution of Debye length κ^{-1} and relative permittivity D_w. A single point charge q' is placed at $\mathbf{r'} = (\rho', \phi', z')$, where $\rho' < a$, in the cylindrical coordinate system (ρ, ϕ, z).

Fourier decomposition of the potential in the periodic angular variable ϕ and the continuous axial coordinate z is given by

$$\psi(\rho; \phi, z) = \frac{1}{2\pi} \sum_{n=-\infty}^{\infty} \int_{-\infty}^{\infty} e^{ikz} e^{in\phi}\, \tilde{\psi}(\rho, n, k)\, dk \tag{A.39}$$

with the inversion

$$\tilde{\psi}(\rho; n, k) = \frac{1}{2\pi} \int_{-\infty}^{\infty} \int_{-\pi}^{\pi} e^{-ikz} e^{-in\phi}\, \psi(\rho; \phi, z) d\phi\, dz \tag{A.40}$$

For $\rho < a$, we proceed in analogy with the planar and spherical cases by adding the Fourier components of the particular and homogeneous solutions of the Poisson equation in the dielectric interior to obtain the full solution. The homogeneous Poisson equation now stipulates

$$\frac{1}{\rho} \frac{\partial}{\partial \rho}\left(\rho \frac{\partial \tilde{\psi}}{\partial \rho} \right) - (k^2\rho^2 + n^2)\tilde{\psi} = 0. \tag{A.41}$$

The solutions of this equation, which are finite at the cylinder axis, are proportional to $I_n(|k|\rho)$, where $I_n(z)$ are modified Bessel functions of the first kind.[160] The particular solution, again the Coulomb potential, is

$$\tilde{\psi}^{(c)} = \frac{q'}{2\pi\varepsilon_0 D_d} e^{-ikz'} e^{-in\phi'}\, I_n(|k|\rho')K_n(|k|\rho) \tag{A.42}$$

where $K_n(z)$ are modified Bessel functions of the second kind,[160] thus $\tilde{\psi} = \tilde{\psi}^{(c)} + A_1 I_n(|k|\rho)$ for $\rho < a$. In the electrolyte, $\rho > a$,

$$\frac{1}{\rho}\frac{\partial}{\partial\rho}\left(\rho\frac{\partial\tilde{\psi}}{\partial\rho}\right) - \left(\alpha^2\rho^2 + n^2\right)\tilde{\psi} = 0 \tag{A.43}$$

where again $\alpha^2 = k^2 + \kappa^2$. The solution must vanish in the $\rho \to \infty$ limit and therefore admits only $\tilde{\psi} = A_2 K_n(\alpha\rho)$.

The matching of boundary conditions is similar to the planar and spherical cases, without the presence of the Stern layer. The Fourier coefficients $\tilde{\psi}_1(\rho; n, k)$ for the potential inside the cylindrical dielectric can be written down, specifically for $\rho' < \rho \leq a$, which in analogy with the previous cases are denoted as

$$\tilde{\psi}_1(\rho; n, k) = \frac{q'}{2\pi\varepsilon_0 D_d} e^{-ikz'} e^{-in\phi'} I_n(|k|\rho')[K_n(|k|\rho) + I_n(|k|\rho)R(k, n)] \tag{A.44}$$

with the definition

$$R(k, n) = \frac{kD_d\overline{K}_n(ka)K_n(\alpha a) - \alpha D_w\overline{K}_n(\alpha a)K_n(ka)}{kD_d\overline{I}_n(ka)K_n(\alpha a) + \alpha D_w\overline{K}_n(\alpha a)I_n(ka)} \tag{A.45}$$

where $\overline{K}_n(x) = K_{n-1}(x) + K_{n+1}(x)$ and $\overline{I}_n(x) = I_{n-1}(x) + I_{n+1}(x)$.

The final expression reads

$$W(\mathbf{r}'', \mathbf{r}') = \frac{e^2}{2\pi^2\varepsilon_0 D_d}\sum_{n=-\infty}^{\infty} e^{in(\phi'' - \phi')}$$

$$\times \int_0^{\infty} dk \cos k(z'' - z')I_n(k\rho')[K_n(k\rho'') + I_n(k\rho'')R(k, n)]. \tag{A.46}$$

The self-energy follows by omitting the first term in the square brackets within the integrand, and letting $\mathbf{r}'' \to \mathbf{r}'$.

Appendix B.   Further Tools in the Ising Model Analysis

In this appendix we summarize some additional techniques which are useful for the analysis of the Ising model. More details on these techniques can be found in the statistical mechanics literature.[58,60]

B.1.   Cluster Expansion

Consider a molecule with N ionizable sites. We parametrize these different states by introducing a state variable s_i for each site i ($i = 1, \ldots, N$) such that $s_i = 1$ if the site is protonated and $s_i = 0$ if the site deprotonated. The set of all site variables $\{s_1, s_2, \ldots, s_N\}$, or briefly $\{s_i\}$, uniquely defines the protonation state of the system. There are 2^N such states. To each of these states $\{s_i\}$ we can associate a free energy $F(\{s_i\})$. This free energy has a particular value for each state and, in principle, could be tabulated. However, since we have only two values $s_i = 0, 1$, the function $F(\{s_i\})$ can be rewritten in a simpler form. Consider first the case $N = 1$. In this case $F(s_1)$ has only two values $F(0)$ and $F(1)$. Thus, we can also write

$$F(s_1) = (1 - s_1)F(0) + s_1 F(1) \tag{B.1}$$

or

$$F(s_1) = F(0) + [F(1) - F(0)]s_1 \tag{B.2}$$

which is a linear function in s_1. Let us repeat the same argument for $N = 2$. The function $F(s_1, s_2)$ depends now on two variables and can have four different values, namely $F(0, 0)$, $F(1, 0)$, $F(0, 1)$, and $F(1, 1)$. Similarly, we can write

$$F(s_1, s_2) = (1 - s_1)(1 - s_2)F(0, 0) + s_1(1 - s_2)F(1, 0)$$

$$+ (1 - s_1)s_2 F(0, 1) + s_1 s_2 F(1, 1) \tag{B.3}$$

or

$$F(s_1, s_2) = F(0, 0)$$

$$+ [F(1, 0) - F(0, 0)]s_1 + [F(0, 1) - F(0, 0)]s_2$$

$$+ [F(1, 1) - F(1, 0) - F(0, 1) + F(0, 0)]s_1 s_2 \tag{B.4}$$

which is a quadratic function in s_1 and s_2. Since $s_i^2 = s_i$ there are no terms proportional to s_i^2. Continuing this kind of argumentation, we observe that the free energy can be written as

$$F(\{s_i\}) = -\sum_i \mu_i s_i + \frac{1}{2!} \sum_{i,j} E_{ij} s_i s_j + \frac{1}{3!} \sum_{i,j,k} L_{ijk} s_i s_j s_k + \cdots \tag{B.5}$$

where the factorials have been introduced for consistency with customary definitions of these coefficients. We have also dropped an additive constant. This free energy is now expanded into linear, quadratic, and cubic contributions in s_i which correspond to singlets, pairs, and triplets of sites. Clearly, one could introduce additional terms, but one expects that this type of expansion converges rather rapidly and consideration of higher-order contributions may not be necessary.

Let us now discuss the different contributions to Eq. (B.5). The first linear term depends on the chemical potential of protons

$$\frac{\beta \mu_i}{\ln 10} = p\hat{K}_i - pH \tag{B.6}$$

where $\hat{\mathrm{p}K}_i$ are the microscopic pK values of the ionizable groups, given that all other groups are deprotonated. This contribution can also be expressed in terms of the free energies, namely

$$\mu_i = F(s_1, \ldots, s_i = 1, \ldots, s_N) - F(s_1, \ldots, s_i = 0, \ldots, s_N) \qquad (B.7)$$

Pairs of sites lead to energy contributions E_{ij}. The pair energy matrix is symmetrical, $E_{ij} = E_{ji}$, and, without loss of generality, we may set $E_{ii} = 0$ as this coefficient can be absorbed in μ_i since $s_i^2 = s_i$. Often it is more useful to introduce the dimensionless pair interaction parameters

$$\varepsilon_{ij} = \frac{\beta E_{ij}}{\ln 10} \qquad (B.8)$$

The pair interaction energies E_{ij} can be expressed in terms of free energies as

$$E_{ij} = F(s_1, \ldots, s_i = 1, \ldots, s_j = 1, \ldots, s_N)$$

$$- F(s_1, \ldots, s_i = 0, \ldots, s_j = 1, \ldots, s_N)$$

$$- F(s_1, \ldots, s_i = 1, \ldots, s_j = 0, \ldots, s_N)$$

$$+ F(s_1, \ldots, s_i = 0, \ldots, s_j = 0, \ldots, s_N) \qquad (B.9)$$

Analogous considerations apply to the triplet energies.

B.2. Other Choices of State Variables

We have mostly focussed on the definition of the free-energy Eq. (B.5), which is called the *lattice gas model*.[57,60,340] Several other equivalent formulations of this model are used. These formulations can all be obtained by a simple transformation of variables. Because we deal with variables that attain only two discrete values, the most general variable transformation $s_i \to \tilde{s}_i$ can be written as

$$\tilde{s}_i = b_i s_i + a_i \qquad (B.10)$$

The transformed variable now has two different values, a_i and $a_i + b_i$. In the physics literature, the most popular choice for these two values is ± 1.[57,60,340] This model is obtained by setting the parameters $a_i = -1$ and $b_i = 2$. The state variables now indicate whether a magnetic spin is aligned in the direction of an external magnetic field or opposite to this direction. As one usually deals with ferromagnets, where spins prefer to align in the same direction, the pair interaction parameters are negative, $\varepsilon_{ij} < 0$. This is equivalent to attractive interactions between sites. The opposite case $\varepsilon_{ij} > 0$, which is of interest here, is relevant for the treatment of antiferromagnets. In protein literature[11,14,16,17,65] one often employs the charge of the group as state variable. This variable is obtained by setting $b_i = 1$ and interpreting

a_i as the charge of the deprotonated group [cf. Eq. (91)]. If we use the new variables, the free energy can be written in the same form as Eq. (B.5), namely

$$F(\tilde{s}_i) = -\sum_i \tilde{\mu}_i \tilde{s}_i + \frac{1}{2!}\sum_{i,j} \tilde{E}_{ij}\tilde{s}_i\tilde{s}_j + \frac{1}{3!}\sum_{i,j,k} \tilde{L}_{ijk}\tilde{s}_i\tilde{s}_j\tilde{s}_k + \cdots \qquad (B.11)$$

but now the coefficients are transformed to[340]

$$\tilde{\mu}_i = \frac{\mu_i}{b_i} + \sum_j \frac{E_{ij}}{b_ib_j}a_j - \frac{1}{2}\sum_{j,k}\frac{L_{ijk}}{b_ib_jb_k}a_ja_k \qquad (B.12)$$

$$\tilde{E}_{ij} = \frac{E_{ij}}{b_ib_j} - \sum_k \frac{L_{ijk}}{b_ib_jb_k}a_k \qquad (B.13)$$

$$\tilde{L}_{ijk} = \frac{L_{ijk}}{b_ib_jb_k} \qquad (B.14)$$

Under such transformations, the functional form of the free energy is invariant but the values of the coefficients change.

B.3. Mean-Field Cluster Expansions

Mean-field cluster expansions represent a systematic way to improve on a mean-field approximation. These methods were introduced in the statistical mechanics literature in the past[264] and were recently used with quite some success for proteins.[265] As discussed in Section 7, the so-called 1-pK and 2-pK models for interfaces also follow from such an expansion.[69]

The method works as follows. We divide the sites into M imaginary clusters $\mathcal{G}_\alpha$ with $\alpha = 1, \ldots, M$. We shall now treat all individual sites within the cluster explicitly as if they were part of a polyprotic molecule. However, we assume that all sites which do not belong to this cluster influence the sites inside the cluster only through their average degree of protonation. The degree of protonation of a site i within a cluster $\mathcal{G}_\alpha$ is

$$\theta_i = \frac{\displaystyle\sum_{s_i \text{ all } i \in \mathcal{G}_\alpha} s_i e^{-\beta \tilde{F}_\alpha(\{s_i\})}}{\displaystyle\sum_{s_i \text{ all } i \in \mathcal{G}_\alpha} e^{-\beta \tilde{F}_\alpha(\{s_i\})}} \qquad (B.15)$$

Taking into account pair interactions only, the free energy for such a cluster is given by

$$\tilde{F}_\alpha = \sum_{i \in \mathcal{G}_\alpha} \tilde{\mu}_i s_i + \frac{1}{2} \sum_{i \in \mathcal{G}_\alpha} \sum_{j \in \mathcal{G}_\alpha} E_{ij} s_i s_j \tag{B.16}$$

Both sums now run over the state variables inside the cluster only. The chemical potential $\tilde{\mu}_i$ now also includes the mean-field contributions and must be replaced by an effective chemical potential

$$\tilde{\mu}_i = \mu_i - \sum_{j \notin \mathcal{G}_\alpha} E_{ij} \theta_j \tag{B.17}$$

The set of equations (B.15)–(B.17) must be solved self-consistently. The advantage of this method is that one can improve successively on the accuracy of the results by increasing the size of the clusters. For weakly interacting systems, this method works well (as any other mean-field treatment). For strongly interacting systems, however, the method converges very slowly and is hardly more practical than a full-scale Monte Carlo simulation.

B.4. Mean-Field Treatment of Nonequivalent Sites

The mean-field model introduced above can be generalized to an arbitrary number of nonequivalent sites. This generalization can be also interesting for homogeneous systems, as the behavior of these systems can be often approximated by splitting the individual sites into several groups. Suppose that we divide N sites into M groups of equivalent sites $\mathcal{G}_\alpha$ $(\alpha = 1, \ldots, M)$ according to some prescription. Within each group there are N_α identical sites with dissociation constants $\mathrm{p}\hat{K}_\alpha$. Equation (183) then generalizes to

$$\log_{10} \frac{1 - \theta_\alpha}{\theta_\alpha} = \mathrm{pH} - \mathrm{p}\hat{K}_\alpha + \sum_\beta \bar{\varepsilon}_{\alpha\beta} \theta_\beta \qquad \text{for } \alpha = 1, \ldots, M \tag{B.18}$$

where θ_α is the degree of protonation of a site of type α and the interaction parameters are given by

$$\bar{\varepsilon}_{\alpha\beta} = \frac{1}{N_\alpha} \sum_{i \in \mathcal{G}_\alpha} \sum_{j \in \mathcal{G}_\beta} \varepsilon_{ij}. \tag{B.19}$$

The overall degree of protonation is obtained from

$$\theta = \sum_\alpha f_\alpha \theta_\alpha \tag{B.20}$$

where $f_\alpha = N_\alpha/N$ is the fraction of sites of type α. If each individual site is viewed as an individual group of sites $(N = M)$, one obtains the so-called Tanford–Roxby approximation[63,67] (see Section 5).

B.5.   Square Lattice with Nearest-Neighbor Interactions

As an example let us briefly discuss the nearest-neighbor pair interaction model on a square lattice. One should stress, however, that due to the short range of the interactions this model is not appropriate for ionizable water–solid interfaces. Nevertheless, it is fitting to mention this model here, as it illustrates the occurrence of second-order phase transition. Such phase transitions are possible in two-dimensional systems and might also occur in some planar ionizable systems. This model has been amply discussed in the statistical mechanics literature.[58,60] The famous analytical solution of this model by Onsager[395] applies only to the ferromagnetic case at $\theta = 1/2$; for an introduction to these aspects of the Ising model see, for example, Ref. 60. The expected protonation behavior is shown in Fig. 112. The Monte Carlo data convey two changes in slope (arrows in Fig. 112) that are indicative of second-order phase transitions. For this system, the locus of this phase transition is known exactly.[396] In the intermediate region, the system prefers to order in a sublattice, where sites are alternatively occupied and empty. This behavior corresponds to alternating ordering of repulsive sites, but in this two-dimensional system this ordering effect is so strong that it leads to a phase transition. Such phase transitions are typical for planar lattice systems dominated by short-range interactions; more extensive discussions are given elsewhere.[340,342]

This type of behavior can be modeled in terms of a mean-field model. However, we have to consider each sublattice as an individual group of sites. The mean-field interaction matrix $\bar{\varepsilon}_{\alpha\beta}$ in Eq. (B.19) has the elements

$$\begin{bmatrix} 0 & 4\varepsilon \\ 4\varepsilon & 0 \end{bmatrix}. \tag{B.21}$$

Equation (B.18) has two solutions. Both solutions are displayed in Fig. 112. One solution represents the familiar one-variable mean-field model discussed earlier [cf. Eq. (183)], while the other corresponds to different occupations of each sublattice.

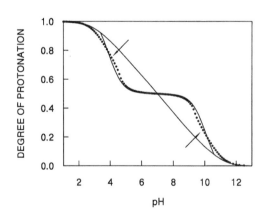

Figure 112. Hypothetical titration curve for a nearest-neighbor pair interaction model on a square lattice with identical sites. Points are Monte Carlo simulation data, and solid lines are the solution of a mean-field model discussed in the text. Arrows denote the locus of the second-order phase transition. The parameters are $p\hat{K} = 10$ and $\varepsilon = 1.5$.

This example also illustrates possible pitfalls associated with mean-field models. While the mean-field model correctly predicts a second-order phase transition in this planar system, there is no reason why the same reasoning should not be applicable to a linear chain. For the linear chain the titration curve is continuous, and no phase transition is present. Nevertheless, the mean-field model will wrongly predict a phase transition in this situation.

Appendix C.   *Affinity Distribution Approach*

In Section 4 we have shown that for noninteracting sites it may be useful to define an affinity distribution $P(K)$ [cf. Eq. (203)]. It turns out that even in the presence of site–site interactions such an affinity distribution is usually a well-defined quantity. The reason for this is that the partition function Ξ is a polynomial in a_H [cf. Eq. (65)] therefore we can always write

$$\Xi = \prod_{n=1}^{N}(1 + \tilde{K}_n a_H)$$ (C.1)

where $-\tilde{K}_n^{-1}$ are the zeros of this polynomial. This expression is precisely identical to a partition function, which describes independent sites [cf. Eq. (196)]. One essential difference in that situation remains, however. While for the system with independent sites the constants $\hat{K}_n$ are always real and positive, for the coefficients $\tilde{K}_n$ in Eq. (C.1) this may not be always the case. Recall that the zeros of a polynomial with real coefficients can be either real or complex. If they are complex, they will occur as complex conjugate pairs. Furthermore, since $\tilde{K}_n > 0$ the zeros of Ξ cannot be real and positive. Therefore, two cases should be investigated separately. (i) All constants $\tilde{K}_n$ are real and positive. (ii) Some of these constants are complex.

Clearly, case (i) is the simpler and more frequent one. Since all $\tilde{K}_i$ are positive and real, the titration curve can be interpreted as a titration curve of a mixture of fictitious monoprotic acids or bases. The corresponding affinity distribution [cf. Eq. (203)] is now given by

$$P(K) = \frac{1}{N}\sum_{i=1}^{N} \delta(K - \tilde{K}_i)$$ (C.2)

where $\delta(x)$ is the Dirac delta function. Since we have assumed that all N zeros are real, the resulting distribution will be necessarily normalized,

$$\int_0^\infty P(K)\,dK = 1$$ (C.3)

Instead of calculating the zeros of Ξ explicitly, we can also obtain the affinity distribution from Eq. (205). Once $P(K)$ is obtained, we can also verify whether all zeros were indeed real. The calculated distribution must namely satisfy the normalization condition Eq. (C.3).

In the case (i) where all zeros of Ξ are real, the titration of interacting sites is exactly equivalent to a titration curve of a mixture of monoprotic acids or bases. In case (ii) some of the zeros of Ξ are complex and the titration curve cannot be interpreted in this fashion. The affinity distribution description is then no longer practical, as one would have to consider two-dimensional distributions of "affinity constants" in the complex plane. It turns out, however, that case (ii) is rather infrequent in practice. This observation is related to the fact that we deal with repulsive (antiferromagnetic) interactions with $\varepsilon_{ij} > 0$.[397] For attractive (ferromagnetic) interactions with $\varepsilon_{ij} < 0$, the Lee–Yang theorem (see, for example, Ref. 57) states that the zeros of the partition function of an Ising model lie on a circle in the complex plane, and thus are complex.

C.1. Derivation from Simple Models

Let us illustrate the above considerations with two simple examples. Figure 113 shows the affinity distributions for models with identical sites and pairwise site–site interactions. The figure illustrates the effect of an increasing number of sites N including infinity. The continuous distributions will be reported as distributions of pK values (or pK spectra). This distribution is proportional to $KP(K)$.

Figure 113a shows the results of the mean-field model, while in Fig. 113b the nearest-neighbor chain model is considered. For the mean-field model, the coefficients of Eq. (65) are given by Eq. (183), and for finite N the zeros can be calculated numerically. For $\varepsilon \geq 0$ the zeros always turn out to be real, and thus an affinity distribution always exists. As $N \to \infty$ the distribution becomes continuous and can be evaluated by inserting Eqs. (223) and (229) into Eq. (205). The calculation leads to the parametric representation of the affinity distribution function[226]

$$K(t^2 + s^2) \sin \alpha s = \hat{K}se^{-\alpha t} \tag{C.4}$$

where $t = 1/2 \pm [1/4 - s(s - 1/\tan \alpha s)]^{1/2}$, $\alpha = \varepsilon \ln 10$, and the distribution is defined by $s = KP(K)/\pi$. This distribution is compared with the case for finite N in Fig. 113a.

Figure 113b shows the results for a linear chain with nearest-neighbor interactions. In this case, the zeros of the partition function can be found analytically.[223] For $\varepsilon > 0$, the zeros again turn out to be real and we can again represent the titration curve in terms of an affinity distribution. In the limit $N \to \infty$, the resulting distribution is found by taking the continuum limit of the analytical expressions of the zeros. The final result reads

$$P(K) = \frac{1}{2\pi K} \frac{1}{\sqrt{4\hat{K}K/(K + u\hat{K})^2 - 1}} \tag{C.5}$$

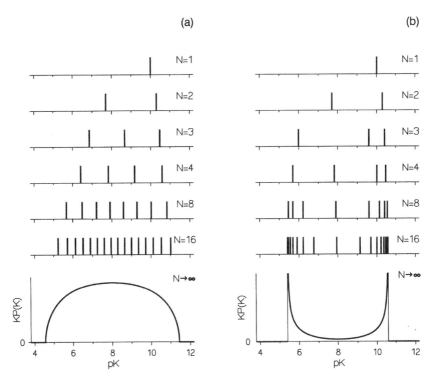

Figure 113. Affinity distributions for the interacting systems with a different number of sites N. All sites are identical and have $\hat{pK} = 10$. (a) Mean-field model with $\bar{\varepsilon} = 4$ and (b) linear-chain model with $\varepsilon = 2$.

in the interval $K_- < K < K_+$ and vanishes elsewhere. The interval boundaries K_- and K_+ are given by the two solutions of the equation $4K\hat{K} = (K + u\hat{K})^2$. This result is displayed in Fig. 113b and compared with the case for finite N.

Figure 114 shows an extension of this analysis to the realistic case of a linear poly(ethylene imine). As discussed in Section 6, in this situation nearest-neighbor triplet interactions are also important.[223] The best fit of such a model of the titration data is shown in Fig. 114a. In this case, the affinity distribution exists as well and is displayed in Fig. 114b.

C.2. Experimental Data Inversion

Given an experimental record of the titration curve θ as a function of pH, one can also try to solve for the affinity distribution $P(K)$ numerically. The integral in Eq. (205) is discretized and one fits the experimental data to a linear superposition of Langmuir isotherms

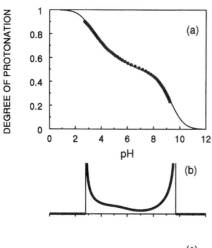

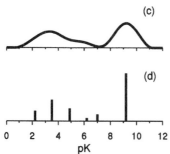

Figure 114. Affinity distribution for linear poly(ethylene imine).[222] (a) Experimental titration curve in 1.0 M NaCl with best fit based on the nearest-neighbor model. The best fit parameters are $p\hat{K}_t \approx 9.10$, $\varepsilon \approx 2.08$, and $\lambda \approx 0.46$ for the pair and triplet parameters. (b) Exact affinity distribution for this model. Distributions obtained from direct inversion of the experimental data are shown below. Regularization (c) for smoothness, and (d) a small number of sites. Reprinted from Ref. 226 with permission from Elsevier Science.

$$\theta = \sum_{m=1}^{M} \rho_m \frac{\tilde{K}_m a_{\mathrm{H}}}{1 + \tilde{K}_m a_{\mathrm{H}}} \qquad (C.6)$$

where $\sum_{m=1}^{M} \rho_m = 1$. To obtain a numerical solution for a set of values ρ_m, the affinity constants $\tilde{K}_m$ are chosen on a fixed, uniformly spaced logarithmic grid. A classical least-squares method would minimize

$$\chi^2 = \sum_{k} \left(\frac{\theta^{(k)} - \theta(a_{\mathrm{H}}^{(k)})}{\sigma^{(k)}} \right)^2 \qquad (C.7)$$

where k runs over all experimental data points while $a_{\mathrm{H}}^{(k)}$, $\theta^{(k)}$, and $\sigma^{(k)}$ are the experimental values for the proton activity, the average degree of protonation, and the experimental error of the average degree of protonation, respectively.

This classical least-squares approach is usually problematic, as it represents a so-called *ill-posed* problem. A given experimental curve can be typically represented by several different affinity distributions equally well. In this situation, χ^2 has an extremely shallow minimum with the consequence that most classical least-squares solution schemes become unstable. The solution of such ill-posed problems has been amply discussed; a good introduction is also available.[398] For the calculation of affinity distributions from experimental data, the most convenient approach to stabilize the solution is based on the assumption of an *a priori* hypothesis about the properties of the distribution. Such a hypothesis could be that the distribution is a smooth function, or that it consists of a few discrete peaks which correspond to a small number of discrete sites.

Once one has decided on such a hypothesis, standard regularized least-squares techniques are available to search for an affinity distribution that is consistent with this hypothesis and the experimental data.[398] Subject to bounds $\rho_m \geq 0$, one minimizes the expression

$$\chi^2 + \eta R \tag{C.8}$$

where we have introduced the regularization function R and the regularization parameter η. The regularization function R represents the mathematical formulation of our *a priori* hypothesis about the affinity distribution and will introduce a penalty for those distributions that are in conflict with it. Distributions consisting of a few isolated, discrete peaks, which correspond to a small number of discrete sites, can be obtained by introducing the regularization function

$$R = \sum_{m>m'} \rho_m \rho_{m'} \tag{C.9}$$

which increases with increasing number of nonzero values of ρ_m. One can bias for smooth distributions with the regularization function

$$R = \sum_m (\rho_{m+1} + \rho_{m-1} - 2\rho_m)^2 \tag{C.10}$$

which approximates the mean square of the curvature of the affinity distribution. The weight of this regularization function is tuned by the regularization parameter η in Eq. (C.8). Increasing η, one introduces an increasingly strong bias toward the properties imposed by the regularizing function. Optimal values of η can be found with various statistical tests.[47,398] The application of these techniques for the interpretation of acid–base titration curves was discussed in detail by Černík *et al.*[47]

In Section 4, we have already shown the applicability of these techniques to the titration curve of a fulvic acid (see Fig. 45). Here we show another example, namely the case of linear poly(ethylene imine). The application of both regularization schemes is shown in Figs. 114c and 114d. None of the regularization techniques is able to extract the actual shape of the affinity distribution accurately.

References

1. J. N. Brønsted, *Z. Phys. Chem.* **169A**, 52 (1934).
2. N. Bjerrum, *Z. Phys. Chem.* **106**, 219 (1923).
3. J. G. Kirkwood and F. H. Westheimer, *J. Chem. Phys.* **6**, 506 (1938).
4. E. J. King, *Acid–Base Equilibria*, Pergamon Press, Oxford, 1965.
5. C. F. Baes and R. E. Mesmer, *The Hydrolysis of Cations*, John Wiley, New York, 1976.
6. D. D. Perrin, B. Dempsey, and E. P. Serjeant, *pK_a Prediction for Organic Acids and Bases*, Chapman & Hall, London, 1981.
7. A. E. Martell and R. J. Motekaitis, *The Determination and Use of Stability Constants*, VCH Publishers, New York, 1988.
8. P. E. Smith and B. M. Pettitt, *J. Phys. Chem.* **98**, 9700 (1994).
9. J. Aragó, A. Bencini, A. Bianchi, E. Garcia-España, M. Micheloni, P. Paoletti, J. A. Ramirez, and P. Paoli, *Inorg. Chem.* **30**, 1843 (1991).
10. W. L. Jorgensen and J. M. Briggs, *J. Am. Chem. Soc.* **111**, 4190 (1989).
11. J. G. Kirkwood, *J. Chem. Phys.* **7**, 351 (1934).
12. K. Linderstrøm-Lang, *C. R. Trav. Lab. Carlsberg* **15**, 1 (1924).
13. A. V. Hill, *J. Physiol. (London)* **40**, iv (1910).
14. C. Tanford, *Physical Chemistry of Macromolecules*, John Wiley, New York, 1960.
15. S. J. Shire, G. I. H. Hanania, and F. R. N. Gurd, *Biochemistry* **13**, 2967 (1974).
16. D. Bashford and M. Karplus, *Biochemistry* **29**, 10219 (1990).
17. A. S. Yang, M. R. Gunner, R. Sampogna, K. Sharp, and B. Honig, *Proteins* **15**, 252 (1993).
18. F. E. Harris and S. A. Rice, *J. Phys. Chem.* **58**, 725 (1954).
19. R. A. Marcus, *J. Phys. Chem.* **58**, 621 (1954).
20. S. Lifson, *J. Chem. Phys.* **26**, 727 (1957).
21. A. Katchalsky, J. Mazur, and P. Spitnik, *J. Polym. Sci.* **23**, 513 (1957).
22. D. Horn, in *Polymeric Amines and Ammonium Salts*, E. J. Goethals (ed.), Pergamon Press, Oxford, 1980.
23. T. Nishio, *Biophys. Chem.* **40**, 19 (1991).
24. C. E. Reed and W. F. Reed, *J. Chem. Phys.* **96**, 1609 (1992).
25. R. G. Smits, G. J. M. Koper, and M. Mandel, *J. Phys. Chem.* **97**, 5745 (1993).
26. A. von Zelewsky, L. Barbosa, and C. W. Schläpfer, *Coord. Chem. Rev.* **123**, 229 (1993).
27. H. Dautzenberg, W. Jaeger, J. Kötz, B. Philipp, C. Seidel, and D. Stscherbina, *Polyelectrolytes*, Hanser Publishers, München, 1994.
28. J. M. Fréchet, *Science* **263**, 1710 (1994).
29. W. Stumm, H. Hohl, and F. Dalang, *Croat. Chem. Acta* **48**, 491 (1976).
30. W. Stumm and J. J. Morgan, *Aquatic Chemistry*, John Wiley, New York, 1981.
31. E. S. Daniels, E. D. Sudol, and M. S. El-Aasser, *Polymer Latices: Preparation, Characterization and Applications*, American Chemical Society, Washington, DC, 1992.
32. O. A. El Seoud, in *Reverse Micelles*, P. L. Luisi and B. E. Straub (eds.), Plenum Press, New York, 1984.
33. B. Lovelock, F. Grieser, and T. W. Healy, *Langmuir* **2**, 443 (1986).
34. B. S. Murray, J. S. Gogfrey, F. Grieser, T. W. Healy, B. Lovelock, and P. J. Scales, *Langmuir* **7**, 3057 (1991).
35. R. Andreu and W. R. Fawcett, *J. Phys. Chem.* **98** 12753 (1994).
36. C. J. Drummond, F. Grieser, and T. W. Healy, *Discuss. Faraday Soc.* **81**, 95 (1986).
37. C. J. Drummond, F. Grieser, and T. W. Healy, *J. Phys. Chem.* **92**, 2604 (1988).

38. R. P. Bell, *The Proton in Chemistry*, Cornell University Press, Ithaca, New York, 1959.
39. D. A. Tomalia, H. Baker, J. Dewald, M. Hall, G. Kallos, S. Martin, J. Roeck, J. Ryder, and P. Smith, *Polym. J.* **17**, 117 (1985).
40. T. Hiemstra, W. H. Riemsdijk, and G. H. Bolt, *J. Colloid Interface Sci.* **133**, 91 (1989).
41. T. Hiemstra, J. C. M. de Wit, and W. H. Riemsdijk, *J. Colloid Interface Sci.* **133**, 105 (1989).
42. H. R. Schulten and M. Schnitzer, *Naturwissenschaften* **80**, 29 (1993).
43. L. Pauling, D. Pressman, and A. L. Grossberg, *J. Am. Chem. Soc.* **66**, 784 (1944).
44. G. Scatchard, J. S. Coleman, and A. L. Shen, *J. Am. Chem. Soc.* **79**, 12 (1957).
45. E. M. Perdue, J. H. Reuter, and R. S. Parrish, *Geochim. Cosmochim. Acta* **48**, 1257 (1984).
46. B. Leuenberger and P. W. Schindler, *Anal. Chem.* **58**, 1471 (1986).
47. M. Černík, M. Borkovec, and J. C. Westall, *Environ. Sci. Technol.* **29**, 413 (1995).
48. P. W. Schindler and H. Gamsjäger, *Kolloid-Z. Z. Polym.* **250**, 759 (1972).
49. D. E. Yates, S. Levine, and T. W. Healy, *Trans. Faraday Soc.* **70**, 1807 (1974).
50. T. W. Healy and L. R. White, *Adv. Colloid Interface Sci.* **9**, 303 (1978).
51. J. Westall and H. Hohl, *Adv. Colloid Interface Sci.* **12**, 265 (1980).
52. J. G. Kirkwood and J. B. Shumaker, *Proc. Natl. Acad. Sci. U.S.A.* **38**, 855 (1952).
53. R. F. Steiner, *J. Chem. Phys.* **22**, 1458 (1954).
54. T. L. Hill, *J. Chem. Phys.* **23**, 623 (1955).
55. E. Ising, *Z. Phys.* **31**, 253 (1925).
56. T. L. Hill, *An Introduction to Statistical Thermodynamics*, Dover Publ. Inc., New York, 1986.
57. T. L. Hill, *Statistical Mechanics*, McGraw-Hill, New York, 1956.
58. R. J. Baxter, *Exactly Solved Models in Statistical Mechanics*, Academic Press, New York, 1982.
59. D. Chandler, *Introduction to Modern Statistical Mechanics*, Oxford University Press, New York, 1987.
60. H. S. Robertson, *Statistical Thermophysics*, Prentice Hall, New Jersey, 1993.
61. C. Tanford and J. C. Kirkwood, *J. Am. Chem. Soc.* **79**, 5333 (1957).
62. W. H. Ortung, *Biochemistry* **9**, 2394 (1970).
63. C. Tanford and R. Roxby, *Biochemistry* **11**, 2192 (1972).
64. M. K. Gilson and B. H. Honig, *Proteins* **3**, 32 (1988).
65. P. Beroza, D. R. Fredkin, M. Y. Okamura, and G. Feher, *Proc. Natl. Acad. Sci. U.S.A.* **88**, 5804 (1991).
66. B. Honig and A. Nicholls, *Science* **268**, 1144 (1995).
67. D. Bashford and M. Karplus, *J. Phys. Chem.* **95**, 9556 (1991).
68. T. Kesvatera, B. Jönsson, E. Thulin, and S. Linse, *J. Mol. Biol.* **259**, 828 (1996).
69. M. Borkovec, *Langmuir* **13**, 2608 (1997).
70. M. Borkovec, J. Daicic, and G. J. M. Koper, *Proc. Natl. Acad. Sci. U.S.A.* **94**, 3499 (1997).
71. T. Gisler, S. F. Schulz, M. Borkovec, P. Schurtenberger, B. D'Aguano, R. Klein, and H. Sticher, *J. Chem. Phys.* **101**, 9924 (1994).
72. B. Jönsson, T. Åkesson, and C. E. Woodward, in *Ordering and Phase Transitions in Charged Colloids*, A. K. Arora and B. V. R. Tata (eds.), VCH Publishers, New York, 1996.
73. Y. M. Chernoberezhskii, in *Surface and Colloid Science*, Vol. 12, E. Matijević (ed.), Plenum Press, New York, 1982.
74. H. S. Harned and B. B. Owen, *Physical Chemistry of Electrolyte Solutions*, Reinhold Publ. Corp., New York, 1958.
75. L. P. Hammett, *Physical Organic Chemistry*, McGraw-Hill, New York, 1940.
76. D. C. Harris, *Quantitative Chemical Analysis*, 5th ed., Freeman, New York, 1998.
77. H. Galster, *pH Measurement*, VCH, Weinheim, 1991.
78. R. G. Bates, *Chem. Rev.* **42**, 1 (1948).
79. S. P. L. Sørensen, *Biochem. Z.* **21**, 131 (1909).
80. F. E. Crane, *J. Chem. Educ.*, **38**, 365 (1961).
81. D. G. Kinniburgh, C. J. Milne, and P. Venema, *Soil Sci. Soc. Am. J.* **59**, 417 (1995).

82. M. L. Macheski, D. J. Wesolowski, D. A. Palmer, and K. Ichiro-Hayashi, *J. Colloid Interface Sci.* **200**, 298 (1998).

83. R. van Duijvenbode, M. Borkovec, and G. J. M. Koper, *Polymer* **39**, 2657 (1997).

84. A. I. Vogel, *A Text-Book of Quantitative Inorganic Analysis*, 3rd ed., Longmans, London, 1961.

85. C. S. Bürgisser, A. M. Scheidegger, M. Borkovec, and H. Sticher, *Langmuir* **10**, 855 (1994).

86. J. C. Eijkel, C. Bosch, W. Olthuis, and P. Bergvelt, *J. Colloid Interface Sci.* **187**, 148 (1997).

87. T. R. Lee, R. I. Carey, H. A. Biebuyck, and G. M. Whitesides, *Langmuir* **10**, 741 (1994).

88. S. R. Holmes-Farley, C. D. Bain, and G. M. Whitesides, *Langmuir* **4**, 921 (1988).

89. R. C. Chatelier, C. J. Drummond, D. Y. C. Chan, Z. R. Vasic, T. R. Gegenbach, and H. J. Griesser, *Langmuir* **11**, 4122 (1997).

90. R. C. Chatelier, A. M. Hodges, C. J. Drummond, D. Y. C. Chan, and H. J. Griesser, *Langmuir* **13**, 3043 (1997).

91. E. McCaffertys and J. P. Wightman, *J. Colloid Interface Sci.* **194**, 344 (1997).

92. S. W. An and R. K. Thomas, *Langmuir* **13**, 6881 (1997).

93. M. Elimelech and C. R. O'Melia, *Colloids Surf.* **44**, 165 (1990).

94. J. Stone-Masui and A. Watillon, *J. Colloid Interface Sci.* **52**, 479 (1975).

95. R. O. James and G. A. Parks, in *Surface and Colloid Science*, Vol. 12, E. Matijevič (ed.), Plenum Press, New York, 1982.

96. R. O. James, J. A. Davis, and J. O. Leckie, *J. Colloid Interface Sci.* **65**, 331 (1978).

97. J. P. H. Zwetsloot and J. C. Leyte, *J. Colloid Interface Sci.* **163**, 362 (1994).

98. R. J. Hunter, *Zeta Potential in Colloid Science*, Academic Press, New York, 1981.

99. J. Lyklema, *Fundamentals of Colloid and Interface Science*, Vol. 2, Academic Press, New York, 1995.

100. G. H. Bolt, *J. Phys. Chem.* **61**, 1166 (1957).

101. R. W. O'Brien and L. R. White, *J. Chem. Soc., Faraday Trans. 2* **77**, 1607 (1978).

102. J. Kilstra, H. P. van Leeuwen, and J. Lyklema, *J. Chem. Soc., Faraday Trans.* **88**, 3441 (1992).

103. W. N. Rolands, R. W. O'Brien, R. J. Hunter, and V. Patrick, *J. Colloid Interface Sci.* **188**, 325 (1997).

104. C. P. Schlichter, *Principles of Magnetic Resonace*, Springer, Berlin, 1980.

105. W. W. Paudler, *Nuclear Magnetic Resonance*, John Wiley, New York, 1987.

106. H. Güther, *NMR Spectroscopy, Basic Principles, Concepts and Applications in Chemistry*, John Wiley, New York, 1995.

107. P. Crews, J. Rodriguez, and M. Jaspars, *Organic Structure Analysis*, Oxford University Press, Oxford, 1998.

108. G. J. M. Koper, M. H. P. van Genderen, C. Elissen-Roman, M. W. P. L. Baars, E. J. Meier, and M. Borkovec, *J. Am. Chem. Soc.* **119**, 6512 (1997).

109. T. Kesvatera, B. Jönsson, E. Thulin, and S. Linse, *Proteins Struct. Func. Genet.* **37**, 106 (1999).

110. D. G. Gorenstein, *Methods Enzymol.* **177**, 295 (1989).

111. K. Mernissi-Arifi, L. Schmitt, G. Schlewer, and B. Spiess, *Anal. Chem.* **67**, 2567 (1995).

112. Y. Takeda, K. Samejima, K. Nagano, M. Watanabe, H. Sugeta, and Y. Kyogoku, *Eur. J. Biochem.* **130**, 383 (1983).

113. W. Schaller and A. D. Robertson, *Biochemistry* **34**, 4714 (1995).

114. M. Delfini, A. L. Segre, F. Conti, R. Barbucci, V. Barone, and P. Ferruti, *J. Chem Soc., Perkin Trans. 2* 900 (1980).

115. D. N. Hague and A. D. Moreton, *J. Chem. Soc., Perkin. Trans. 2* 265 (1994).

116. O. Zhang, E. K. Lewis, J. P. Olivier, and J. D. Forman-Kay, *J. Biomolecular NMR* **4**, 845 (1994).

117. R. Richarz and K. Wütrich, *Biochemistry* **17**, 2263 (1978).

118. Y. Oda, M. Yoshida, and S. Kanaya, *J. Biol. Chem.* **268**, 88 (1993).

119. W. R. Baker and A. Kintanar, *Arch. Biochem. Biophys.* **327**, 189 (1996).

120. M. D. Sørensen and J. J. Led, *Biochemistry* **33**, 13727 (1994).

121. Y. Oda, T. Yamazaki, K. Nagayama, S. Kanaya, Y. Kuroda, and H. Nakamura, *Biochemistry* **33**, 5275 (1994).
122. L. Braunschweiler and R. R. Ernst, *J. Magn. Res.* **53**, 521 (1983).
123. G. Bodenhausen and D. J. Ruben, *Chem. Phys. Lett.* **69**, 185 (1980).
124. M. Zhang, E. Thulin, and H. J. Vogel, *J. Protein Chem.* **13**, 527 (1994).
125. J. L. Markley and W. M. Westler, *Biochemistry* **35**, 11092 (1996).
126. F. Abildgaard, A. M. Munk-Jorgensen, J. J. Led, T. Christensen, E. B. Jensen, F. Junker, and H. Dalboge, *Biochemistry* **31**, 8587 (1992).
127. J. Schleucher, M. Schwendinger, M. Sattler, P. Schmidt, O. Schedletsky, S. J. Glaser, O. W. Sorensen, and C. Griesinger, *J. Biomolecular NMR* **4**, 301 (1994).
128. J. G. Petrov and D. Möbius, *Langmuir* **6**, 746 (1990).
129. P. Fromherz and B. Masters, *Biochim. Biophys. Acta* **356**, 270 (1974).
130. R. A. Hall, P. J. Thistlethwaite, F. Grieser, N. Kimizuka, and T. Kunitake, *Langmuir* **10**, 3743 (1994).
131. J. G. Petrov, E. E. Polymeropoulos, and H. Möhwald, *J. Phys. Chem.* **100**, 9860 (1996).
132. X. Zhao, S. Ong, H. Wang, and K. B. Eisenthal, *Chem. Phys. Lett.* **214**, 203 (1993).
133. K. Nakamoto, Y. Morimoto, and A. E. Martell, *J. Am. Chem. Soc.* **85**, 309 (1963).
134. C. Contescu, V. Popa, and J. A. Schwarz, *J. Colloid Interface Sci.* **180**, 149 (1996).
135. N. J. Harrick, *Internal Reflection Spectroscopy*, John Wiley, New York, 1967.
136. N. J. Harrick, *Internal Reflection Spectroscopy, Review and Supplement*, Harrick Scientific Corp., Ossining, New York, 1985.
137. S. J. Hug and B. Sulzberger, *Langmuir* **10**, 3587 (1994).
138. J. T. Edsall, R. B. Martin, and B. R. Hollingworth, *Proc. Natl. Acad. Sci. U.S.A.* **44**, 505 (1958).
139. Y. Nozaki and C. Tanford, *J. Am. Chem. Soc.* **89**, 742 (1967).
140. F. Figueirido, G. S. Del Buono, and R. M. Levy, *J. Phys. Chem.* **100**, 6389 (1996).
141. M. Kinoshita, S. Iba, and M. Harada, *J. Chem. Phys.* **105**, 2487–2499.
142. H. Friedman, *Ionic Solution Theory*, Interscience, New York, 1962.
143. J. C. Rasaiah and H. L. Friedman, *J. Chem. Phys.* **48**, 2742 (1968).
144. G. M. Torrie and J. P. Valleau, *J. Phys. Chem.* **86**, 3251 (1982).
145. D. J. Tannor, B. Marten, R. Murphy, R. A. Friesner, D. Sitkoff, A. Nicholls, M. Ringnalda, W. A. Goddard III, and B. Honig, *J. Am. Chem. Soc.* **116**, 11875 (1994).
146. J. L. Chen, L. Noodleman, D. A. Case, and D. Bashford, *J. Chem. Phys.* **98**, 11059 (1994).
147. B. Kallies and R. Mitzner, *J. Phys. Chem. B* **101**, 2959 (1997).
148. R. L. Cleland, *Macromolecules* **17**, 634 (1984).
149. D. F. Evans and H. Wennerström, *The Colloidal Domain*, VCH, New York, 1994.
150. L. Kotin and M. Nagasawa, *J. Chem. Phys.* **36**, 873 (1962).
151. J. Warwicker and H. C. Watson, *J. Mol. Biol.* **157**, 671 (1982).
152. M. Ullner, B. Jönsson, and P. O. Widmark, *J. Chem. Phys.* **100**, 3365 (1994).
153. J. de Groot, G. J. M. Koper, M. Borkovec, and J. de Bleijser, *Macromolecules* **31**, 4182 (1998).
154. M. P. Allen and D. J. Tildesley, *Computer Simulation of Liquids*, Oxford Science Publications, Oxford, 1991.
155. D. Frenkel and B. Smit, *Understanding Molecular Simulation, from Algorithms to Applications*, Academic Press, San Diego, 1996.
156. N. A. Metropolis, A. W. Rosenbluth, M. N. Rosenbluth, A. Teller, and E. Teller, *J. Chem. Phys.* **21**, 1087 (1953).
157. J. N. Israelachvili, *Intermolecular and Surface Forces*, 2nd ed., Academic Press, London, 1991.
158. D. A. McQuarrie, *Statistical Mechanics*, Harper & Row, New York, 1976.
159. B. Jönsson, H. Wennerström, and B. Halle, *J. Phys. Chem.* **84**, 2179 (1980).
160. J. D. Jackson, *Classical Electrodynamics*, 2nd ed., John Wiley, New York, 1975.
161. R. A. Marcus, *J. Chem. Phys.* **23**, 1057 (1955).
162. Y. N. Vorobjev and H. A. Scheraga, *J. Phys. Chem.* **97**, 4855 (1993).

163. H. X. Zhou, *J. Chem. Phys.* **100**, 3152 (1994).
164. K. A. Sharp and B. Honig, *J. Phys. Chem.* **94**, 7684 (1990).
165. B. Beresford-Smith, D. Y. C. Chan, and D. J. Mitchell, *J. Colloid Interface Sci.* **105**, 216 (1986).
166. C. E. Woodward and B. Jönsson, *J. Phys. Chem.* **92**, 2000 (1988).
167. R. D. Groot, *J. Chem. Phys.* **95**, 9191–9203.
168. R. Kjellander, *J. Chem. Phys.* **99**, 10392 (1995).
169. L. Guldbrand, B. Jönsson, H. Wennerström, and P. Linse, *J. Chem. Phys.* **80**, 2221 (1984).
170. J. C. Rasaiah and H. L. Friedman, *J. Chem. Phys.* **50**, 3965 (1969).
171. D. N. Card and J. P. Valleau, *J. Chem. Phys.* **52**, 6232 (1970).
172. C. N. Patra and S. K. Ghosh, *J. Chem. Phys.* **100**, 5219 (1994).
173. S. Nordholm and A. D. J. Haymet, *Aust. J. Chem.* **33**, 2013 (1980).
174. S. Nordholm, *Chem. Phys. Lett.* **105**, 302 (1984).
175. B. R. Svensson, B. Jönsson, M. Fushiki, and S. Linse, *J. Phys. Chem.* **96**, 3135 (1992).
176. R. Kjellander, *J. Phys. Chem.* **99**, 10392 (1995).
177. J. C. Rassiah, *J. Chem. Phys.* **52**, 704 (1970).
178. J. P. Valleau and L. K. Cohen, *J. Chem. Phys.* **72**, 5935 (1980).
179. K. S. Pitzer, *Acc. Chem. Res.* **10**, 371 (1977).
180. K. S. Pitzer, in *Activity Coefficients in Electrolyte Solutions*, R. M. Pytkowitz (ed.), CRC Press, Boca Raton, 1979.
181. P. Richmond, *J. Chem. Soc., Faraday Trans. 2* **70**, 1067 (1974).
182. M. Medina-Noyola and B. I. Ivlev, *Phys. Rev. E* **52**, 6281 (1995).
183. F. H. Stillinger, *J. Chem. Phys.* **35**, 1584 (1961).
184. J. W. Adamson, *Physical Chemistry of Surfaces*, 5th ed., John Wiley, New York, 1990.
185. S. J. Miklavic, *J. Chem. Phys.* **103**, 4794 (1995).
186. B. Honig, K. Sharp, and A.-S. Yang, *J. Phys. Chem.* **97**, 1101 (1995).
187. M. Dubois, T. Zemb, L. Belloni, A. Deville, P. Levitz, and R. Setton, *J. Chem. Phys.* **96**, 2277 (1992).
188. R. Kjellander and S. Marcelja, *J. Chem. Phys.* **82**, 2122 (1985).
189. R. Kjellander, T. Åkesson, B. Jönsson, and S. Marcelja, *J. Chem. Phys.* **97**, 1424 (1985).
190. H. Greberg, R. Kjellander, and T. Åkesson, *Mol. Phys.* **87**, 407 (1996).
191. A. Kitao, F. Hirata, and N. Go, *J. Phys. Chem.* **97**, 10231 (1993).
192. M. Ullner, C. E. Woodward, and B. Jönsson, *J. Chem. Phys.* **105**, 2056 (1996).
193. B. R. Brooks, R. E. Bruccoleri, B. D. Olafson, D. J. States, S. Swaminathan, and M. Karplus, *J. Comput. Chem.* **4**, 187 (1983).
194. A. D. Mackerell, J. Wiorkiewicz, and M. Karplus, *J. Am. Chem. Soc.* **117**, 11946 (1995).
195. P. Flory, *Statistical Mechanics of Chain Molecules*, Interscience, New York, 1969.
196. B. Widom, *J. Chem. Phys.* **39**, 2808 (1963).
197. B. Svensson, T. Åkesson, and C. E. Woodward, *J. Chem. Phys.* **95**, 2717 (1991).
198. Y. Zhou, *J. Phys. Chem. B* **102**, 10615 (1998).
199. B. Noszál, in *Biocoordination Chemistry*, K. Burger (ed.), Ellis Horwood, New York, 1990.
200. J. Wyman, *J. Mol. Biol.* **11**, 631 (1965).
201. H. S. Simms, *J. Am. Chem. Soc.* **48**, 1239 (1926).
202. M. Borkovec and G. J. M. Koper, *J. Phys. Chem.* **98**, 6038 (1994).
203. M. Ullner and B. Jönsson, *Macromolecules* **29**, 6645 (1996).
204. L. J. Henderson, *Am. J. Physiol.* **21**, 427 (1908).
205. K. A. Hasselbalch, *Biochem. Z.* **78**, 112 (1916).
206. J. March, *Advanced Organic Chemistry*, McGraw-Hill, New York, 1977.
207. D. D. Perrin, *Dissociation Constants of Organic Bases in Aqueous Solutions*, Butterworth, London, 1965.
208. D. D. Perrin, *Dissociation Constants of Organic Bases in Aqueous Solutions, Supplement 1972*, Butterworth, London, 1972.

209. D. D. Perrin, *Dissociation Constants of Inorganic Acids and Bases in Aqueous Solutions*, Butterworth, London, 1969.
210. E. P. Serjeant and B. Dempsey, *Dissociation Constants of Organic Acids in Aqueous Solutions*, Pergamon Press, Oxford, 1979.
211. R. M. Smith and A. E. Martell, *NIST Critically Selected Stability Constants of Metal Complexes Database, Version 5.0*, U.S. Department of Commerce, Gaithersburg, 1998.
212. R. M. Smith and A. E. Martell, *Critical Stability Constants*, Vol. 6, Plenum Press, New York, 1989.
213. R. M. Smith and A. E. Martell, *Critical Stability Constants*, Vol. 5, Plenum Press, New York, 1986.
214. I. Brandariz, F. Arce, X. L. Armesto, F. Penedo, and M. S. de Vicente, *Monatsh. Chem.* **124**, 249 (1993).
215. A. Albert and E. P. Serjeant, *Determination of Ionization Constants*, 2nd ed., Chapman and Hall, London, 1971.
216. T. Ishimitsu, S. Hirose, and H. Sakurai, *Talanta* **24**, 555 (1977).
217. D. Signer and K. A. Dill, *J. Phys. Chem.* **93**, 6737 (1989).
218. D. L. Hunston, *Anal. Biochem.* **63**, 99 (1975).
219. C. Contescu, J. Jaginello, and J. A. Schwarz, *Langmuir* **9**, 1754 (1993).
220. R. J. Sips, *J. Chem. Phys.* **16**, 490 (1948).
221. C. J. F. Bötcher and P. Bordewijk, *Theory of Electric Polarization*, 2nd ed., Vol. II, Elsevier Scientific, Amsterdam, 1978.
222. M. Borkovec and G. J. M. Koper, *Langmuir* **10**, 2863 (1994).
223. G. J. M. Koper and M. Borkovec, *J. Chem. Phys.* **104**, 4204 (1995).
224. S. W. Provencher, *Comput. Phys. Commun.* **27**, 213 (1982).
225. B. A. Dempsey and C. R. O'Melia, in *Aquatic and Terrestrial Humic Materials*, R. F. Christman and E. T. Gjissing (eds.), Ann Arbor Science, Ann Arbor, 1983.
226. M. Borkovec, U. Rusch, M. Černík, G. J. M. Koper, and J. C. Westall, *Colloids Surf. A* **107**, 285 (1996).
227. J. C. M. De Wit, W. H. Riemsdijk, M. M. Nederlof, D. G. Kinniburgh, and L. K. Koopal, *Anal. Chim. Acta* **232**, 189 (1990).
228. D. G. Kinniburgh, W. H. van Riemsdijk, L. K. Koopal, M. Borkovec, M. H. Benedetti, and M. J. Avena, *Colloids Surf. A* **151**, 147 (1999).
229. S. W. Karickhoff, V. K. McDaniel, C. Melton, A. N. Vellino, D. E. Nute, and L. Carriera, *Environ. Toxicol. Chem.* **10**, 1405 (1991).
230. I. D. Brown, *Structure and Bonding* **2**, 1 (1981).
231. A. Kossiakoff and D. Harker, *J. Am. Chem. Soc.* **60**, 2047 (1938).
232. J. E. Ricci, *J. Am. Chem. Soc.* **70**, 109 (1948).
233. W. L. Bleam, *J. Colloid Interface Sci.* **159**, 312 (1993).
234. I. D. Brown, *Phys. Chem. Minerals* **15**, 30 (1987).
235. J. Sauer and R. Ahlrichs, *J. Chem. Phys.* **93**, 2575 (1990).
236. P. Nagy, K. Novak, and G. Szasz, *J. Mol. Struct.* **201**, 257 (1989).
237. T. Brinck, J. S. Murray, and P. Politzer, *J. Org. Chem.* **56**, 5012 (1991).
238. J. R. Rustad, B. P. Hay, and J. W. Halley, *J. Chem. Phys.* **102**, 427 (1995).
239. K. H. Kim and Y. C. Martin, *J. Med. Chem.* **34**, 2056 (1991).
240. M. J. Potter, M. K. Gilson, and J. A. McCammon, *J. Am. Chem. Soc.* **116**, 10298 (1994).
241. F. H. Westheimer and J. G. Kirkwood, *J. Chem. Phys.* **6**, 513 (1938).
242. F. H. Westheimer and M. W. Shookhoff, *J. Am. Chem. Soc.* **61**, 555 (1939).
243. E. Rajasekaran, B. Jayaram, and B. Honig, *J. Am. Chem. Soc.* **116**, 8238 (1994).
244. B. Noszál and P. Sándor, *Anal. Chem.* **61**, 2631 (1989).
245. C. Tanford and J. D. Hauenstein, *J. Am. Chem. Soc.* **78**, 5287.
246. C. Tanford, J. D. Hauenstein, and D. G. Rands, *J. Am. Chem. Soc.* **77**, 6409 (1955).
247. F. C. Kokesh and F. H. Westheimer, *J. Am. Chem. Soc.* **93**, 7270 (1971).
248. K. Bartik, C. Redfield, and C. M. Dobson, *Biophys. J.* **66**, 1180 (1994).

249. M. Inoue, H. Yamada, T. Yasukochi, R. Kuroki, T. Miki, T. Horiuchi, and T. Imoto, *Biochemistry* **31**, 5545 (1992).
250. W. R. Forsyth, M. K. Gilson, J. Antosiewicz, O. R. Jaren, and A. D. Robertson, *Biochemistry* **37**, 8643 (1998).
251. M. E. Huque and H. J. Vogel, *J. Protein Chem.* **12**, 695 (1993).
252. S. D. Lewis, F. A. Johnson, and J. A. Shafer, *Biochemistry* **20**, 48 (1981).
253. J. D. Forman-Kay, G. M. Clore, and A. M. Gronenborn, *Biochemistry* **31**, 3442 (1992).
254. N. A. Wilson, E. Barbar, J. A. Fuchs, and C. Woodward, *Biochemistry* **34**, 8932 (1995).
255. M. F. Jeng, A. Holmgren, and H. J. Dyson, *Biochemistry* **34**, 10101 (1995).
256. J. Qin, G. M. Clore, and A. M. Gronenborn, *Biochemistry* **35**, 7 (1996).
257. L. P. McIntosh, G. Hand, P. E. Johnson, M. D. Joshi, M. Körner, L. A. Plesniak, L. Ziser, W. W. Wakarchuk, and S. G. Withers, *Biochemistry* **35**, 9958 (1996).
258. T. P. O'Connell and J. P. G. Malthouse, *Biochem. J.* **317**, 35 (1996).
259. M. Oliveberg, V. L. Arcus, and A. R. Fersht, *Biochemistry* **34**, 9424 (1995).
260. J. Antosiewicz, J. A. McCammon, and M. K. Gilson, *J. Mol. Biol.* **238**, 415 (1994).
261. A. H. Juffer, *Biochem. Cell. Biol.* **76**, 198 (1998).
262. M. K. Gilson, *Curr. Opinion Struct. Biol.* **5**, 216 (1995).
263. J. Antosiewicz, J. A. McCammon, and M. K. Gilson, *Biochemistry* **35**, 7819 (1996).
264. D. M. Burley, in *Phase Transitions and Critical Phenomena*, Vol. 2, C. Domb and M. S. Green (eds.), Academic Press, New York, 1975, p. 329.
265. M. K. Gilson, *Proteins* **15**, 266 (1993).
266. M. Fernández and P. Fromherz, *J. Phys. Chem.* **81**, 1755 (1977).
267. D. Bashford, M. Karplus, and G. W. Canters, *J. Mol. Biol.* **203**, 507 (1988).
268. L. Sandberg and O. Edholm, *Biophys. Chem.* **65**, 189 (1997).
269. P. Beroza and D. A. Case, *J. Phys. Chem.* **100**, 20156 (1996).
270. Y. Y. Sham, Z. T. Chu, and A. Warshel, *J. Phys. Chem.* **101**, 4458 (1997).
271. M. Schaefer, M. Sommer, and M. Karplus, *J. Phys. Chem. B* **101**, 1663 (1997).
272. A. H. Juffer, P. Argos, and H. J. Vogel, *J. Phys. Chem.* **101**, 7664 (1997).
273. A. Warshel and J. Åqvist, *Annu. Rev. Biophys. Biophys. Chem.* **20**, 267 (1991).
274. A. Warshel, S. T. Russel, and A. K. Churg, *Proc. Natl. Acad. Sci. U.S.A.* **81**, 4785 (1984).
275. T. Kesvatera, B. Jönsson, E. Thulin, and S. Linse, *Biochemistry* **33**, 14170 (1994).
276. G. King, F. S. Lee, and A. Warshel, *J. Chem. Phys.* **95**, 4366 (1991).
277. T. Simonson and D. Perahia, *J. Am. Chem. Soc.* **117**, 7987 (1995).
278. T. Simonson and C. L. Brooks III, *J. Am. Chem. Soc.* **118**, 8452 (1996).
279. G. Löffler, H. Schreiber, and O. Steinhauser, *J. Mol. Biol.* **270**, 520 (1997).
280. M. K. Gilson, A. Rashin, R. Fine, and B. Honig, *J. Mol. Biol.* **183**, 503 (1985).
281. M. Fushiki, B. Svensson, B. Jönsson, and C. E. Woodward, *Biopolymers* **31**, 1149 (1991).
282. J. Warwicker, *J. Mol. Biol.* **236**, 887 (1994).
283. A.-S. Yang and B. Honig, *J. Mol. Biol.* **231**, 459 (1993).
284. S. Kuramitsu and K. Hamaguchi, *J. Biochem.* **87**, 1215 (1980).
285. M. Rico, J. Santoro J., C. Gonzales, M. Bruix, and J. L. Neira, in *Structure, Mechanism and Function of Ribonucleases*, C. M. Cuchillo, R. de Llorens, M. V. Nogués, and X. Parés (eds.), Departament de Bioquímica i Biologia Molecular, Universita Autònoma di Barcelona, Bellaterra, Spain, 1991.
286. T. J. You and D. Bashford, *Biophys. J.* **69**, 1721 (1995).
287. B. R. Svensson, B. Jönsson, C. E. Woodward, and S. Linse, *Biochemistry* **30**, 5209 (1991).
288. M. Ullner, B. Jönsson, B. Söderberg, and C. Peterson, *J. Chem. Phys.* **105**, 2056 (1996).
289. F. L. B. daSilva, B. Svensson, T. Åkesson, and B. Jönsson, *J. Chem. Phys.* **109**, 2624 (1998).
290. M. K. Gilson and B. H. Honig, *Nature* **330**, 84 (1987).
291. C. M. Groeneveld, M. C. Ouwerling, C. Erkelens, and G. W. Canters, *J. Mol. Biol.* **200**, 189 (1988).

292. A. H. Juffer, E. F. F. Botta, B. A. M. van Keulen, A. van der Ploeg, and H. J. C. Berendsen, *J. Comput. Phys.* **97**, 144 (1991).
293. B. Svensson and B. Jönsson, *J. Comput. Chem.* **16**, 370 (1995).
294. C. N. Pace, D. V. Laurents, and J. A. Thompson, *Biochemistry* **29**, 2564 (1990).
295. T. E. Creighton, *Proteins*, Freeman, New York, 1984.
296. C. R. Matthews, *Annu. Rev. Biochem.* **62**, 653 (1993).
297. J. M. Dabora, J. G. Pelton, and S. Marqusee, *Biochemistry* **35**, 11951 (1996).
298. M. Nagasawa and A. Holtzer, *J. Am. Chem. Soc.* **86**, 531 (1964).
299. M. Mandel, *Eur. Polym. J.* **6**, 807 (1970).
300. T. Kitano, S. Kawaguchi, K. Ito, and A. Minakata, *Macromolecules* **20**, 1598 (1987).
301. T. L. Hill, *Arch. Biochem. Biophys.* **57**, 229 (1955).
302. M. Borkovec and G. J. M. Koper, *Macromolecules* **30**, 2158 (1997).
303. D. Soumpasis, *J. Chem. Phys.* **69**, 3190 (1978).
304. G. S. Manning, *J. Phys. Chem.* **85**, 870 (1981).
305. R. L. Cleland, J. L. Wang, and D. M. Detweiler, *Macromolecules* **17**, 634 (1984).
306. A. P. Sassi, S. Beltrán, H. H. Hooper, H. W. Blanch, J. Prausnitz, and R. A. Siegel, *J. Chem. Phys.* **97**, 8767 (1992).
307. N. Madras and A. D. Sokal, *J. Stat. Phys.* **50**, 109 (1988).
308. P. Flory, *Principles of Polymer Chemistry*, Cornell University Press, Ithaca, 1992.
309. M. Ullner, B. Jönsson, B. Söderberg, and C. Peterson, *J. Chem. Phys.* **104**, 3048 (1996).
310. P. G. de Gennes, P. Pincus, R. M. Velasco, and F. Brochard, *J. Phys.* **37**, 1461 (1976).
311. P. G. de Gennes, *Scaling Concepts in Polymer Physics*, Cornell University Press, Ithaca, 1979.
312. M. Borkovec and G. J. M. Koper, *Prog. Colloid Polym. Sci.* **109**, 142 (1998).
313. B. C. Bonekamp, *Colloids Surf.* **41**, 267 (1989).
314. J. B. Baliff, C. Lerf, and C. W. Schläpfer, *Chimia* **48**, 336 (1994).
315. R. L. Cleland, J. L. Wang, and D. M. Detweiler, *Macromolecules* **15**, 386 (1982).
316. S. Nilsson and W. Zhang, *Macromolecules* **23**, 5234 (1990).
317. D. S. Olander and A. Holtzer, *J. Am. Chem. Soc.* **90**, 4549 (1968).
318. Y. Muroga, K. Suzuki, Y. Kawaguchi, and M. Nagasawa, *Biopolymers* **11**, 137 (1972).
319. Y. Kawaguchi and M. Nagasawa, *J. Phys. Chem.* **73**, 4382 (1969).
320. M. Nagasawa, T. Murase, and K. Kondo, *J. Phys. Chem.* **69**, 4005 (1965).
321. A. Katchalsky, *J. Polym. Sci.* **7**, 393 (1951).
322. F. Oosawa, *Polyelectrolytes*, Marcel Dekker, New York, 1971.
323. J. de Groot, J. G. Hollander, and J. de Bleijser, *Macromolecules* **30**, 6884 (1997).
324. X. Wang, T. Komoto, I. Ando, and T. Otsu, *Macromol. Chem.* **189**, 1845 (1988).
325. J. B. Christensen, E. Tipping, D. G. Kinniburgh, C. Gron, and T. H. Christensen, *Environ. Sci. Technol.* **32**, 3346 (1998).
326. M. J. Avena, L. K. Koopal, and W. H. van Riemsdijk, *J. Colloid Interface Sci.* **217**, 37 (1999).
327. J. Westall, in *Geochemical Processes at Mineral Surfaces*, J. A. Davis and K. F. Hayes (eds.), American Chemical Society, Washington, DC, 1986.
328. J. Sonnefeld, *J. Colloid Interface Sci.* **155**, 191 (1993).
329. H. Ohshima, W. T. Healy, and L. R. White, *J. Colloid Interface Sci.* **90**, 17 (1982).
330. T. Hiemstra and W. H. van Riemsdijk, *Colloids Surf.* **59**, 7 (1991).
331. J. R. Bargar, S. N. Towle, G. E. Brown Jr., and G. A. Parks, *J. Colloid Interface Sci.* **185**, 473 (1997).
332. M. Schudel, H. Behrens, H. Holthoff, R. Kretschmar, and M. Borkovec, *J. Colloid Interface Sci.* **196**, 241 (1997).
333. T. Hiemstra, P. Venema, and W. H. van Riemsdijk, *J. Colloid Interface Sci.* **184**, 680 (1996).
334. J. A. Davis, R. A. James, and J. O. Leckie, *J. Colloid Interface Sci.* **63**, 480 (1978).
335. T. Hiemstra and W. H. van Riemsdijk, *J. Colloid Interface Sci.* **179**, 488 (1996).
336. R. Charmas and W. Piasecki, *Langmuir* **12**, 5458 (1996).

337. R. Charmas, W. Piasecki, and W. Rudzinski, *Langmuir* **11**, 3199 (1995).

338. S. Levine, *J. Colloid Interface Sci.* **37**, 619 (1971).

339. M. T. M. Koper, *J. Electroanal. Chem.* **450**, 189 (1998).

340. K. Binder and D. P. Landau, *Adv. Chem. Phys.* **76**, 91 (1989).

341. B. Dünweg, A. Michev, and P. A. Rikvold, *J. Chem. Phys.* **94**, 3958 (1991).

342. I. Jäger, *Surface Sci.* **254**, 300 (1991).

343. C. A. Barlow and J. R. Macdonald, *Adv. Electrochem. Electrochem. Eng.* **6**, 1 (1967).

344. D. A. Sverjensky, *Geochim. Cosmochim. Acta* **58**, 3133 (1994).

345. F. J. Hingston, A. M. Posner, and J. P. Quirk, *J. Soil Sci.* **23**, 177 (1972).

346. K. Pulver, P. W. Schindler, J. C. Westall, and R. Grauer, *J. Colloid Interface Sci.* **101**, 554 (1984).

347. C. Ludwig and P. W. Schindler, *J. Colloid Interface Sci.* **169**, 284 (1995).

348. J. Lützenkirchen, *Environ. Sci. Technol.* **32**, 3149 (1998).

349. I. H. Harding and T. W. Healy, *J. Colloid Interface Sci.* **107**, 382 (1985).

350. V. E. Shubin, I. V. Isakova, M. P. Sidorova, A. Y. Men'shikova, and T. G. Evseeva, *Kolloidn. Zh.* **52**, 935 (1990).

351. S. H. Behrens, D. I. Christl, R. Emmerzael, P. Schurtenberger, and M. Borkovec, *Langmuir* **16**, 2566 (2000).

352. A. S. Russel, P. J. Scales, C. S. Mangelsdorf, and L. R. White, *Langmuir* **11**, 1112 (1995).

353. V. E. Shubin, R. J. Hunter, and R. W. O'Brien, *J. Colloid Interface Sci.* **159**, 174 (1993).

354. M. R. Gittings and D. A. Saville, *J. Colloid Interface Sci.* **11**, 798 (1995).

355. M. Abe, A. Kuwabara, K. Yoshihara, K. Ogino, M. J. Kim, T. Kondo, and H. Ohshima, *J. Phys. Chem.* **98**, 2991 (1994).

356. T. D. Evans, J. R. Leal, and P. W. Arnold, *J. Electroanal. Chem.* **105**, 161 (1979).

357. D. A. Sverjensky and N. Sahai, *Geochim. Cosmochim. Acta* **60**, 3773 (1996).

358. P. Venema, T. Hiemstra, P. G. Weidler, and W. H. van Riemsdijk, *J. Colloid Interface Sci.* **198**, 282 (1998).

359. C. Contescu, A. Contescu, and J. A. Schwarz, *J. Phys. Chem.* **98**, 4327 (1994).

360. J. Lyklema, G. J. Fokkink, and A. de Keizer, *Prog. Colloid Polym. Sci.* **83**, 46 (1990).

361. A. de Keizer, L. G. J. Fokkink, and J. Lyklema, *Colloids Surf.* **49**, 149 (1990).

362. L. G. J. Fokkink, A. de Keizer, and J. Lyklema, *J. Colloid Interface Sci.* **127**, 116 (1989).

363. W. Rudzinski, R. Charmas, and S. Partyka, *Colloids Surf. A* **70**, 111 (1993).

364. J. R. Rustad, A. R. Felmy, and B. P. Hay, *Geochim. Cosmochim. Acta* **60**, 1563 (1996).

365. H. Kawakami and S. Yoshida, *J. Chem. Soc., Faraday Trans. 2* **80**, 921 (1984).

366. D. G. Hall, *Langmuir* **13**, 91 (1997).

367. M. Blesa, A. G. J. Maroto, and A. E. Regazzoni, *J. Colloid Interface Sci.* **140**, 287 (1990).

368. V. E. Heinrich, *Prog. Surf. Sci.* **14**, 175 (1983).

369. J. B. Peri and A. L. Hensley, *J. Phys. Chem.* **70**, 2926 (1968).

370. J. B. Peri, *J. Phys. Chem.* **69**, 220 (1965).

371. H. Knözinger and P. Ratnasamy, *Catal. Rev. Sci. Eng.* **17**, 31 (1978).

372. V. Barrón and J. Torrent, *J. Colloid Interface Sci.* **177**, 407 (1996).

373. R. H. Yoon, T. Salman, and G. Donnay, *J. Colloid Interface Sci.* **70**, 483 (1979).

374. R. J. Atkinson, A. M. Posner, and J. P. Quirk, *J. Phys. Chem.* **71**, 550 (1967).

375. L. Sigg and W. Stumm, *Colloids Surf.* **2**, 101 (1981).

376. W. A. Zeltner and M. A. Anderson, *Langmuir* **4**, 469 (1988).

377. P. G. Weidler, S. J. Hug, T. P. Wetche, and T. Hiemstra, *Geochim. Cosmochim. Acta* **62**, 3407 (1998).

378. G. A. Parks and P. L. de Bruyn, *J. Phys. Chem.* **66**, 967 (1962).

379. P. Hesleiter, D. Babic, N. Kallay, and E. Matijević, *Langmuir* **3**, 815 (1987).

380. N. H. G. Penners, L. K. Koopal, and J. Lyklema, *Colloids Surf.* **21**, 457 (1986).

381. A. W. M. Gibb and L. K. Koopal, *J. Colloid Interface Sci.* **134**, 122 (1990).

382. T. Hiemstra, H. Yong, and W. H. van Riemsdijk, *Langmuir* **15**, 5942 (1999).

383. M. J. G. Janssen and H. N. Stein, *J. Colloid Interface Sci.* **109**, 508 (1986).
384. M. J. G. Janssen and H. N. Stein, *J. Colloid Interface Sci.* **111**, 112 (1986).
385. Y. G. Bérubeé and P. L. de Bruyn, *J. Colloid Interface Sci.* **27**, 305 (1968).
386. R. Sprycha, *J. Colloid Interface Sci.* **102**, 173 (1984).
387. L. Bousse, N. F. De Rooij, and P. Bergfeld, *Surf. Sci.* **135**, 479 (1983).
388. T. Hiemstra, W. H. van Riemsdijk, and M. G. M. Bruggenwert, *Neth. J. Agric. Sci.* **35**, 281 (1987).
389. T. Hiemstra and W. H. van Riemsdijk, *Langmuir* **15**, 8045 (1999).
390. R. P. Abendroth, *J. Colloid Interface Sci.* **34**, 591.
391. G. Vigil, Z. Xu, S. Steinberg, and J. Israelachvili, *J. Colloid Interface Sci.* **165**, 367 (1996).
392. T. F. Tadros and J. Lyklema, *J. Electroanal. Chem.* **17**, 267 (1968).
393. B. N. J. Persson, *Surf. Sci.* **258**, 451 (1991).
394. K. Binder and A. P. Young, *Rev. Mod. Phys.* **58**, 801 (1986).
395. L. Onsager, *Phys. Rev.* **65**, 117 (1944).
396. X. N. Wu and F. Y. Wu, *Phys. Lett.* **144**, 123 (1990).
397. M. Suzuki, C. Kawabata, S. Ono, Y. Karaki, and M. Ikeda, *J. Phys. Soc. Jpn.* **29**, 837 (1970).
398. W. H. Press, S. A. Teukolsky, W. T. Vetterling, and B. P. Flannery, *Numerical Recipes*, Cambridge University Press, Cambridge, 1992.

3

Combined Application of Radiochemical and Electrochemical Methods for the Investigation of Solid/Liquid Interfaces

Kálmán Varga, Gábor Hirschberg, Pál Baradlai, and Melinda Nagy

1. Introduction

1.1. On the Significance of Radiotracer Methods in Sorption Studies

The entire array of charged species and oriented dipoles existing at the electrode/electrolyte interface, i.e., the structure of the electrical double layer, can have predominant effects on electrode processes. Several, now classic, books (see e.g., Refs. 1–3) have presented a description and analysis of the structure of the electrical double layer, emphasizing the central role of two types of adsorption (such as *specific* and *nonspecific*) in the understanding and mechanistic interpretation of various interfacial phenomena. Owing to the potential relevance of this field, comprehensive investigations of the sorption processes occurring at electrode/electrolyte (especially at metal/solution) heterogeneous systems have comprised a substantial part of electrochemical studies for many decades. Extended fundamental and applied research has been carried out on various, more or less closely related topics. In addition to an examination of the adsorption, electrosorption, electrocatalytic, etc., behavior of different metal (mainly noble metal) electrodes, studies into the kinetics and mechanisms of corrosion, corrosion inhibition, and radioactive contamination processes on metallic constructional materials of industrial importance have also entered the spotlight of scientific interest (see Refs. 3–8 and references cited therein).

An adequate explanation of a given interfacial phenomenon requires a thorough and wide-ranging examination of a metal/solution heterogeneous system. In order

---

Kálmán Varga, Gábor Hirschberg, Pál Baradlai, and Melinda Nagy • Department of Radiochemistry, University of Veszprém, Veszprém, Hungary.

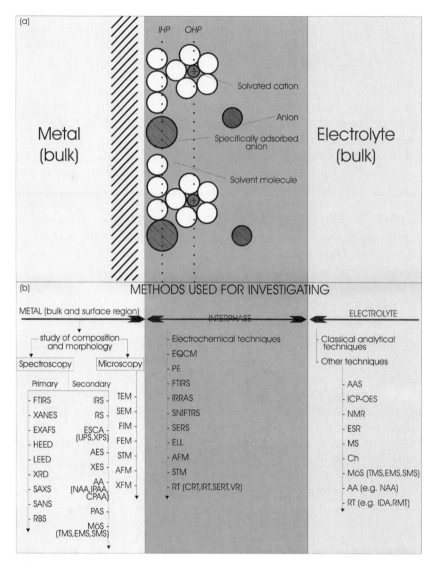

Figure 1. (a) A proposed model of the electrode–solution interface (double-layer region). Abbreviations: IHP, inner Helmholtz plane; OHP, outer Helmholtz plane. (b). A survey of available experimental techniques. Abbreviations: FTIRS, Fourier Transform Infrared Spectroscopy; XANES, X-ray Absorption Near Edge Structure; EXAFS, Extended X-ray Absorption Fine Structure; HEED, High Energy Electron Diffraction; LEED, Low Energy Electron Diffraction; XRD, X-ray Diffraction; SAXS, Small Angle X-ray Scattering; SANS, Small Angle Neutron Scattering; RBS, Rutherford Ion Backscattering Spectroscopy; IRS, Infrared Spectroscopy; RS, Raman Spectroscopy; ESCA, Electron Spectroscopy for Chemical Analysis; UPS, Ultraviolet Photoelectron Spectroscopy; XPS, X-ray Photoelectron Spectroscopy; AES, Auger Electron Spectroscopy; XES, X-ray Emission Spectroscopy; AA, Activation Analysis; NAA, Neutron Activation Analysis; IPAA, Instrumental Photon Activation Analysis; CPAA, Charged Particle Activation Analysis; PAS, Positron Annihilation Spectroscopy; MöS, Mössbauer

to gain a more complete insight into a problem, parallel or simultaneous applications of independent experimental techniques are generally needed, which yield information not only regarding the interface between the electrode and the solution, but also about both phases. Figure 1 gives a schematic presentation of the metal/solution double-layer structure (Fig. 1a) and a brief survey of the investigation methods (Fig. 1b). Figure 1b reveals that a wide variety of experimental techniques is available for the study of sorption phenomena and for the characterization of the structure and the state of the surface and interface. It may also be seen in this figure that besides the classical electrochemical, spectroscopic, optical, and microscopic techniques, application of radioactive tracers (RT) represents an important subgroup of methods used in such studies.

Although the radiotracer technique in electrochemical research has been utilized for seven decades, the wider spread of radioactive labeling methods has been employed only in the last 35 years. Owing to the intensive methodological and technical developments, *in situ* radiotracer methods have been elaborated which, coupled with traditional electrochemical techniques (e.g., voltammetry, chronoamperometry), enable us to study the surface properties of single- and polycrystalline metals. The *in situ* radiotracer methods have a number of advantages: (i) They provide *direct* (*and in situ*) information about the electrode processes and are extremely useful for the determination of the surface excess concentration (Γ) of adsorbed species as a function of various experimental parameters (time, potential, pH, solution concentration, temperature, etc.). (ii) They are *molecule specific* and ensure a relatively easy way for data processing and the calculation of Γ values by using the *straightforward relationship* between the surface counting rate and surface concentration. (iii) They have *high sensitivity* due to the fact that the latter is primarily limited by the type of measured radiations, molar activity of radiotracers, real surface area of the electrode, and counting efficiency of the nuclear detector. (iv) By using upgraded versions of the *in situ* radiotracer methods alternatively, *no limitation* on the electrode materials and their crystallographic orientations is imposed. (v) Numerous radioisotopes can be used as tracers, ensuring an extension of the applicability of the radiotracer technique for the investigation of substances

Figure 1. (Continued) Spectroscopy; TMS, Transmission Mössbauer Spectrometry; EMS, Emission Mössbauer Spectrometry; SMS, Scattering Mössbauer Spectrometry; TEM, Transmission Electron Microscopy; SEM, Scanning Electron Microscopy; FIM, Field Ion Microscopy; FEM, Field Emission Microscopy; STM, Scanning Tunneling Microscopy; AFM, Atomic Force Microscopy; XFM, X-ray Fluorescence Microscopy; EQCM, Electrochemical Quartz Crystal Microbalance; PE, Piezoelectric Technique; IRRAS, Infrared Reflection–Absorption Spectroscopy; SNIFTRS, Substractively Interfacial Fourier Transform Infrared Spectroscopy; SERS, Surface Enhanced Raman Spectroscopy; ELL, Ellipsometry; RT, Radiotracer Technique; CRT, Common Radiotracer Technique; IRT, Indirect Radiotracer Technique; SERT, Secondary Effect Radiotracer Technique; VR, Voltradiometry; AAS, Atomic Absorption Spectroscopy; ICP-OES, Inductively Coupled Plasma Optical Emission Spectroscopy; NMR, Nuclear Magnetic Resonance; ESR, Electron Spin Resonance; MS, Mass Spectroscopy; CH, Chromatography; IDA, Isotope Dilution Analysis; RMT, Radiometric Titration.

labeled with radioisotopes emitting *not only pure* β-, *but* γ- *(or X-rays) or* β-*and* γ-*radiations*, simultaneously. (We shall return to a more detailed discussion of the latter issue in Section 2.1.)

In the last 15 years several monographs[5,9-12] and review articles[13-18] have been published on this topic, covering in detail experimental techniques, important subfields, and research trends.

1.2. General Problems of Radioactive Contamination Studies

The corrosion/corrosion prevention and radioactive contamination/decontamination problems, appearing in connection with the safe operation of the cooling circuits of water-cooled nuclear reactors, perpetuate interest in the study of fundamental aspects of contamination and corrosion phenomena of constructional materials used in nuclear power stations. It is well known that primary circuits of water-cooled nuclear reactors may become contaminated with radioactive isotopes during their normal operation.[8] The most significant contributors to the radioactivity of corrosion products are ^{58}Co, ^{60}Co, ^{54}Mn, ^{51}Cr, and ^{59}Fe, as also in the case of Soviet-type pressurized water reactors (PWRs) ^{110m}Ag.[8,19] It is of further interest to note that several aggressive anions such as Cl^- and SO_4^{2-} accumulated from the secondary side coolant of PWRs (e.g., so-called hide-out at ´steam generators) may cause significant corrosion failure of various surfaces of special importance. Knowledge of the fundamental aspects of these processes may contribute to the development of more efficient corrosion protection, and primarily enables us to ascertain the best ways for the decontamination of structural materials.[8,19c-d,20]

Establishing the inevitable links between corrosion and contamination phenomena should lead to a new approach in handling problems of contamination/decontamination and corrosion/corrosion prevention. A schematic presentation of this approach is shown in Fig. 2, in which it is essential to emphasize the importance of the electrochemical aspects of the contamination and decontamination processes. An inspection of relevant literature data reveals that the significance of electrochemical aspects has not been taken into account in the evaluation of standardized investigation methods of contamination–decontamination for a long time. Recently, industrial practice has proven that a number of contamination issues cannot be satisfactorily interpreted, even semiempirically, if the adequate explanation of important electrochemical aspects is disregarded.

It is also seen in Fig. 2 that, in response to the need for a better understanding of the kinetics and mechanism of the above-mentioned processes, the first step is the application of laboratory measuring systems. (The next step required is the investigation of the phenomena in simulated model systems.) The methods used on the laboratory scale should be able to combine advantages of the experimental techniques of radiochemistry, electrochemistry, and corrosion science to explain the relevant problems. Therefore, one has to choose investigation methods from the potentially great number of possibilities compiled in Fig.1b that can answer most of the questions posed. Among these methods, the simultaneous use (coupling) of *in situ* radiotracer and electrochemical techniques is considered to be one of the most powerful tools.

The aim of the present article is to give a brief overview on the methodology and applicability of radiotracer methods to be utilized in an investigation of interfacial phenomena at solid/liquid heterogeneous systems. In addition, we shall highlight several significant aspects of recent progress in this field. An important objective of this chapter is to present upgraded versions of two *in situ* radiotracer methods used alternatively, and to demonstrate—by selected experimental results—their applicability for the extensive study of adsorption (electrosorption), contamination, and corrosion processes.

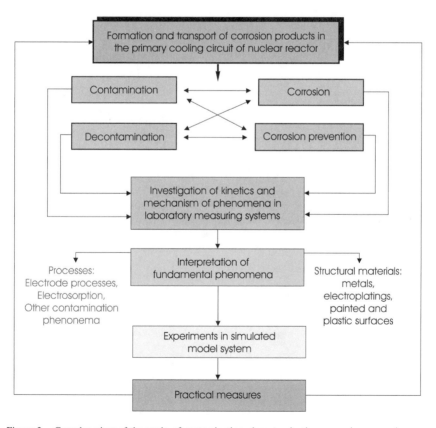

Figure 2.   Complex view of the study of contamination–decontamination, corrosion–corrosion prevention problems that occur in nuclear power plants.

2. An Overview of Radiotracer Methods

2.1. Methods Used for the Investigation of Interfacial Phenomena

The application of radiotracer methods has played a key role in studies of electrode/electrolyte interfaces for many decades. For instance, the very idea of the method worked out by Joliot[21] in 1930 is still used in the modern *in situ* "foil" radiolabeling technique. Despite the difficulties arising from the application of radioactive materials, it should be noted that there appears to be continuous progress in the field of radioelectrochemistry or even a "renaissance" of radioactive labeling.

In accordance with considerations outlined elsewhere,[9,15] Fig. 3 offers a classification of radiotracer methods which can be employed to investigate a large number of interfacial phenomena belonging to more or less interrelated branches

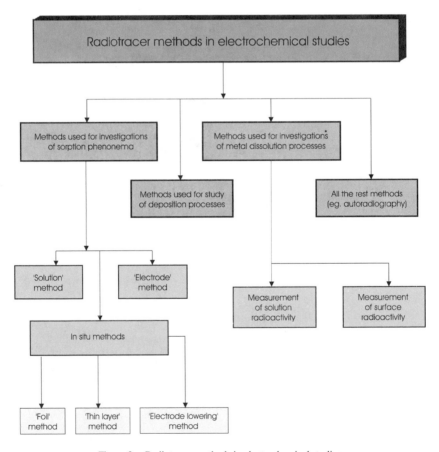

Figure 3.   Radiotracer methods in electrochemical studies.

of electrochemistry. This section deals in detail with radioactive labeling techniques used for studying *sorption processes*, because they are of great importance from both methodological and application aspects. The majority of these methods can be directly applied to the investigation of dissolution and deposition of metals and enables us to study the mechanism of other phenomena of applied research such as corrosion and corrosion inhibition, heterogeneous catalytic processes, contamination and decontamination, etc.

For the purpose of this review a brief survey on the history and methodology of radiotracer sorption studies will be presented first.

2.1.1. Experimental Techniques: A Historical Survey

The first significant progress in the application of radiotracer methods for the study of electrosorption phenomena occurred some forty years ago due to the pioneering work of groups headed by Hackermann,[22] Schwabe,[23] Bockris,[3,24] and Balashova and Kazarinov.[9,25,26] In the early 70s two more laboratories joined (Horányi,[13,14,17,27] and Sobkowski and Wieckowski[28]) and achieved important results in improving investigation methods, and widening their field of application by studying misinterpreted or unexplored electrosorption processes. During the last 15 years further centers were formed in the USA (Wieckowski and co-workers in Urbana,[10–12,29] Schlenoff and co-workers in Tallahassee[30]), France (Marcus and co-workers[31]), and Hungary (Varga and co-workers[18,32–35] in Veszprém).

In considering the methodology of experimental techniques applied for sorption studies, *ex situ* ("electrode" and "solution") and *in situ* methods could be distinguished. As shown in Fig. 3, the *in situ* (i.e., the most up-to-date) techniques can be divided into three main subgroups, namely "foil," "thin layer," and "electrode lowering" methods. Further subdivision could be introduced on the basis of:

(i) the studied phenomena: simultaneous, competitive, induced or enhanced adsorption of anions and cations, or electrosorption (adsorption or chemisorption) of organic species, etc.;

(ii) the use of information conveyed by radiation measurements (direct or indirect radioactive labeling);

(iii) the quality, state, and surface structure of the adsorbent, such as smooth surfaces of well-defined single- and polycrystals, or rough surfaces or powdered metals;

(iv) measured radiation: primary radiation [β or low-energy γ (X-ray)], or secondary radiation (backscattered β, induced X-ray, or electrons).[5,13,17]

In the application of radioactive tracers to sorption studies, the oldest but still popular method is measuring the surface activity of samples after being in contact with the solution containing the radiolabeled substances for a given period of time. The surface excess concentration of the studied species can be determined from the net counting rate originating from the surface, the net counting rate due to the solution layer wetting the adsorbent (if any), the molar (or specific) activity of the labeled species, the real surface area of the adsorbent, and the proportionality factor between the activity of adsorbed species and the measured counting rate. The methods of this group are called "measuring after immersion" (by Reinhard[15]) or "electrode" methods (by Kazarinov[9]).

The "electrode" method is widely used due to its methodological and technical simplicity. Further advantages are its high sensitivity (it is mainly defined by the specific activity of the adsorbate) and that any kind (α, β, and γ) of radiation emitted by radiotracers can be used for nuclear detection. However, there are fundamental shortcomings which cause obstacles in its application. The adsorption measurements are carried out with the interruption of the sorption process, and several important parameters (such as electrode potential or state of the surface) cannot be controlled after removing the electrode from the solution. It is also difficult to handle the wetting liquid film. If the amount of adsorbate in the solution layer wetting the surface is not negligible as compared to the adsorbed amount, the net counting rate originating from this layer has to be taken into account with a correction factor or one should remove the wetting solution layer from the surface. In the first case the error in the determination of the correction factor, and in the second case the decrease in the adsorbed amount due to rinsing, diminishes the accuracy of the method, which usually exceeds $\pm 10\%$.

In view of the mentioned disadvantages, the "electrode" method yields more or less acceptable results only in studies of electrosorption of organic molecules, strong chemisorption of anions, and deposition of cations. Despite its limitations, this method was formerly frequently used by numerous research groups all over the world (see Ref. 9 and references cited therein). (It is noteworthy that the contribution of several laboratories to the development and application of the "electrode" method in the past few decades seems to be considerable[36–38].)

Radiotracer techniques that are based on the measurement of the concentration change of the solution in contact with the adsorbent represent an important group in the study of sorption phenomena. They are called "solution" methods.

By applying radiolabeled substances, a change in the radioactivity of the solution yields information about the decreasing solution concentration of investigated species and, consequently, the amount of such species accumulated at the surface. If the total volume of the solution, the real surface area of the adsorbent, the solution concentration, and the specific activity of the adsorbate are known, then by measuring the net counting rate of a fixed volume of solution before and at given intervals during the sorption process, the surface excess concentration can easily be calculated.

The simplest "solution" method can be applied together with the "electrode" method, i.e., the radioactive concentration of the solution is determined before and after the immersion of the sample. In the intermittent (step-by-step) method, aliquot volumes of the solution are taken and measured. Finally, in the continuous "solution" method the radioactive concentration of the solution is measured continuously throughout the sorption process.[9,37b–d]

Since the "solution" method is an indirect means to determine the surface excess concentration, its application is strongly limited by the adsorptive capacity of the adsorbent. The error in the determination of Γ is usually greater than 15%.

The *in situ* techniques (Fig. 3) constitute a third group of radiolabeling methods which enable the continuous study of surface excess concentration (radioactivity) of adsorbed species on the electrode surface immersed in a solution. All these methods utilize the "thin layer" principle put forward by Aniansson[39] and are considered to be the most advanced technical solutions for the investigation of interfacial phenomena in solid/liquid heterogeneous systems from both a radiochemical and an electrochemical standpoint.

The very basis of the "thin layer" principle is that application of radioisotopes emitting β-particles or γ- (X-ray) photons of energy not higher than 20 keV allows one to measure a significant intensity surplus related to the sorption excess on the electrode surface above a small solution background. Methodologically, this means that owing to the absorption and self-absorption of β- and low-energy ($E_\gamma \leq 20$ keV) γ-particles the detector can "see" radiation coming from a thin solution layer only, making the intensity of the solution background small. When labeled species accumulate on the surface in the course of a sorption process, they arrive in the "visual field" of the detector. Therefore, an intensity excess can be measured which is proportional to the surface excess concentration of the species studied. [It should be noted that in case of labeling with γ-emitting isotopes of medium energy, the favorable ratio of sorption excess/solution background intensity can only be achieved if the background intensity comes from a thin (*ca* 0.5 mm), homogeneous solution layer. The latter is created by suitable shielding in a specially designed cell, while the radiolabeled species can accumulate on the surface from the total volume of the circulating solution.]

In situ radiotracer methods can be divided into three subgroups according to the technical realization of the "thin layer" principle (Fig. 3).

The so-called "foil" methods constitute the oldest subgroup of *in situ* radiotracer techniques. The original principle underlying these methods, proposed by Frederick Joliot,[21] involves a thin foil covered with an adsorbent (or an adsorbent foil), which forms part of the measuring cell and separates the detector from the solution containing the labeled adsorbate. The detector measures the sum of the intensity of the radiation emitted by the labeled adsorbate on the adsorbent foil and the intensity originating from a thin layer of bulk solution due to the self-absorption of β-particles.

Advanced designs of this radiotracer technique for the investigation of adsorption phenomena on metal surfaces were described by Schwabe and Weissmantel[23b] and Bockris and Bolmgren[24] in the early 60s. At the beginning of the 70s, a more improved version of the "foil" method was worked out by Horányi.[13,17,27] The schematic presentation of his radioelectrochemical instrument is shown in Fig. 4, which consists of a traditional three-compartment electrochemical cell. A gold-

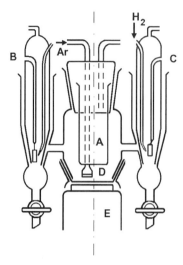

Figure 4. Measuring setup for adsorption studies by a version of the "foil" method[27]: (A) central compartment, (B) counter electrode, (C) reference electrode, (D) main electrode, (E) scintillation detector.

plated plastic foil covered with the adsorbent forms the main electrode, which is the bottom of the cell, while the scintillation detector is placed beneath the main electrode. In this setup the detector and working electrode are separated, and the electrode can be readily replaced, providing a simple and rapid way to carry out experiments. During the past 25 years this latter version of "foil" methods has been a powerful tool with which to study the adsorption, electrosorption, and electrocatalytic behaviors of several inorganic and organic species. A brief summary of the most important results obtained by Horányi and co-workers can be found elsewhere.[5,13,14,17] In 1972, working independently, a similar cell construction was described by Sobkowski et al[28a] in Poland. In 1975, Wieckowski[28b] modified the Sobkowski cell: a glass scintillator was used instead of the plastic one. A layer of metal deposited onto one face of the glass scintillation disk formed the working electrode. Wieckowski's measuring cell is illustrated schematically in Fig. 5.

In accordance with the basic principles outlined above, the "foil" method was originally developed for labeling with radioisotopes emitting low or medium energy β-radiations. (As we shall discuss later on, it is possible to use isotopes emitting γ-photons, or characteristic X-rays, of energy lower than 20 keV as well.) The lack of suitable β-emitting isotopes (13 pure β-emitting isotopes are available only) strongly reduced the applicability of the "foil" method and inspired researchers to find methods for the *in situ* investigation of species labeled with γ-emitting isotopes.

The number of radioisotopes suitable as tracers could be increased by using the so-called "thin layer" method, based on the "thin layer" theory for labeling with isotopes emitting γ- or high energy β-radiations. The fundamental idea and the first cell design were published by Kafalas and Gatos in 1958,[40] according

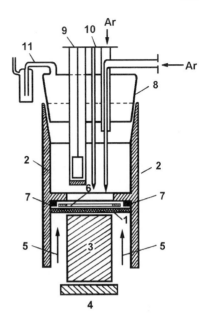

Figure 5. The instrument constructed by Wieck-owski[28b]: (1) TRTE (tritium radiation transparent electrode), (2) Teflon frame, (3) light pipe, (4) photomultiplier (EMI 9514 S), (5) tightening, (6) Teflon gasket, (7) rubber gasket, (8) Kel-F sealed glass (Pyrex) ground joint, (9) counter electrode (closed by fritted disk), (10) hydrogen reference electrode (Luggin capillary), (11) argon bubbler, inlets of argon.

to which the solution of the labeled adsorbate circulates continuously through the cell in a thin layer (*ca* 0.5 mm) between the electrode and the detector. Using an appropriate cell design and shielding (Fig. 6), the detector measures the intensity of the radiation emitted by species being accumulated on the smooth surface of the electrode (*ca* 1 cm^2) and the intensity coming from a small volume (*ca* 0.04 cm^3) of solution (solution background). The advantages of this technique include the use of a host of available radioactive tracers, and the ability to study smooth, well-defined electrode surfaces in solutions of low concentrations ($\leq 10^{-6}$ mol dm^{-3}). The disadvantages of the "thin layer" method include a complicated measuring system, the radioactive contamination of the elements of the circulation loop, difficulties in the measurement and control of the electrode potential, and finally, insufficient sensitivity toward the sorption of weakly bonded species. For these reasons the "thin layer" method has not been

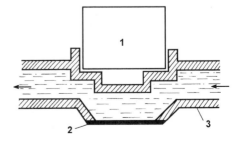

Figure 6. Cell for adsorption measurement by the "thin layer" method: (1) radiation counter, (2) metal specimen (working electrode), (3) glass cell body.[40]

applied widely for the study of electrosorption phenomena during the past four decades. However, several efforts to upgrade this technique for the investigation of various adsorption processes have been described.[9,23c,41]

The third subgroup of radiotracer procedures utilizing the "thin layer" principle consists of the so-called "electrode lowering" method. This technique was established in 1966 and later refined by Kazarinov *et al.*[9,26] In the measuring cell illustrated in Fig. 7, the intensity of radiation is detected alternatively throughout

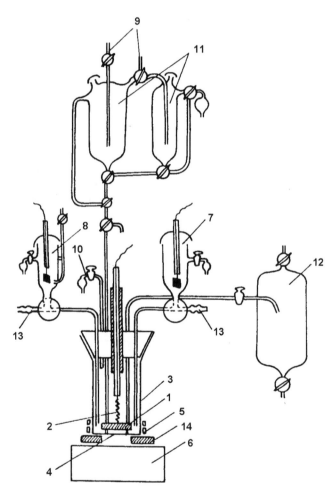

Figure 7. Cell for adsorption measurement by the "electrode lowering" method: (1) working electrode, (2) flexible contact, (3) cell body, (4) membrane, (5) membrane mounting, (6) radiation counter, (7) auxiliary electrode, (8) reference electrode, (9) gas inlet, (10) gas outlet, (11) vessel for preparation of the solution, (12) solution sink, (13) connecting tubes, (14) shielding.[26]

the sorption process. To measure the intensity from the labeled substances adsorbed on the electrode surface, the disk-shaped electrode is lowered to a thin membrane serving as the bottom of the central compartment of an electrochemical cell, as well as the window of a gas-flow detector. Thus, the electrode disk shields the detector against the radiation from the bulk solution, i.e., only the sum of the net adsorbate counting rate and the counting rate originating from a thin solution layer (*ca* 10 μm) beneath the lowered electrode is measured. Positioning the electrode at some distance over the range of the β-radiation of the radioisotope used for labeling, the counter measures the β-particles originating from the bulk solution while the sorption process proceeds. If several parameters, such as the absorption coefficient for β-radiation, the real surface area of the adsorbent, molar activity, and solution concentration of labeled species, are known, the Γ values can be determined from the counting rates measured with the electrode in the two positions (raised and lowered) .

The "electrode lowering" technique attempts to unify the advantages of both the "foil" and "thin layer" methods. According to the constructors, it is suitable for investigating sorption phenomena on smooth and microrough surfaces using both β- and γ-emitting isotopes for labeling. However, the present authors have experienced difficulties in applying the original version of this method, especially in the case of γ-emitting isotopes. Some aspects of the methodology need to be reconsidered and its sensitivity improved. In spite of these drawbacks, both the original and refined versions of the "electrode lowering" method are still in use in different laboratories (see references cited in Refs. 9–17 and 42, and Section 2.1.2).

2.1.2. Recent Progress

The most important developments in radiotracer studies of sorption phenomena since the middle 80s can be summarized as follows[5,10–12,17,18]:
   (1) Progress in the use of radiotracer techniques, such as
      (a) extension, completion, and generalization of the basic principles;
      (b) amplification of the host of isotopes (and radiations) used for the sorption experiments;
      (c) simultaneous use (coupling) of the radiotracer techniques with electrochemical methods;
      (d) demand for combined application of radiotracer methods with surface analytical (XPS, AES, MöS, FTIRS, SERS) and other techniques (ICP-OES, etc.).
   (2) Broadening the application field, and the occurrence of special areas of application.
In the case of the "foil" method, special designs of radioelectrochemical cells, and a significant improvement in their methodology and applicability have been observed.

The extension and generalization of the basic principles of the "foil" method apply to pure β- and γ- (X-ray) emitting isotopes,[32, 33, 43] to isotopes emitting both β- and γ-particles,[32,33,44–47] as well as to the measurement of secondary radiation [backscattered β-radiations,[48] X-rays (characteristic X-rays or/and bremsstrahlung) induced by β-radiation,[49] and electrons induced by γ- (X-ray) radiations[47,50]]. For the first time in radiotracer studies, energy spectra of primary and secondary radiation originating from the radioelectrochemical cell have been analyzed, providing important information on the following topics:

(1) the role of changes (various deformations) in the energy spectrum of primary β-radiation emitted by radiolabeled species during sorption phenomena[47,50b,51];

(2) the potential for the application of low energy ($E < 15$ keV) γ- (X-ray) photons in the determination of Γ values[43];

(3) the possibilities for the detection of β- and γ-emitting isotopes ([110m]Ag, ^{60}Co, etc.)[32,33,47,50b];

(4) the simultaneous use (coupling) of electrochemical methods and radiotracer techniques based on the measurement of secondary radiations.[48, 49]

New cell constructions were developed for the study of ion exchange processes,[30] as well as sorption and corrosion behaviors of poly- and single crystalline metal electrodes.[31] An upgraded version of Horányi's[13,27] cell was applied for *in situ* investigations of metals, paintings, and plastic coatings of industrial importance (Fig. 8).[18,32–34] In the radioelectrochemical cell shown in Fig. 8 a fourth compartment (F) is attached to the main compartment (A) for measurement of pH by a combined glass electrode. The auxiliary electrode (C) is of a noble metal (gold or platinum), and the reference electrode (B) is the reference hydrogen electrode (RHE). The bottom of the main compartment is a thin plastic (PVC or polyethylene) foil (of thickness 13–15 μm). A gold conductive layer (0.3–0.8 μm; 1–1.5 mg cm^{-2}), deposited on the foil by Ar$^+$ ion sputtering, holds the main electrode (D) of the cell.

The formation of the main electrode depends on the material to be studied:

(A) *In the case of metals there are three possibilities for the adsorbent formation, as illustrated by the model of the main compartment in Fig. 8, as follows:*

   (i) a structured metal layer formed by electrodeposition in order to obtain electrodes with high real surface area (e.g., platinized platinum[13, 27];

   (ii) thin metal layers (of thickness 0.3–1.0 μm) formed by Ar$^+$ ion sputtering on a plastic foil (e.g., gold[34,45,47,52–54] or aluminum[55];

   (iii) layers obtained by sedimentation of powdered metal samples (e.g., stainless steel[52,56–59] or low-alloyed steel[60]).

(B) *In the case of other inorganic or organic materials the main electrode can be:*

   (i) a plastic foil forming the bottom of the cell (e.g., polyethylene[33,44,46]);

   (ii) a polymer film on the gold-plated plastic foil (e.g., polymer film electrode[14,17]);

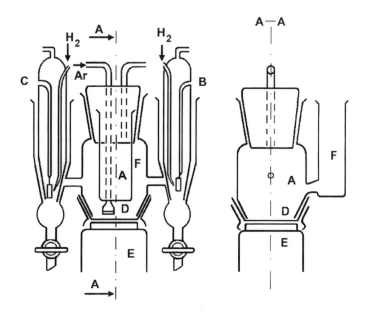

MEASURING CELL

A - main compartment   B - reference electrode   C - auxiliary electrode
D - main electrode   E - detector   F - pH measurement

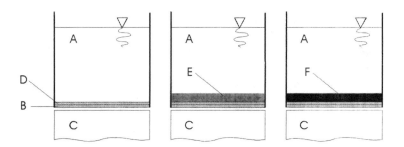

THE MODEL OF THE MAIN COMPARTMENT FOR DIFFERENT TYPES
OF ADSORBENT FORMATIONS

A - solution   B - foil   C - detector
D - gold layer   E - steel powder   F - coating

Figure 8. Measuring setup for *in situ* investigations of metals, paintings, and plastic coatings by an upgraded version of the "foil" method.[34]

(iii) a protective coating created on the gold-plated plastic foil (e.g., epoxide resins[32,61]).

It should be noted that the study of industrial metals by the "foil" method is usually carried out with powdered samples. The use of powdered adsorbents of high real surface areas increases the sensitivity of the method without substantial modification of the original properties of the sample surface. With this kind of main electrode it is possible to detect species at even less than 0.01 monolayer coverage (with respect to the geometric surface area). For instance, the detection limit of the "foil" method for some ions important in corrosion (e.g., Cl⁻ ions labeled with ^{36}Cl) or for inhibitor molecules (e.g., ^{14}C-labeled organic substances, or CrO_4^{2-} labeled with ^{51}Cr) is substantially higher than that of the "electrode lowering" technique.

As regards the methodological aspect, we should emphasize the dynamic "*voltradiometric*" method worked out by Horányi, which combines cyclic voltammetry with the radiotracer "foil" technique.[14,17,45] During measurements in a cell designed by Horányi, cyclic voltammetric and cyclic voltradiometric curves are recorded simultaneously to give information about possible links between redox processes and the chemical nature of the adsorbed or desorbed species.

In several cases direct radiotracer methods cannot be applied owing to some technical restrictions, such as (i) no radioisotopes emitting suitable radiation are available for labeling, (ii) preparation of labeled species is too difficult (and/or expensive) and their applicability is strongly limited, (iii) limitation in the available concentration range, and (iv) no possibility to distinguish the adsorption of labeled compound from that of a product formed from it. In such instances, the species to be studied can be examined by the *indirect radiotracer method*[5,14,17] in which we follow the adsorption of a simple, well-known molecule (a suitably selected labeled indicator species) that has some relation with the adsorption of the studied species. The latter could be based on competitive, enhanced, or induced adsorption.

Since 1987, significant progress in the methodology and applicability of the "electrode lowering" method has been made by Wieckowski and co-workers[10–12, 16,29] with the so-called *in situ* "*thin gap*" radiotracer technique. The instrument reported by Krauskopf and Wieckowski[29] is shown Fig. 9. The most important novelty of this technique is the use of smooth electrodes and a glass scintillator polished to optical quality, making it possible to reduce the gap distance between the electrode and scintillator to a value as low as 1–2 μm. As the application of polished surfaces improves the sensitivity of "electrode lowering" method by one order of magnitude, it is evident that method worked out by Wieckowski *et al.* will play an important role in the interpretation of fundamental phenomena at well-defined electrode surfaces. Published experimental results[10–12,16,29] reveal that the "thin gap" method is extremely useful for the study of sorption properties of well-defined, smooth surfaces of single crystals and polycrystalline electrodes.

There are possibilities for the further development of the methodology and technical solutions for the "thin gap" method,[34] but the expressions used for quantitative evaluation of data will have to be refined in some aspects.[35] The main

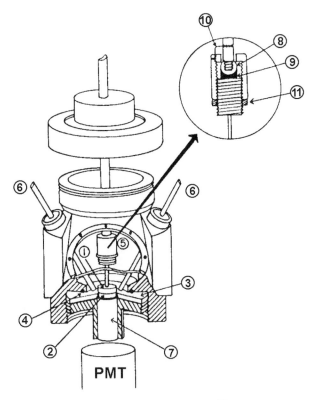

Figure 9. Diagram of the cell constructed by Krauskopf *et al.*[(29)]: (1) platinum electrode, (2) glass scintillator, (3) Macor ceramic disk cell bottom, (4) Teflon O-ring, (5) flexible elbow (see inset), (6) cell ports (six), (7) light pipe, (8) stainless steel sphere, (9) concave Teflon spacer, (10) platinum wire for electrical contact, (11) lock nut.

components of an upgraded version of the "thin gap" technique[(34–35)] is depicted schematically in Fig. 10A–C, while a photograph of the entire measuring instrument is shown in Fig. 10D. The central part of the measuring setup is a glass–Teflon–austenitic stainless steel measuring cell (Fig. 10A,B), which is designed for simultaneous radiochemical and electrochemical measurements and is equipped with a reference hydrogen electrode and an auxiliary electrode of a noble metal (gold or platinum). Li-glass scintillators of thickness varying between 0.2–4 mm are mounted into the ceramic bottom of the cell and polished to optical quality. For reference and auxiliary experiments, a 0.1–5-mm-thick β-plastic scintillator and silicon semiconductor detector can also be used. With the present design of the "thin gap" method any species can be used, if labeled with isotopes that emit pure β-radiation of maximum energy ($E_{\beta max}$) higher than 18 keV ($H_2^{35}SO_4$, $H^{36}Cl$, organic and inorganic inhibitors labeled with ^{14}C, ^{35}S, and ^{32}P). Furthermore, the methodological developments[(35c,d)] make it possible to detect also β- and γ- (e.g.,

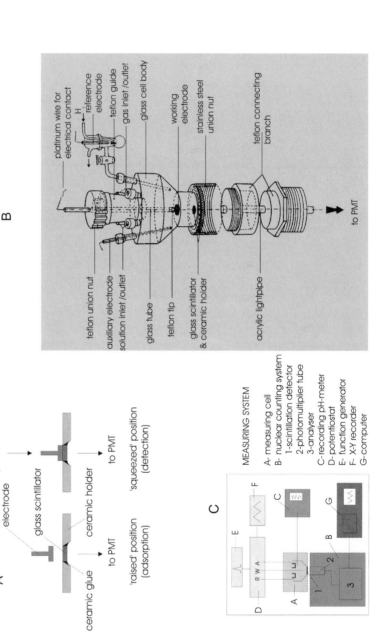

A

pressure

electrode

glass scintillator

ceramic holder

to PMT

'raised' position
(adsorption)

to PMT

'squeezed' position
(detection)

ceramic glue

B

platinum wire for
electrical contact

H

reference
electrode

teflon guide

gas inlet /outlet

glass cell body

working
electrode

stainless steel
union nut

teflon connecting
branch

to PMT

teflon union nut

auxiliary electrode

solution inlet /outlet

glass tube

teflon tip

glass scintillator
& ceramic holder

acrylic lightpipe

C

MEASURING SYSTEM

A- measuring cell
B- nuclear counting system
1-scintillation detector
2-photomultiplier tube
3-analyser
C-recording pH-meter
D-potentiostat
E- function generator
F- X-Y recorder
G-computer

D

E

F

R W A

C

A

1

2

B

3

G

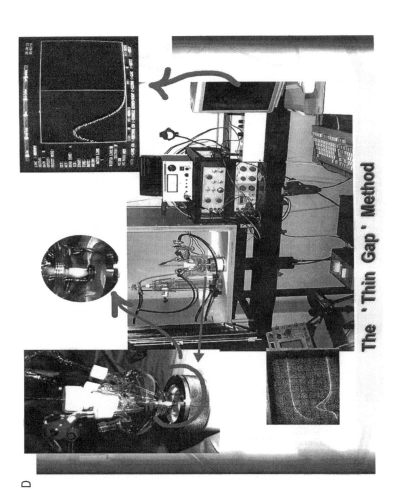

Figure 10. An upgraded version of the "thin gap" method[34,35]: (A) schematic of the experiments; (B) simplified view of the radioelectrochemical cell; (C) measuring system; (D) photograph of the whole measuring instrument.

^{110m}Ag or ^{60}Co), as well as low energy ($E < 20$ keV) γ- (X-ray) (e.g. ^{51}Cr or ^{55}Fe) emitting radioisotopes. A nuclear data processing system based on dual- and multichannel (8k) pulse-height analyzers and auxiliary electrochemical devices (Fig. 10C,D) enables one to characterize the surface structure and state via sorption processes.

Finally, the research trends and fields of application of radiotracer methods in which significant results have recently been achieved are as follows[5,10–12,17]:

(i) Deeper insights in fundamental issues with regard to interfacial phenomena in electrode/electrolyte systems have been gained by making use of:

- well-defined surfaces of single crystals and polycrystalline metals,
- compact electrodes with high real surface area or mainly rough surfaces,
- powdered adsorbents.

(ii) Investigations of induced and enhanced adsorption phenomena; studies of modified electrode surfaces.

(iii) Studies of sorption phenomena on various constructional materials of industrial importance (iron, alloys, aluminum, plastics, etc.) in connection with corrosion, corrosion inhibition, and contamination processes.

(iv) Investigation of the properties of polymer film electrodes.

2.2. Methods Used for the Investigation of Radioactive Contamination–Decontamination of Constructional Materials

Due to an increase in the number of nuclear power stations and other institutions (companies) using or processing radionuclides, radioactive contamination has become a problem of practical importance. In the nuclear industry, the vast majority of contamination takes place on the surface of constructional materials (metals, plastics, painted surfaces, and galvanic coatings) in contact with the radioactive solution. Knowledge of the kinetics and the mechanism of the accumulation of various radioactive contaminants is of essential importance not only for environmental protection and safety, but also for the normal operation of nuclear reactors.

The contamination–decontamination phenomena occurring at the solid/liquid interfaces have been studied extensively for decades and the number of relevant publications is enormous. With the exception of a few papers dealing with all aspects of contamination–decontamination phenomena,[8] most of the publications report new findings on industrial contamination–decontamination problems or decontamination methods. In the qualification of the contamination–decontamination behavior of surfaces of industrial importance, standardized investigation methods have been playing a predominant role. Twelve standardized procedures, laboratory and pilot plant methods, as well as results of the investigation of several radioactive contaminants are summarized elsewhere.[52]

Figure 11 is compiled from available literature data and gives an overview on the methods used to study radioactive contamination–decontamination processes in solid/liquid heterogeneous systems.

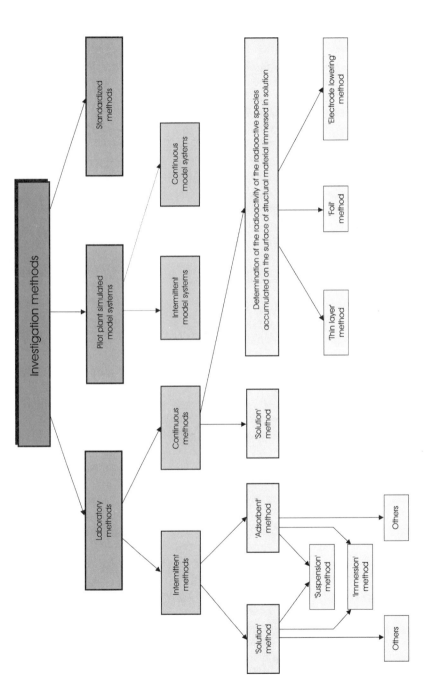

Figure 11. Investigation methods used to study radioactive contamination–decontamination processes.

3. Selected Results

It is well known[1-3] that two types of ion adsorption (such as specific and nonspecific) at electrified interphases should be distinguished. Although the specific adsorption processes could be monitored by a number of independent experimental methods (see Fig. 1), evaluation of the surface excess concentration of specifically adsorbed ions sometimes requires troublesome experiments.

The *in situ* radiotracer techniques described earlier offer a unique possibility to investigate the specific sorption of various ions in the presence of a great excess of other ions (supporting electrolyte) by labeling them. In this section, recent findings obtained in our laboratory concerning three selected areas of specific sorption phenomena are reviewed in order to demonstrate the potentialities of the combined application of radiochemical and electrochemical methods.

3.1. Adsorption of Anions and Cations on Polycrystalline Gold

3.1.1. Comparative Study of Specific Adsorption of Cl^-, HSO_4^-/SO_4^{2-} and HSO_3^-/SO_3^{2-} Ions

The main part of the following radiotracer and voltammetric experiments[53] was carried out by using a version of the "foil" method described in Section 2.1.2 (see Fig. 8). In these studies, the geometric surface area of the gold electrode layer was 4.9 cm^2, and the roughness factor varied between 2.6 and 6.2. In addition, some adsorption measurements were performed on a disk-shaped polycrystalline gold electrode (1.00 cm in diameter and 0.30 cm in height) in solutions containing H_2SO_4 labeled with ^{35}S by a version of the "thin gap" method presented in Fig. 10. The geometric surface area of the compact Au electrode was 0.78 cm^2, and the roughness factor of the surface varied between 2.7 and 3.5.

The dependence of the adsorption of anions on the potential was measured in two ("continuous" and "interrupted") polarization modes. The potential of the working electrode was continuously shifted in the "continuous" adsorption studies, while several voltammetric scans (3–5) at a sweep rate of 100 mV s^{-1} over the studied potential range were applied between two consecutive adsorption measurements in the "interrupted" polarization mode, in order to renew the electrode surface and to remove trace impurities. [Potential values quoted in this section are given on the reference hydrogen electrode (RHE) scale.]

Figures 12a and 13a show the voltammograms of a polycrystalline gold electrode in 1 mol dm^{-3} and 0.1 mol dm^{-3} $HClO_4$ supporting electrolyte, respectively, before (solid line) and after (dashed line) the addition of 2×10^{-4} mol dm^{-3} H_2SO_4 labeled with ^{35}S. The voltammetric behavior of the gold electrode is basically identical in both $HClO_4$ solutions and correlates well with the literature data)[62] demonstrating that the purity of the polycrystalline gold/$HClO_4$ system fulfills the requirements of electrochemical practice. A comparison of the dashed and solid lines in Figs. 12a and 13a shows that the presence of HSO_4^-/SO_4^{2-} ions in the electrolyte causes the onset of electrooxidation of the gold surface to shift toward more positive potential values by *ca* 80 mV, which is indicative of site-blocking by surface interacting anions.

In spite of the similarity of the voltammograms in Figs. 12a and 13a, the Γ vs. *E* curves of Figs. 12b and 13b indicate that the adsorption of HSO_4^-/SO_4^{2-} ions on the gold surface is highly dependent upon the experimental conditions and the applied polarization modes. As a consequence of the dominant effects of the above-mentioned parameters, a difference as high as one order of magnitude in Γ

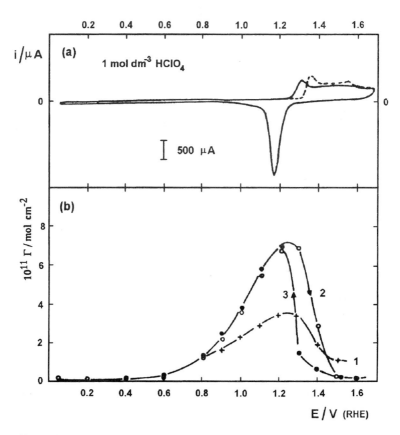

Figure 12.   (a) Cyclic voltammetry of the polycrystalline gold electrode (formed by Ar^+ ion sputtering) in 1 mol dm^{-3} $HClO_4$ in the absence (solid line) and presence (dashed line) of 2×10^{-4} mol dm^{-3} H_2SO_4 labeled with ^{35}S. Scan rate, 25 mV s^{-1}. (b) Potential dependence of bisulfate/sulfate adsorption on gold in the interrupted (curves 2 and 3, positive- and negative-going plots, respectively) polarization mode under the experimental conditions as above. For comparison, curve 1 shows the potential dependence of bisulfate/sulfate adsorption measured in 1 mol dm^{-3} $HClO_4$ in the continuous (positive-going) polarization mode.

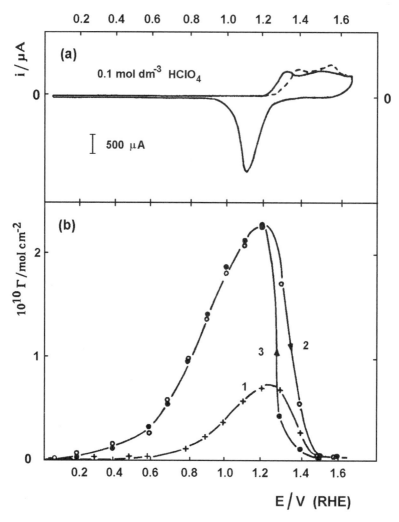

Figure 13. (a) Cyclic voltammetry of the polycrystalline gold electrode (formed by Ar^+ ion sputtering) in 0.1 mol dm^{-3} $HClO_4$ in the absence (solid line) and presence (dashed line) of 2×10^4 mol dm^{-3} H_2SO_4, labeled with ^{35}S. Scan rate, 25 mV s^{-1}. (b) Potential dependence of bisulfate/sulfate adsorption on gold in the interrupted (curves 2 and 3, positive- and negative-going plots, respectively) polarization mode under the experimental conditions as above. For comparison, curve 1 shows the potential dependence of bisulfate/sulfate adsorption measured in 1 mol dm^{-3} $HClO_4$ in the interrupted (positive-going) polarization mode.

values at the same solution (bulk) concentration of H_2SO_4 can be observed. However, it should be emphasized that the source of the significantly different Γ values is not due to a random error of measurement. The detected behavior is reproducible and the characteristic Γ vs. E curves (curve 1 in Fig. 12b as well as curves 2 and 3 in Fig. 13b) are very similar to observations made by Horányi *et*

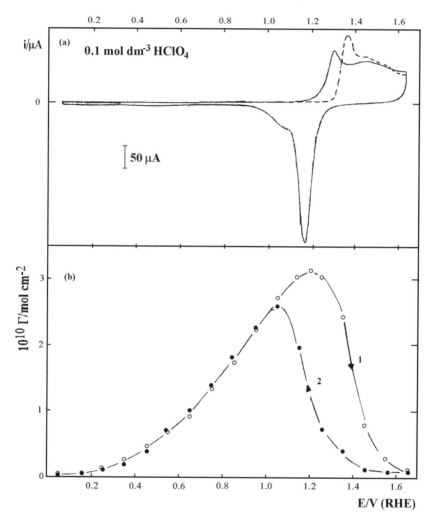

Figure 14. (a) Cyclic voltammetry of a disk-shaped Au_{poly} electrode in 0.1 mol dm^{-3} $HClO_4$ in the absence (solid line) and presence (dashed line) of 2×10^{-4} mol dm^{-3} H_2SO_4. Scan rate, 25 mV s^{-1}. (b) Potential dependence of bisulfate/sulfate adsorption on gold in the interrupted (curves 1 and 2, positive- and negative-going plots, respectively) polarization mode, under the experimental conditions as above.

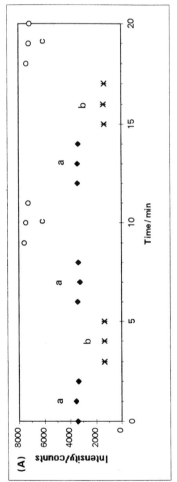

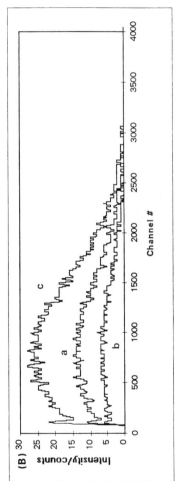

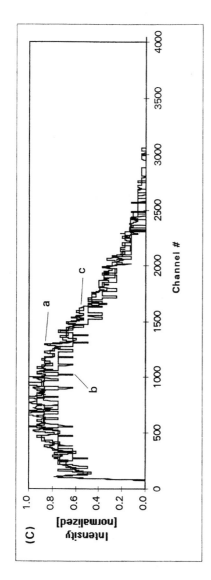

Figure 15. (A) Study of the reproducibility of the counting rates obtained in the "raised" (adsorption) position (section a) and in the "squeezed" (detection) position (sections b and c). The counting rates were detected upon adsorption of SO_4^{2-} ions labeled with ^{35}S on polycrystalline gold in 0.1 mol dm^{-3} $HClO_4 + 2 \times 10^{-4}$ mol dm^{-3} H_2SO_4. Potential values: (b) $E = 0.05$ V (no sulfate adsorption occurs); (c) $E = 1.20$ V. (B) Multichannel pulse-height spectra of ^{35}S originating from the radioelectrochemical cell under experimental conditions given in A. (C) Multichannel pulse-height spectra of ^{35}S normalized for maximum intensity. Experimental conditions as detailed above.

al.[63] and Wieckowski *et al.*[64,65] A comparison of plots "a" and "b" in Figs. 12 and 13 reveals that the hysteresis—which is obtained between the positive- and negative-going plots in Figs. 12b and 13b, and found to be independent of the absolute values of the surface excess concentrations—may most likely be due to the irreversibility of the surface oxidation of the gold electrode.

It is interesting to note that adsorption of bisulfate/sulfate ions was also studied on the smooth surface of a compact, polycrystalline Au electrode (Au_{poly}) by a combination of the "thin gap" method (Fig. 10) and voltammetry, in order to gain a deeper insight into the reliability and simultaneous applicability of the two independent radiotracer techniques. Figure 14a shows cyclic voltammograms of the Au_{poly} electrode taken in 0.1 mol dm^{-3} $HClO_4$ in the absence (solid line) and presence (dashed line) of 2×10^{-4} mol dm^{-3} H_2SO_4 labeled with ^{35}S. The corresponding surface excess concentration vs. electrode potential profiles are displayed in Fig. 14b. In addition, attempts were made to study the reproducibility of the Γ values (counting rates) and the energy distribution of β-radiation upon adsorption of HSO_4^-/SO_4^{2-} ions. Results shown in Fig. 15 reveal that the reproducibility of counting rates measured at various experimental conditions is excellent (Fig. 15a) and no measurable alteration in the energy spectra of β-radiation coming from the radioelectrochemical cell can be detected (Fig. 15b,c). A comparison between Figs. 13a and 14a demonstrates that there are no fundamental differences in the voltammetric behavior of the Ar^+ ion sputtered gold layer and the compact polycrystalline gold electrode. Although the surface excess concentration of HSO_4^-/SO_4^{2-} ions on the disk-shaped gold (Fig. 14b) is higher than that measured by the "foil" method (Fig. 13b), differences in the Γ values shown in Figs. 13b and 14b are less than 25%. All these results, in accordance with published studies,[63–65] provide evidence that the adsorption of HSO_4^-/SO_4^{2-} ions takes place on the oxide-free surface of gold and that it is reversible with respect to both their bulk solution concentration and electrode potential.

Moreover, a careful inspection of Fig. 12b reveals that the significant difference in the Γ values obtained by "continuous" and "interrupted" polarization techniques is obviously related to differences in the state (activity) of the gold surface. (The presence of trace impurities in the solution at extremely low concentration does not necessarily exert any effect on the shape of the voltammetric curves, but their competitive sorption may influence the surface excess concentration of HSO_4^-/SO_4^{2-} ions to be adsorbed weakly on the gold electrode.) These assumptions, however, do not offer any information as to the reason for the significantly higher adsorption of HSO_4^-/SO_4^{2-} ions in 0.1 mol dm^{-3} than in 1 mol dm^{-3} $HClO_4$ supporting electrolyte, using the same polarization procedure (see Fig. 13b).

In order to analyze the above mentioned issue, four factors must be taken into account[53]:

    (i) competition between ClO_4^- and HSO_4^-/SO_4^{2-} ions during the adsorption;
    (ii) pH dependence of the adsorption of HSO_4^-/SO_4^{2-} ions;
    (iii) presence and competitive adsorption of trace impurities (e.g., Cl^-);

(iv) roughness factor of the electrode surface.

Published data in Refs. 17 and 62 and references cited therein clearly show that adsorption of ClO_4^- ions on various noble metals is very low, i.e., no significant specific adsorption of ClO_4^- ions on a gold surface probably occurs. The relative adsorption strength of the anions on gold in increasing order is as follows[17, 62]

$$F^-, ClO_4^- \ll HSO_4^-/SO_4^{2-} < Cl^- \tag{1}$$

Therefore, it is immediately clear that the active competition between ClO_4^- and HSO_4^-/SO_4^{2-} ions—even at ten times higher perchloric acid concentration—can be ruled out. Naturally, an increase of one order of magnitude in the concentration of the supporting electrolyte causes changes in the pH of the solution, as well as in the concentration of trace impurities. Therefore, taking into consideration the weak and pH-independent electrosorption of HSO_4^-/SO_4^{2-} ions on gold surfaces,[53,63–65] it follows that the Γ values are predominantly influenced by the electrochemical environment, i.e., in the present case by the competitive sorption of solution impurities (mostly Cl^- ions). (It is well known that the trace amount of Cl^- in the $HClO_4$ supporting electrolyte, using even ultrapure $HClO_4$, is not negligible.) The latter assumption is strongly supported by the results of competitive adsorption processes in a solution containing Cl^- labeled with ^{36}Cl and HSO_4^-/SO_4^{2-} ions (Fig. 16). Obviously, the surface excess of Cl^- ions on the gold electrode is relatively high and the adsorbed species cannot be removed by the addition of a large excess of nonlabeled HSO_4^-/SO_4^{2-} ions into the solution. All these experimental data, in accordance with published material,[53,63,65] attest to the much higher adsorption strength and substantially stronger interaction of Cl^- ions with the gold surface.

The *in situ* radiotracer technique coupled with electrochemical measurements allows us to compare the adsorption behavior of species formed from aqueous SO_2 (HSO_3^-/SO_3^{2-} ions or/and molecular SO_2 as well as their oxidation and reduction products) with that of HSO_4^-/SO_4^{2-} ions in order to obtain more information about the nature and sorption properties of the above-mentioned species.[54]

Figure 17a shows the voltammetry of the polycrystalline gold, which was created by Ar^+ ion sputtering, in $HClO_4$ supporting electrolyte in the absence (dotted line) and presence (solid line) of 2×10^{-4} mol dm^{-3} HSO_3^- ions. Γ vs. E curves of the accumulation of HSO_3^- labeled with ^{35}S on gold in 1 mol dm^{-3} $HClO_4$ (at pH = 0) are shown in Fig. 17b (curves 1 and 2). For comparison, curves 3 and 3′ in Fig. 17b illustrate the potential dependence of the adsorption of radiolabeled HSO_4^- ions (or/and molecular SO_2 as well as their oxidation and reduction products) under the above-mentioned experimental conditions. As seen in Fig. 17b, at least four zones of bisulfite sorption can be distinguished in the potential range of 0.05 V to 1.65 V, namely at $E \leq 0.30$ V aqueous SO_2 is reduced to yield a gold surface covered with reduction products (zone I in curve 1). Under the assumption that at 0.05 V only reduced adspecies are present on the electrode surface, the number of electrons required for the reduction of one sorbed molecule containing radiolabeled sulfur

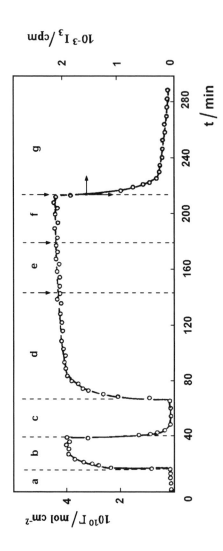

Figure 16. A study of the mobility of adsorbed Cl⁻ ions. Γ vs. time curves of chloride adsorption on polycrystalline gold in 0.1 mol dm^{-3} HClO₄ + 2 × 10^{-5} mol dm^{-3} labeled HCl under various experimental conditions: (a) and (c) 0.05 V; (b) and (d) 1.20 V; (e) addition of 2 × 10^{-5} mol dm^{-3} H₂SO₄ at 1.20 V; (f) addition of 2 × 10^{-5} mol dm^{-3} H₂SO₄ at 1.20 V; (g) addition of 1 × 10^{-3} mol dm^{-3} unlabeled HCl at 1.20 V.

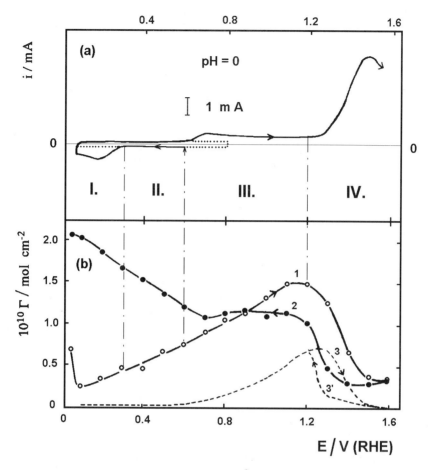

Figure 17. (a) Voltammetric curves of gold in 1 mol dm^{-3} HClO$_4$ before (dotted line) and after (solid line) the addition of 2×10^{-4} mol dm^{-3} HSO$_3^-$ labeled with ^{35}S. Initial polarization potential (0.60 V) as well as the direction of polarization are indicated by arrows. Scan rate, 100 mV s^{-1}. (b) Potential dependence of the sorption of bisulfite ions (and their oxidation and reduction products) on gold (curves 1 and 2, positive- and negative-going plots. respectively). Experimental conditions as above. For comparison, curves 3 and 3' (positive- and negative-going plots, respectively) show the potential dependence of bisulfate adsorption on gold in 1 mol dm^{-3} HClO$_4$ containing 2×10^{-4} mol dm^{-3} Na$_2$SO$_4$ (labeled with ^{35}S).

(epm) could be calculated, using the charge ($Q = 45$ μC cm^{-2}) involved in the adsorption as well as the stationary surface excess concentration measured at 0.05 V ($\Gamma = 7 \times 10^{-11}$ mol cm^{-2}). (The reduction charge Q was obtained by integrating the cathodic-going i–E curve over the potential range 0.05–0.30 V. The Q value was corrected for the double-layer contribution.) The epm value at 0.05 V was found to be 6.6. Thus, taking into consideration possible errors in the estimation of data used for the epm calculation, it is possible that the reduction products adsorbed at

0.05 V are sulfides. Naturally, the reduction of bisulfite to elemental sulfur in the first reduction step cannot be excluded. It is noteworthy that the generation of sulfides (and polysulfides) as possible adspecies on gold and platinum upon electroreduction of aqueous SO_2 has been pointed out by Weaver et al.[66] and Szklarczyk et al.[67] Furthermore, both a decrease and an increase in the Γ values at potentials below 0.30 V can be observed (zone I in curve 1, Fig. 17b), which may imply that accumulation of molecular SO_2 (or bisulfite) is superimposed on the potential-dependent adsorption of reduction products.

At potentials between 0.30 V and 0.60 V (zone II in curve 1, Fig. 17b), where the aqueous SO_2 is essentially stable to oxidation and reduction, surface concentration of adspecies (bisulfite or molecular SO_2) increases with increasing potential. A further increase in Γ values can be detected in the potential region of 0.60 V to 1.20 V, which is indicative of the continuous formation and adsorption of oxidation products (zone III in curve 1, Fig. 17b). At $E = 1.20$ V and above (zone IV, curve 1), in accordance with the voltammetric data shown in Fig. 17a, it can be concluded that (i) reduction products are oxidized and desorbed readily at the early stages of the electrooxidation of gold surface, and (ii) an intensive decrease in the adsorption of oxidation products begins along with the electrooxidation of gold, and indicates that surface oxides do not provide active sites for sorption of SO-type species.

The apparent hysteresis observed between the positive- and negative-direction Γ vs. E plots (zones III and IV in curves 1 and 2, Fig. 17b) firmly supports the contention that no accumulation of oxidation products takes place on oxide-covered gold. On the negative-direction polarization (zones III and IV, curve 2), however, the adsorption of oxidation products commences at potentials at which the free metal sites are created and consistently follows the irreversible surface redox behavior of the substrate. It is of special interest to note that at potentials below 0.60 V, where no oxidation of aqueous SO_2 occurs, a significant accumulation of unidentified S-containing species on the gold surface covered with oxidation products can be observed (zones I and II, curve 2). (A plausible explanation of the above-mentioned phenomenon is given elsewhere.[54]) Moreover, a comparison of the accumulation of species created from aqueous SO_2 (curves 1 and 2) with the adsorption of bisulfate ions (curves 3 and 3′ in Fig.17b) reveals that (i) in contrast to bisulfite sorption, a reversible adsorption of HSO_4^- species proceeds on the oxide-free surface of gold (see also Figs. 13b–15b), and (ii) HSO_4^-/SO_4^{2-} ions are weakly bonded to the substrate (as detailed above, competitive adsorption of the trace amounts of solution contaminating Cl^- ions suppresses the bisulfate adsorption), i.e., the relative adsorbability of bisulfate/sulfate is significantly lower than that of the adspecies formed from aqueous SO_2.

In order to obtain a better insight into the complex sorption phenomena, the dependence of bisulfite and bisulfate accumulations on potential was studied in ClO_4^--containing test solutions at pH values of 3.5 and 6.5. (Figs. 18a and b). Curves 1 and 2 in Figs. 17b and 18a show that the hysteresis of the positive- and negative-going plots of bisulfite/sulfite adsorption increases with increasing pH. This effect can most likely be attributed to the enhanced irreversibility of the formation and reduction of the surface oxide at higher pH values. The increase in surface excess concentrations with increasing pH values reflected by Figs. 17 and 18 implies a higher accumulation of bisulfite/sulfite ions (as well as their oxidation products) on gold not covered with reduced adspecies.[54] (In this context, it should be noted that the relative proportion of bulk species created from aqueous SO_2 as well as the nature of surface active sites also change with pH, which might exert an influence on the saturation Γ values.[54]) By considering the potential dependence of bisulfate/sulfate adsorption at the pH values studied (dashed lines in Figs. 17 and 18), it is clear that the surface concentration of adsorbed anions diminishes with increasing pH. This effect can rather be ascribed to the competitive adsorption of solution contaminants than anything else.[53,63,64]

In summary, it can be concluded that the radiotracer and voltammetric results shown in Figs. 12–18 provide direct confirmation of the essentially different sorption character of chloride, bisulfate/sulfate, and bisulfite/sulfite ions as well as of SO-type species formed (and adsorbed on gold) in the course of bisulfite/sulfite electrooxidation. (An explanation regarding the possible nature of the latter oxidized adspecies is also given in Ref. 54.)

3.1.2. Electrosorption of Ag and Co Species

Accumulation of Ag and Co species on gold electrodes was studied by the simultaneous use of the "foil" method and cyclic voltammetry, in order to demonstrate the applicability of the *in situ* radiotracer technique in studies of cation adsorption processes.

3.1.2.1. Underpotential Deposition of Silver. Underpotential deposition (UPD) of silver on gold, i.e., the formation of submonolayer or monolayer coverages of Ag deposits occurring at more positive potentials than the reversible Nernst potential in the same electrolyte, has been extensively investigated during the past two decades due to its practical and theoretical importance (see Refs. 47 and 68–71, and references cited therein). The refined *in situ* radiotracer "foil" method (Fig. 8) combined with voltammetric measurements is a powerful tool for the comprehensive investigation of the above-mentioned deposition phenomena. The work presented in this section is focused on the extent and character of UPD of silver labeled with [110m]Ag. The potential dependence of Ag electrosorption was measured in two ("continuous" and "interrupted") polarization modes. All experiments were conducted at room temperature in $0.1 \text{ mol dm}^{-3} \text{ HClO}_4 + 2 \times 10^{-4} \text{ mol dm}^{-3} \text{ Ag(I)}$ under oxygen-free (99.999 v% Ar) atmosphere. Other experimental conditions are as detailed in Section 3.1.1.

A special (dE/dX) β-plastic scintillation detector (Thorn Emi Ltd.; thickness, 100 μm) was developed for the *in situ* determination of the surface excess of Ag adatoms labeled with [110m]Ag ($E_{\beta 1 max} = 83 \text{ keV}$; $E_{\beta 2 max} = 530 \text{ keV}$) and Co species

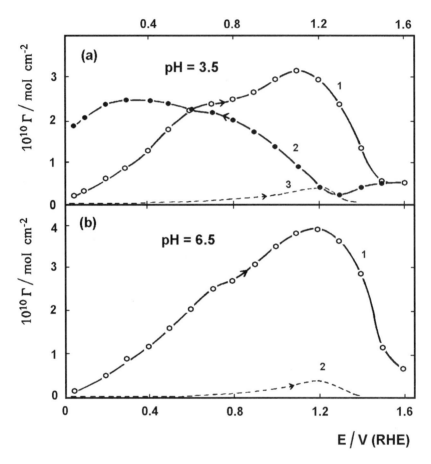

Figure 18. (a) Potential dependence of the sorption of bisulfite/sulfite ions (and their oxidation and reduction products) on gold in ClO_4^- solution (c = 1 mol dm^{-3}, pH = 3.5) containing 2×10^{-4} mol dm^{-3} HSO_3^-/SO_3^{2-} labeled with ^{35}S (curves 1 and 2, positive and negative-going plots, respectively). For comparison, curve 3 (positive-going plot) shows the potential dependence of SO_4^{2-} adsorption on gold in ClO_4^- solution (c = 1 mol dm^{-3}, pH = 3.5) containing 2×10^{-4} mol dm^{-3} Na_2SO_4 labeled with ^{35}S. (b) Potential dependence of the sorption of bisulfite/sulfite ions (and their oxidation products) on gold in ClO_4^- solution (c = 1 mol dm^{-3}, pH = 6.5) containing 2×10^{-4} mol dm^{-3} HSO_3^-/SO_3^{2-} labeled with ^{35}S (curve 1, positive-going plot). For comparison, curve 2 (positive-going plot) shows the potential dependence of SO_4^{2-} adsorption on gold in ClO_4^- solution (c = 1 mol dm^{-3}, pH = 6.5) containing 2×10^{-4} mol dm^{-3} Na_2SO_4 labeled with ^{35}S.

labeled with ^{60}Co ($E_{\beta max}$ = 312 keV). The main features of the pulse–height spectra of ^{110m}Ag, originating from the radioelectrochemical cell, are shown in Fig. 19. The experimental setup used allowed us to study the sorption phenomena by detecting β-particles emitted by ^{110m}Ag radionuclides. Furthermore, the *dE/dX* detector placed beneath the cell is suitable for investigating the sorption behavior of species

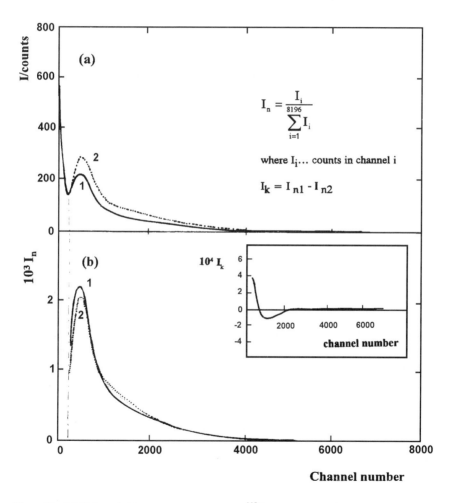

Figure 19. Multichannel (8K) pulse–height spectra of ^{110m}Ag measured by β-plastic *dE/dX* detector at various experimental conditions (measuring live time, 300 s). (a) Spectra of ^{110m}Ag source originating from the radioelectrochemical cell at different potential values of polycrystalline gold: (1) *E* = 1.60 V/RHE (solution [background]); (2) *E* = 0.70 V/RHE (^{110m}Ag accumulation). (b) Normalized spectra of ^{110m}Ag. Experimental conditions as detailed for Fig. 19 a; difference between spectra 1 and 2 is shown in inset.

labeled with radioisotopes emitting either low (or medium) energy β-particles or β- and γ-radiation, simultaneously (e.g., ^{110m}Ag and ^{60}Co). Disturbing effects of the secondary radiations (γ-ray, characteristic X-ray or bremsstrahlung, Compton and photoelectrons) on the detection are negligible or can be eliminated (see Fig. 19b). To our knowledge, *this is the first attempt to measure the surface concentration of Ag submonolayers* using an *in situ* radiotracer method.[47]

Possible links between the voltammetric curves and Γ vs. E profiles obtained on polycrystalline gold in a solution containing 0.1 mol dm^{-3} HClO$_4$ and 2×10^{-4} mol dm^{-3} AgNO$_3$ labeled with ^{110m}Ag are shown in Fig. 20. The potential window-opening study (into cathodic direction) of the voltammetric behavior of gold in a solution containing Ag$^+$ ions reveals the redox features of the system (Fig. 20a). Two well-resolved reduction (underpotential deposition) peaks at 0.88 and 1.24 V as well as two oxidation (stripping) peaks at 0.91 and 1.26 V can be distinguished in the UPD range (the reversible Nernst potential for bulk silver formation is 0.652 V/RHE). Also, it is clear from Fig. 20a that the underpotential deposition of Ag$^+$ ions takes place in at least two potential zones (UPD I and II). In the zone of 0.78–1.30 V formation of silver adatoms probably occurs, while between 0.65 V and 0.78 V (just before the bulk deposition) the mechanism is different, i.e., three-dimensional clusters are most likely formed. The latter assumption is strongly supported by results gained from studies of HSO$_4^-$/SO$_4^{2-}$ adsorption enhanced by Ag-adatoms.[63,68] Deposition of bulk silver commences at 0.65 V, which is identical with the calculated value of the reversible Nernst potential.

The radiotracer results in Figs. 20b,c show that the deposition of silver on polycrystalline gold is significant over the entire potential range that is more negative than the onset of surface electrooxidation. No measurable surface excess values were found above 1.30 V, i.e., in this region Γ_{Ag} is less than 5×10^{-12} mol cm^{-2}. In the range of 0.65–1.30 V, Γ values not higher than 3.5×10^{-10} mol cm^{-2} were detected by both polarization techniques. This result provides evidence that the fraction coverage of the surface (Θ) in the UPD region is no more than 0.20, assuming that a surface concentration of about 2×10^{-9} mol cm^{-2} corresponds to one monolayer.[69] The $\Theta = 0.20$ value is significantly smaller than that evaluated from the microcoulometric data.[70]

Furthermore, the "interrupted" polarization technique does not cause any differences in the Γ values obtained during the anodic- and cathodic-going polarizations (Fig. 20b), while in the case of the "continuous" polarization mode an apparent hysteresis of Γ vs. E curves occurs (Fig. 20c). These observations indicate strongly that silver deposition is essentially influenced by the state and pretreatment of the gold surface, which is obviously related to the applied polarization technique. Specifically, higher Γ values are detected on an oxide-free than on an oxide-covered surface of gold (curves 2 and 1 in Fig. 20c, respectively) in the course of the negative-going polarization. However, the hysteresis between curves 1, 1′ and 2, 2′ in Fig. 20c is independent of the initial state of the gold surface. It appears that this hysteresis is due to the larger amount of Ag deposited by the "continuous"

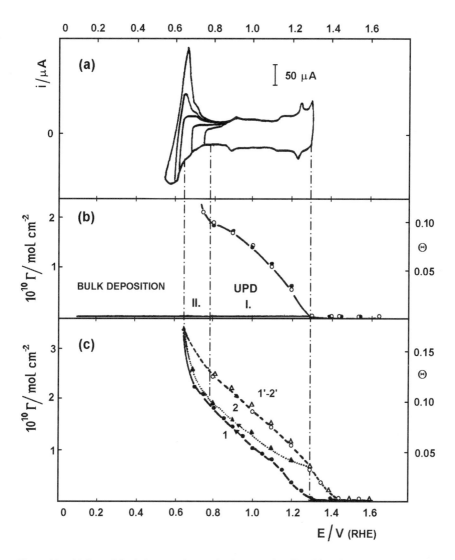

Figure 20. (a) Potential window-opening study (into negative direction) of silver deposition on a polycrystalline gold layer in 0.1 mol dm^{-3} HClO$_4$ + 2 × 10^{-4} mol dm^{-3} AgNO$_3$ labeled with ^{110m}Ag. Scan rate, 10 mV s^{-1}. (b) Potential dependence of silver deposition on polycrystalline gold layer in the interrupted polarization mode (empty and filled circles represent negative- and positive-going plots, respectively). Experimental conditions as above. (c) Potential dependence of silver deposition on polycrystalline gold layer in the continuous polarization mode (curves 1 and 2, and 1′ and 2′ represent negative- and positive-going plots, respectively). The measurements of curves 1 and 1′, and curves 2 and 2′ were started on oxide-free and partially oxide-covered gold surfaces, respectively. Experimental conditions as above.

polarization technique (Fig. 20c, zone UPD II), which requires substantially higher overpotential to dissolve (see Ref. 71 and references cited therein).

The cyclic voltammetric and radiotracer results in Fig. 20 lead to the conclusion that the underpotential deposition of Ag^+ ions on an oxide-free surface of gold and dissolution of Ag adatoms are quasi-reversible processes. At underpotential values, silver is probably deposited as a clean metal or silver oxide (Ag_2O and AgO).[47] The relative amount of the latter phase is higher in the deposit formed on "interrupted" polarization. In addition, slow structural changes of the gold surface, being associated with strong interactions between Ag and Au, presumably also take place due to the cyclic voltammetric treatment.

3.1.2.2. Cobalt Accumulation: Evidence of Colloid Formation. A radiochemical and electrochemical study of cobalt accumulation on polycrystalline gold electrode in $NaClO_4$ and borate buffer solution was carried out by means of the *in situ* radiotracer "foil" method (Fig. 8) and voltammetry.[45] Figures 21a and b indicate that the extent and kinetics of cobalt accumulation are primarily affected by the pH of the solution. The prolonged adsorption time and the shape of Γ vs. time curves at higher pH (> 7) are probably due to the simultaneous formation and sorption of the hydrolysis products of cobalt. Curves 1 and 2 in Fig. 21a show that the saturation Γ values exceed the monolayer coverage calculated for cobalt ions ($\Gamma = 4.3 \times 10^{-10}$ mol cm^{-2} for hydrated cobalt and 3.36×10^{-9} mol cm^{-2} for metallic cobalt), indicating the effect of colloid formation.[44–46] The presence of borate species suppresses the cobalt accumulation under open circuit conditions and during cathodic polarization (curves 6 and 7 in Fig. 21a). The reason for this effect can be attributed not only to the competitive sorption of borate species, but also to borate species decreasing the colloid formation of cobalt.[44–46] Furthermore, Fig. 21b attests to the relative adsorbability of the hydrolysis products of cobalt [$CoOH^+$, $Co(OH)_2$, etc.] being higher than that of Co^{2+}.

Figure 22a shows the cyclic voltammograms of a gold electrode in the borate buffer of pH = 7.6 in the absence (dashed line) and presence (solid lines) of cobalt ions. In certain cases, radiotracer methods are applicable to follow accumulation processes under potentiodynamic conditions. Voltradiometric curves can only yield unambiguous information on accumulation phenomena, if these processes are accompanied by high-intensity changes in comparison with the solution background. Figures 22b and c confirm that the accumulation of Co species during potential cycling can be investigated by voltradiometry. These plots correspond to continuous thickening of the oxide layer. The increase of cobalt accumulation measured during anodic polarization is due to the change of the oxidation state of Co(II). Previous results[45] reveal that at potentials more positive than 1.05 V, oxide layers containing CoOOH, Co_3O_4, and CoO_2 are formed.

The value of Q_{anodic} in Fig. 22a is always larger than $Q_{cathodic}$, giving a strong indication that during the negative-going potential sweep only a part of the previously formed oxide is reduced. This result means that the cobalt species accumulated during anodic polarization in borate buffer cannot be completely removed by

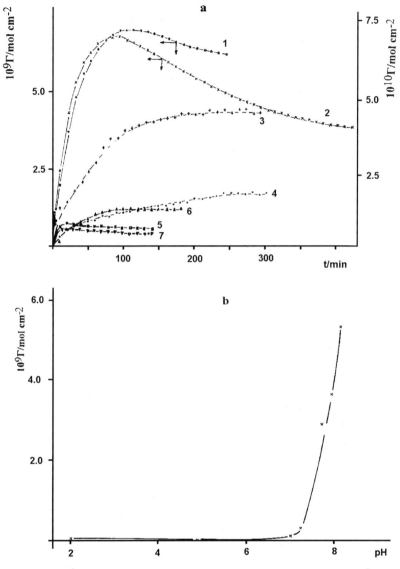

Figure 21. (a) Time dependence of cobalt accumulation on gold electrode in 0.5 mol dm^{-3} NaClO$_4$: 1, initial pH = 8.8, which decreases to 8.1 during measurement; 2, initial pH = 8.4, which decreases to 7.9 during measurement; 3, pH = 7.5; 4, pH = 7.0; 5, pH = 4.9 and in borate buffer ($c_{boric\ acid}$ = 0.1 mol dm^{-3}, $c_{perchlorate}$ = 0.5 mol dm^{-3}); 6, pH = 7.6; 7, pH = 4.9. ($c_{Co(II)}$ = 2 × 10^{-4} mol dm^{-3}). (b) pH dependence of cobalt accumulation on gold electrode in 0.5 mol dm^{-3} NaClO$_4$.

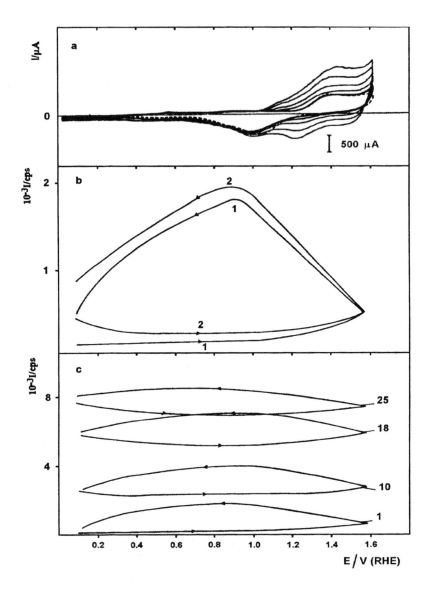

Figure 22.   (a) Cyclic voltammograms of gold electrode in borate buffer before (dashed line) and after (solid line) the addition of 2×10^{-4} mol dm^{-3} Co(II) into the solution at pH = 7.6 (scan rate, 25 mV s^{-1}). (b) Voltradiometric curves of cobalt (labeled with ^{60}Co) absorption on gold electrode in borate buffer solution at pH = 7.6 (scan rate, 25 mV s^{-1}, $c_{Co(II)} = 2 \times 10^{-4}$ mol dm^{-3}); first (1) and second (2) cycles. (c) Voltradiometric curves of cobalt (labeled with ^{60}Co) sorption on gold electrode in borate buffer solution at pH = 7.6 (scan rate, 25 mV s^{-1}, $c_{Co(II)} = 2 \times 10^{-4}$ mol dm^{-3}; Arabic numbers indicate which cycle is presented by the corresponding curve).

cathodic potentiostatic steps. The accumulation of cobalt at higher potentials during a cathodic-going potential sweep (Fig. 22b and c) may be the result of the build-up of Co^{2+} ions into the cobalt-oxide layer through electrochemical reaction.

All these results give additional information with regard to not only the kinetics and mechanism of the accumulation of Co-containing species on metal surfaces, but also the decrease or release of trace cobalt from industrial solutions contaminated with ^{60}Co radionuclide.

3.2. Accumulation of ^{110m}Ag on an Austenitic Stainless Steel

Knowledge of the state and sorption behavior of corrosion products adhering to metal surfaces in cooling circuits of water-cooled nuclear reactors is of great importance for a number of practical reasons. For instance, the accumulation of radioactive cations (such as $^{60}Co^{2+}$, $^{110m}Ag^{+}$, $^{51}Cr^{3+}$, $^{54}Mn^{2+}$, etc.) from the reactor coolant onto the surface layer of corrosion products is a very important factor in the elucidation of corrosion and radioactive contamination processes occurring in nuclear power stations. Owing to the significance of this problem, a large number of studies have been devoted to examining the sorption of various radioactive cations on metallic (and nonmetallic) constructional materials used in the nuclear industry.[8,19,52]

The contamination caused by ^{110m}Ag activated in the core appears to be a significant problem in some Soviet-made PWRs, and may lead to troubles in the continuous measurement of other contaminants in the primary side coolant. Laboratory model systems (such as the "foil" and "thin gap" methods presented in Section 2.1.2) make it possible to study the kinetics and mechanism of silver accumulation on the surfaces of, e.g., austenitic stainless steel type 08X18H10T (GOST 5632-61), used as the structural material of primary circuits. In order to demonstrate the usefulness of *in situ* radiotracer studies in this field, this section covers some radiochemical research carried out in a model solution of the primary side coolant of PWRs, containing silver labeled with ^{110m}Ag. In addition, the sorption of silver and the possibility of measuring the corresponding radiation are studied in the presence of other radioactive species (samples from a nuclear power plant). Some data characterizing the steel, as well as the composition of solution samples originating from a nuclear power station have been published.[32] The composition of the model solution used is described in the legend of Fig. 23. Other experimental conditions are as detailed in Section 3.1.

Figure 23 shows the time dependence of silver accumulation on a disk-shaped sample of austenitic stainless steel measured by the "thin gap" method in a model solution of primary side coolant.[35c,d] A highly time- and potential-dependent sorption of silver species takes place on the smooth electrode surface; no quasi-equilibrium surface excess of deposited silver is reached even after a period of 18 h at a given potential value. The fact that a surface excess of $\Gamma = 2 \times 10^{-9}$ mol dm^{-3} corresponds to one monolayer coverage of Ag^{+} ions[69] provides evidence that only

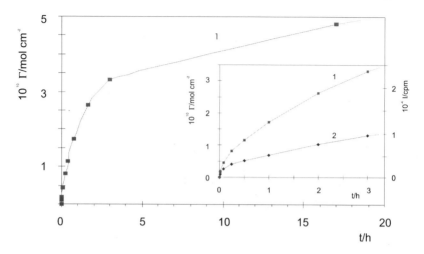

Figure 23.   Time dependence of the accumulation of silver labeled with ^{110m}Ag on austenitic stainless steel type 08X18H10T (GOST 5632-61) measured by the "electrode lowering" method. The supporting electrolyte is a model solution of the primary circuit coolant of NPPs, containing H_3BO_3 (3 g dm^{-3}), K^+ (3.5 × 10^{-3} dm^{-3}), Na$^+$ (2 × 10^{-4} g dm^{-3}), Li$^+$ (8 × 10^{-4} g dm^{-3}), NH$_3$ (3 × 10^{-2} g dm^{-3}), N$_2$H$_4$ (2 × 10^{-5} g dm^{-3}), Ag$^+$ (1 × 10^{-6} mol dm^{-3}); pH$_{25\ °c}$ = 7.7. In the insert, curves 1 and 2 show the accumulation at an open circuit potential (E = 0.75 V) and at 1.10 V, respectively.

a limited part of the real surface area of the steel sample (less than 25%) is occupied by silver ions. The surface excess concentration values at potentials more positive than the open circuit potential are smaller (see insert in Fig. 23), indicating the significant effect of the electrode potential on the silver deposition processes. This observation is highly supported by the Γ vs. *E* profiles presented in Fig. 24. Figure 24a shows the cyclic voltammetry of the stainless steel in a borate buffer solution with and without silver ions. This voltammogram demonstrates that the studied steel exhibits passive features over a wide potential range (between –0.10 and 1.10 V) in both the absence and presence of Ag$^+$ ions labeled with ^{110m}Ag. The Γ vs. *E* curves in Fig. 24b and c reveal a considerable decrease in Γ values at more anodic potentials than the open-circuit corrosion potential (curve 1, Fig. 24b). On the other hand, a potential shift into cathodic direction results in a significant increase in the surface excess concentration of silver (curve 1′ in Fig. 24b, and Fig. 24c). All these results indicate that the extent of contamination caused by ^{110m}Ag$^+$ ions is strongly influenced by the potential values to be chosen within the passive region of the steel surface.

Figure 25 shows that the ^{110m}Ag accumulation on powdered stainless steel from solution samples, originating from a nuclear power plant, can be studied by measuring the different radiation of the same isotope. Moreover, the b and c branches of curves 2 and 3 in Fig. 25, in accordance with the results presented in Figs. 23 and 24, reveal the effects of potential shifts on the sorption of ionic silver.[32]

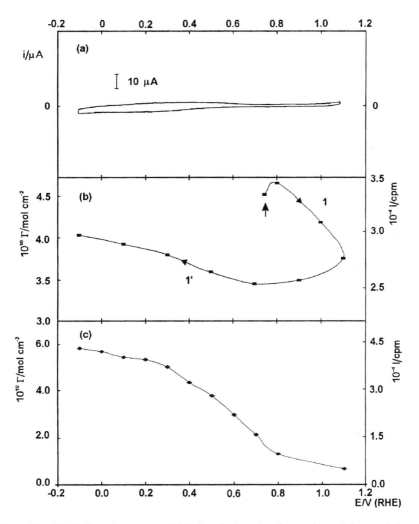

Figure 24. (a) Cyclic voltammetry of disk-shaped electrode of austenitic stainless steel [type: 08X18H10T (GOST 5632-61)] in the model solution (for details see legend to Fig. 23). Scan rate, 10 mV s^{-1}. (b) Potential dependence of the adsorption of silver labeled with ^{110m}Ag on stainless steel in the model solution (curves 1 and 1′, positive- and negative-going plots, respectively). Arrow pointing upward indicates open-circuit potential. The experiment was carried out in continuous polarization mode, waiting for 30 min at each potential before obtaining intensity values. (c) Negative-going plot of silver deposition on stainless steel starting from $E = 1.10$ V. Experimental conditions are as above.

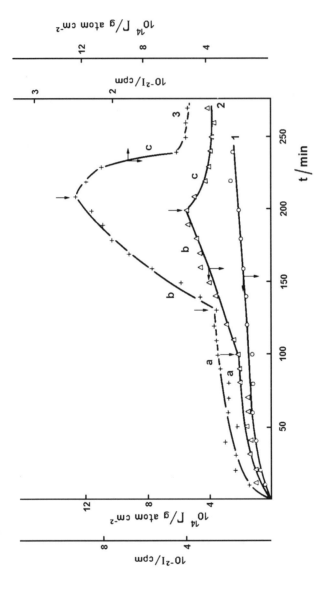

Figure 25. Accumulation of silver species labeled with ^{110m}Ag from different solutions, originating from NPP, on austenitic stainless steel type 08X18H10T (GOST 5632–61) under different experimental conditions: (1) β-radiation of ^{110m}Ag was measured under open-circuit conditions. (2) $K\alpha$-radiation of ^{110m}Ag was measured under: (a) open-circuit conditions; (b) cathodic potential shift; (c) anodic potential shift. (3) $K\alpha$-radiation of ^{110m}Ag was measured under: (a) open-circuit conditions; (b) cathodic potential shift; (c) anodic potential shift. Other experimental conditions are as detailed in Ref. 32.

From these illustrative examples several implications arise, as follows:

(i) Accumulation of ^{110m}Ag on austenitic stainless steel is most likely due to the cementation (underpotential deposition) of radioactive silver ions.[35d]

(ii) Both the deposition (contamination) and dissolution of deposited silver (decontamination and/or surface prevention) can be influenced by the electrochemical parameters of the system studied.

3.3. Sorption Behavior of Duplex Stainless Steels in HCl and H₂SO₄ Solutions

This section reports new findings obtained by the combined radiochemical and electrochemical studies of the sorption of Cl^- and HSO_4^-/SO_4^{2-} ions on surfaces of Ru-alloyed and Ru-free duplex stainless steels.[58,59] Chemical compositions of these experimental alloys are given in Table 1. By using the results of ICP-optical emission spectroscopy (ICP-OES) and surface analytical techniques, such as Auger-electron spectroscopy (AES) and X-ray photoelectron spectroscopy (XPS),[58, 72] some possible links between sorption phenomena and the structure and chemical composition of passive layers are also presented and discussed.

3.3.1. Time and Concentration Dependence of Cl^- and HSO_4^-/SO_4^{2-} Accumulations

The passivation phenomena of duplex stainless steels modified with ruthenium was investigated by measuring the HSO_4^-/SO_4^{2-} and Cl^- accumulations by the *in situ* radiotracer "foil" method. Powdered stainless steel samples were exposed to dilute reducing acid solutions in which they passivated spontaneously, in order to gain a better understanding of the role of different species in the entire corrosion process. Figure 26 displays the time dependence of Cl^- sorption on two duplex stainless steels. The surface concentration of Cl^- ions on both steels is small ($\Gamma < 1 \times 10^{-10}$ mol cm^{-2}) and the study of the mobility of labeled chloride ions

Table 1. Chemical Composition of the Duplex Stainless Steels Studied (wt%)

No.	Alloy Type	Fe	Cr	Ni	Mo	Ru	Mn	Si	S	P	O	N	C
377	Fe-22%Cr 9%Ni 3%Mo 0%Ru	rest	22.0	9.07	2.81	—	<0.1	0.070	0.01	<0.01	0.021	0.006	0.03
380	Fe-22%Cr 9%Ni 3%Mo 0.3%Ru	rest	22.4	9.24	2.92	0.28	<0.1	0.030	0.01	<0.01	0.037	0.005	0.02

accumulated on these substrates (Fig. 26b) indicates that no strong embedding of Cl^- occurs. The bisulfate/sulfate accumulation on both duplex stainless steels is depicted graphically in Fig. 27 and indicates a saturation surface concentration of about 4.5×10^{-10} mol cm^{-2}. The results of the exchange of labeled HSO_4^-/SO_4^{2-} ions sorbed on duplex stainless steels (curve 3b in Fig. 27a and curve 1b in Fig. 27b) reveal that a part of the sulfate/bisulfate species is strongly bonded to the spontaneously formed passive layers.

To gain further information on the competitive sorption processes in solutions containing Cl^- and HSO_4^-/SO_4^{2-} ions, attempts were made to study the mobility of species accumulated on both duplex stainless steels. Results obtained with the steel containing 0.28% Ru are shown in Fig. 28. At quasi-equilibrium conditions the surface excesses of bisulfate/sulfate ions on the surfaces of both steels are not influenced by the large excess of Cl^- ions (Fig. 28a). However, addition of a small amount of HSO_4^-/SO_4^{2-} ions to the solution phase results in a significant decrease in sorbed labeled Cl^- (Fig. 28b). These phenomena give a strong indication that: (i) only a very limited part of the real surface area of steel samples is occupied by Cl^- ions (pitting corrosion sites); (ii) sorption of HSO_4^-/SO_4^{2-} ions most likely takes place on the total real surface of steel samples (large surface coverage and build-up);

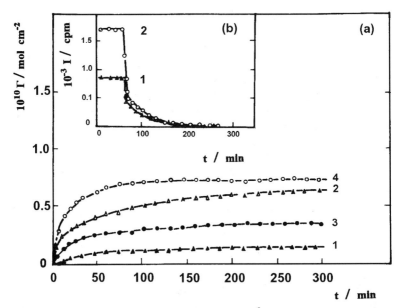

Figure 26. Γ vs. time curves of Cl^- accumulation in 0.1 mol dm^{-3} HClO$_4$ under open-circuit conditions: (a) Curves (1) and (2) on duplex stainless steel No. 377 at Cl^- concentrations of 2×10^{-5} and 1×10^{-4} mol dm^{-3}, respectively; (3) and (4) on duplex stainless steel No. 380 at Cl^- concentrations of 2×10^{-5} and 1×10^{-4} mol dm^{-3}, respectively. (b) Study of the mobility of labeled Cl^- accumulated on duplex steels Nos. 377 and 380 (curves 1 and 2, respectively) by the addition of a large excess of nonlabeled HCl ($c = 1 \times 10^{-2}$ mol dm^{-3}).

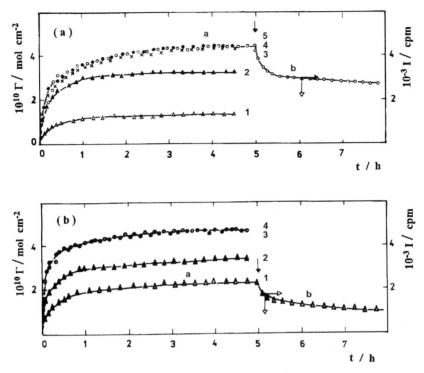

Figure 27. Γ vs. time curves of H_2SO_4 accumulation in 0.1 mol dm^{-3} HClO$_4$ at open-circuit potentials: (a) On duplex stainless steel No. 377. (1) 5×10^{-6}; (2) 1×10^{-5}; (3 a) 2×10^{-5}; (4) 1×10^{-4}; (5) 1×10^{-3} mol dm^{-3} labeled H_2SO_4; (3 b) addition of 1×10^{-2} mol dm^{-3} nonlabeled H_2SO_4 (open-circuit potentials were shifted over the range of -50 to 700 mV). (b) On duplex stainless steel No. 380. (1 a) 5×10^{-6}; (2) 2×10^{-5}; (3) 1×10^{-4}; (4) 5×10^{-4} mol dm^{-3} labeled H_2SO_4; (1 b) addition of 1×10^{-2} mol dm^{-3} nonlabeled H_2SO_4 (open-circuit potentials were shifted over the range of -80 to 720 mV.)

(iii) adsorbability of HSO_4^-/SO_4^{2-} ions is significantly higher than that of Cl$^-$ ions. It is also obvious that the Ru content does not exert significant effects on the sorption behavior of the duplex stainless steel surfaces in either type of acid solution.

In accordance with the results of XPS and AES studies,[58,72] it is possible that passive layers of excellent corrosion resistance are formed on both duplex stainless steels at the above mentioned experimental conditions. This assumption is strongly supported by the apparently low rate of the dissolution processes,[59] indicating that redistribution of the main alloying elements of the surface layers are presumably modest in the course of the spontaneous transformation of duplex steel surfaces.

3.3.2. Potential Dependence of Cl$^-$ and HSO_4^-/SO_4^{2-} Accumulations

The Γ vs. E curves obtained with the powered samples of the duplex steels which passivated spontaneously, starting from the open-circuit potential, show that

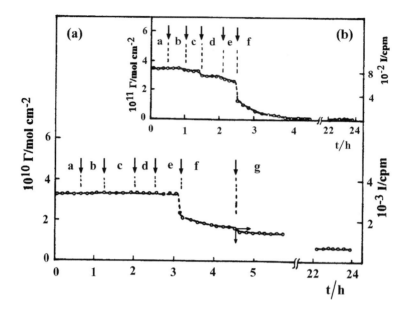

Figure 28.   (a) Effects of HCl additions on the surface concentration of bisulfate/sulfate ions accumulated on duplex stainless steel with 0.28% under Ru open-circuit conditions. ($c = 2 \times 10^{-5}$ mol dm^{-3}). Concentrations of HCl: (a) 0; (b) 1×10^{-5}; (c) 2×10^{-5}; (d) 2×10^{-4}; (e) 1×10^{-3} mol dm^{-3}; (f) and (g) additions of 1×10^{-2} mol dm^{-3} Na$_2$CrO$_4$ and nonlabeled H$_2$SO$_4$, respectively. (b) Effects of H$_2$SO$_4$ additions on the surface concentration of chloride ions accumulated on duplex stainless steel with 0.28% Ru at open-circuit conditions. ($C_{HCl} = 2 \times 10^{-5}$ mol dm^{-3}). Concentrations of H$_2$SO$_4$: (a) 0; (b) 5×10^{-6}; (c) 2×10^{-5}; (d) 2×10^{-4}; (e) 2×10^{-3} mol dm^{-3}; (f) addition of 1×10^{-2} mol dm^{-3} nonlabeled HCl.

there is no considerable potential dependence of HSO$_4^-$/SO$_4^{2-}$ accumulation on the surfaces over the potential range of 0 to 1.20 V, as depicted by curves 2 and 2′, and 3 and 3′ in Fig. 29. This finding correlates well with the fact that both studied steels exhibit passive features over a wide potential region in 0.1 mol dm^{-3} HClO$_4$ in the absence and presence of H$_2$SO$_4$ up to a concentration of 1×10^{-3} mol dm^{-3} (see Fig. 30). The potential dependence of Cl$^-$ sorption shown by curves 1 and 1′ in Fig. 29 does not differ significantly from that of bisulfate/sulfate ions. In this case, the cathodic polarization curve shows a potential shift towards more positive values, resulting in a slight increase in the surface excess of Cl$^-$ ions (curve 1′). All these results supply further evidence that the extent of the accumulation of aggressive anions depends decisively on the structure and probably on the composition of the passive layer formed on the steel surface.

   Comparative studies were carried out with surface analytical techniques such as AES and XPS,[58,72] in order to characterize the spontaneously formed passive films on these duplex steels. The results show that, with both materials, HCl and H$_2$SO$_4$ cause a moderate enrichment of Cr in the passive layers, mostly in the form

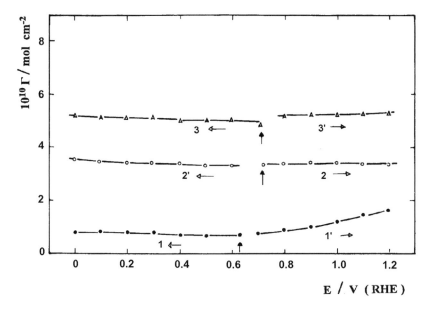

Figure 29. Potential dependence of HSO_4^-/SO_4^{2-} and Cl^- accumulation in 0.1 mol dm^{-3} HClO$_4$ under various experimental conditions: (1 and 1') at Cl$^-$ concentration of 1×10^{-4} mol dm^{-3} on duplex steel No. 380; (2 and 2') at H$_2$SO$_4$ concentration of 2×10^{-5} mol dm^{-3} on duplex steel No. 380; (3 and 3') at H$_2$SO$_4$ concentration of 1×10^{-4} mol dm^{-3} on duplex steel No. 377. The open-circuit potentials are indicated by arrows.

of Cr_2O_3 and $Cr(OH)_3$, which is normally found in stainless steels. The relative amount of $Cr(OH)_3$ dominates in the outer layer, while Cr_2O_3 is present deeper in the layer. XPS analysis on both steels indicates the existence of various types of iron oxides [Fe_3O_4, Fe_2O_3, and $FeO(OH)$]. It appears that Fe_3O_4 and Cr_2O_3 contents are higher in the Ru-containing samples. A very slight enrichment of the Ni (in metallic form) can also be detected in the sample's surface region. An

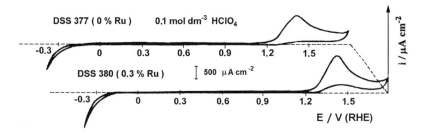

Figure 30. Cyclic voltammetric curves of duplex stainless steels in 0.1 mol dm^{-3} HClO$_4$. Scan rate, 20 mV s^{-1}.

interesting fact is that only in the case of duplex steel with 0.28% Ru, spontaneously passivated in HCl, was a significant enrichment of Mo detected. However, no Ru enrichment of the passive layers was detected after exposure to either acid solution.

These results, in accordance with data presented elsewhere,[58,59,72] suggest that during selective dissolution of Fe the majority of the surface defect sites are initially occupied by Ni, Mo, and Cr. It is possible that the surface concentration of Ru also reaches a critical value, which may be below the detection limit of the AES. Both the extent and strong character of bisulfate/sulfate accumulation are most likely related to the redistribution of these alloying components.

4. Conclusions

The very aim of this chapter was to highlight the significance and potentialities of the combined application of radiotracer and electrochemical methods in studies of sorption phenomena occurring at solid/liquid interfaces. Bearing in mind that most of the prospective readers are not very familiar with radiochemistry, an attempt was made to discuss the theoretical and technical details of the methodology and applicability of the radiotracer methods used not only for the examination of the electrosorption, electrocatalytic, etc., behaviors of different metal electrodes, but also for the investigation of corrosion, corrosion inhibition, and radioactive contamination processes on constructional materials of industrial importance. Owing to the fact that results of radiotracer research in several areas of electrochemistry have been reviewed periodically,[5,9-18] mostly the new and significant developments in this field were taken into consideration.

To demonstrate the usefulness of the upgraded versions of two *in situ* radiotracer methods applied alternatively, selected experimental situations have been described in some detail, which, in some cases, were combined with other (electrochemical, surface analytical) methods.

ACKNOWLEDGMENTS. This work was supported by the Hungarian Science Foundation (OTKA Grant No F015695), the Ministry of Education of Hungary (MKM Grant Nos. 112/96 and 0789/97), and Paks Nuclear Power Plant.

References

1. A. J. Bard and L. R. Faulkner, *Electrochemical Methods,* John Wiley and Sons, New York, 1980.
2. E. Gileadi, *Electrode Kinetics,* VCH Publishers, New York, 1993.
3. J. O'M. Bockris and S. U. M. Khan, *Surface Electrochemistry*, Plenum Press, New York, 1993.
4. J. Lipkowski and P. N. Ross (eds.), "Adsorption of Molecules at Metal Electrodes," in *Frontiers of Electrochemistry*, Vol. 1, VCH Publishers, New York, 1992.
5. G. Horányi, in *Catalysis*, J. J. Spivey (ed.), Vol. 12, The Royal Society of Chemistry, Thomas Graham House, Cambridge, 1996, p. 254.

6. E. B. Budevski, G. T. Staikov, and W. J. Lorenz, *Electrochemical Phase Formation and Growth*, VCH Publishers, New York, 1996.

7. P. Marcus and J. Oudar (eds.), *Corrosion Mechanism in Theory and Practice*, Marcel Dekker, New York, 1995.

8. a. D. H. Lister, "Activity transport and corrosion processes in PWRs," in *Water Chemistry of Nuclear Reactor Systems 6*, BNES, London, 1992 p. 49. b. A. P. Murray, *Nucl. Technol.* **74**, 324 (1986). c. G. C. W. Comley, *Prog. Nucl. Energy* **16**, 41 (1985).

9. V. E. Kazarinov and V. N. Andreev, in *Comprehensive Treatise of Electrochemistry*, E. Yeager, J. O'M. Bockris, B. Conway, and S. Sarangapani (eds.), Vol. 9, Plenum Press, New York, 1984, p. 393.

10. A. Wieckowski, in *Modern Aspects of Electrochemistry*, R. E. White, J. O'M. Bockris, and B. Conway (eds.), Vol. 21, Plenum Press, New York, 1990, p. 65.

11. E. K. Krauskopf and A. Wieckowski, in *Frontiers of Electrochemistry*, J. Lipkowski and R. P. Ross (eds.), Vol. 1, VCH Publishers, New York, 1992, p. 119.

12. P. Zelenay and A. Wieckowski, in *Modern Techniques for In Situ Surface Characterization*, H. D. Abruna (ed.), VCH Publishers, New York, 1991, p. 479.

13. G. Horányi, *Electrochim. Acta* **25**, 43 (1980).

14. G. Horányi, *B. Electrochemistry* **5**, 235 (1989).

15. G. Reinhard, *Isotopenpraxis*, **18**, 41 (1982); **18**, 157 (1982).

16. M. E. Gamboa-Aldeco, K. Franaszczuk, and A. Wieckowski, in *The Handbook of Surface Imaging and Visualization*, A. T. Hubbard (ed.), CRC Press, New York, 1995, p. 635.

17. G. Horányi, *Rev. Anal. Chem.* **14**, 1 (1995).

18. K. Varga, *Kémiai Közlemények* **83**, 77 (1996) (Hungarian).

19. a. G. L. Horváth, P. Ormai, T. Pintér, and I. C. Szabó, *Kernenergie* **30**, 38 (1987). b. Von H. Hepp, H. Generlich, and E. Jaensch, *Kraftwerkstechnik* **59**, 158 (1979). c. *Second International Seminar on Primary and Secondary Side Water Chemistry of Nuclear Power Plants*, Proceedings, Balaton-füred, Hungary, Sept 19–23, 1995. d. *Third International Seminar on Primary and Secondary Side Water Chemistry of Nuclear Power Plants*, Proceedings, Balatonfüred, Hungary, Sept 16–20, 1997.

20. F. P. Ford and P. L. Andresen, in *Corrosion Mechanism in Theory and Practice*, P. Marcus and J. Oudar (eds.), Marcel Dekker, New York, 1995, p. 501.

21. F. Joliot, *J. Chim. Phys. Phys.-Chim. Biol.* **27**, 119 (1930).

22. a. R. A. Powers and N. Hackerman, *J. Electrochem. Soc.* **100**, 314 (1953). b. N. Hackerman and S. Stephens, *J. Phys. Chem.* **58**, 904 (1954).

23. a. K. Schwabe, *Chem. Technol.* **10**, 469 (1958). b. K. Schwabe and Ch. Weissmantel, *Z. Phys. Chem.* **215**, 48 (1960). c. K. Schwabe and W. Schwenke, *Electrochim. Acta* **9**, 1003 (1964).

24. E. A. Bolmgren and J. O'M. Bockris, *Nature* **186**, 305 (1960).

25. N. A. Balashova and V. E. Kazarinov, in *Electroanalytical Chemistry*, A. Bard (ed.), Vol. 3, Marcel Dekker, New York, 1969, p. 135.

26. a. V. E. Kazarinov, *Electrohimiya* **2**, 1170 (1966). b. V. E. Kazarinov, G. J. Tysyachnaya, and V. N. Andreev, *J. Electroanal. Chem.* **65**, 391 (1975).

27. a. G. Horányi, J. Solt, and F. Nagy, *J. Electroanal. Chem.* **31**, 87 (1971). b. J. Solt, G. Horányi, and F. Nagy, *Acta Chim. Acad. Sci. Hung.* **63**, 385 (1970).

28. a. J. Sobkowski and A. Wieckowski, *J. Electroanal. Chem.* **34**, 185 (1972). b. A. Wieckowski, *J. Electrochem. Soc.* **122**, 252 (1975).

29. E. K. Krauskopf, K. Chan, and A. Wieckowski, *J. Phys. Chem.* **91**, 2327 (1987).

30. a. Ming Li and J. B. Schlenoff, *Anal. Chem.* **66**, 824 (1994). b. J. B. Schlenoff and Ming Li, *Ber. Bunsenges Phys. Chem.* **100**, 943 (1996).

31. a. P. Marcus, in *Corrosion Mechanisms in Theory and Practice*, P. Marcus and J. Oudar (eds.), Marcel Dekker, New York, 1995, p. 239. b. J. M. Herbelin, N. Barbouth, and P. Marcus, *J. Electrochem. Soc.* **137**, 3410 (1990).

32. K. Varga, E. Maleczki, E. Házi, and G. Horányi, *Electrochim. Acta* **35**, 817 (1990).
33. A. Kolics, E. Maleczki, K. Varga, and G. Horányi, *J. Radioanal. Nucl. Chem., Articles* **158**, 121 (1992).
34. M. Nagy, P. Baradlai, L. Tomcsányi, and K. Varga, *ACH—Models in Chem.* **132**, 561 (1995).
35. a. G. Hirschberg, Z. Németh, and K. Varga, *J. Electroanal. Chem.* **456**, 171 (1998). b. G. Hirschberg and K. Varga, in *Meeting Abstract Volume, Joint International Meeting of the ECS and ISE*, Paris (France), August 30–September 5, 1997, pp. 1175–1176. c. G. Hirschberg, K. Varga, P. Baradlai, Z. Németh, J. Schunk, and P. Tilky, in *Proceedings of the Third International Seminar on Primary and Secondary Side Water Chemistry of Nuclear Power Plants*, Proceedings, Balatonfüred, Hungary, Sept 16–20, 1997. d. G. Hirschberg, P. Baradlai, K. Varga, G. Myburg, J. Schunk, and P. Tilky, *J. Nucl. Mater.* **265**, 273 (1999).
36. a. G. Tóth and J. Miller, *Magy. Kém. Folyóirat* **78**, 282 (1972); **78**, 523 (1972). b. J. Miller and G. Tóth, *Isotopenpraxis* **3**, 19 (1967).
37. a. J. Kónya and A. Bába, *J. Electroanal. Chem.* **109**, 125 (1980). b. L. Várallyai, J. Kónya, F. H. Kármán, E. Kálmán, and J. Telegdi, *Electrochim. Acta* **36**, 981 (1991). c. F. H. Kármán, E. Kálmán, L. Várallyai, and J. Kónya, *Z. Naturforsch.* **46a**, 183 (1991). d. L. Várallyai, J. Kónya, E. Kálmán, and F. H. Kármán, *ACH—Models in Chem.* **132**, 551 (1995).
38. a. T. Drozda and E. Maleczki, *J. Radioanal. Nucl. Chem., Letters* **95**, 339 (1985). b. T. Drozda and E. Maleczki, *J. Radioanal. Nucl Chem., Articles* **152**, 321 (1991).
39. G. Aniansson, *J. Phys. Chem.* **55**, 1286 (1951).
40. J. A. Kafalas and H. C. Gatos, *Rev. Sci. Instrum.* **29**, 47 (1958).
41. J. G. N. Thomas and A. D. Mercer, *Mater. Chem. Phys.* **10**, 1 (1984).
42. D. Poskus and G. Agafonovas, *J. Electroanal. Chem.* **393**, 105 (1995).
43. A. Kolics and G. Horányi, *Appl. Radiat. Isot.* **47**, 551 (1996).
44. A. Kolics and K. Varga, *J. Colloid Interface Sci.* **168**, 451 (1994).
45. A. Kolics and K. Varga, *Electrochim. Acta* **40**, 1835 (1995).
46. A. Kolics, K. Varga, E. Maleczki, and G. Horányi, *J. Radioanal. Nucl. Chem., Articles*, **170**, 457 (1993).
47. a. K. Varga and M. Nagy, *Magy. Kém. Folyóirat* **103**, 297 (1997). b. K. Varga, M. Nagy, and G. Hirschberg, *J. Electroanal. Chem.* (to be published).
48. a. G. Horányi, *J. Electroanal. Chem.* **354**, 319 (1993). b. G. Horányi, *J. Electroanal. Chem.* **370**, 67 (1994). c. A. Kolics and G. Horányi, *J. Electroanal. Chem.* **374**, 101 (1994).
49. a. A. Kolics and G. Horányi, *J. Electroanal. Chem.* **376**, 167 (1994). b. A. Kolics and G. Horányi, *Electrochim. Acta* **40**, 2465 (1995).
50. a. A. Kolics and G. Horányi, *J. Electroanal. Chem.* **372**, 261 (1994). b. A. Kolics, E. Maleczki, and G. Horányi, *J. Radioanal. Nucl. Chem., Articles* **170**, 443 (1993).
51. A. Kolics and G. Horányi, *Electrochim. Acta* **41**, 791 (1996).
52. K. Varga, Ph.D. thesis, Veszprém, 1990 (Hungarian).
53. a. M. Nagy and K. Varga, *Magy. Kém. Folyóirat* **100**, 174 (1994). b. K. Varga, M. Nagy, and P. Baradlai, *J. Electroanal. Chem.* (to be published).
54. K. Varga, P. Baradlai, and A. Vértes, *Electrochim. Acta* **42**, 1143 (1997).
55. L. Tomcsányi, K. Varga, I. Bartik, G. Horányi, and E. Maleczki, *Electrochim. Acta* **34**, 855 (1989).
56. K. Varga, E. Maleczki, and G. Horányi, *Electrochim. Acta* **33**, 25 (1988); **33**, 1167 (1988); **33**, 1775 (1988); **31**, 1667 (1986).
57. K. Varga, P. Baradlai, W. O. Barnard, G. Myburg, P. Halmos, and J. H. Potgieter, *Electrochim. Acta* **42**, 25 (1997).
58. P. Baradlai, J. H. Potgieter, W. O. Barnard, L. Tomcsányi, and K. Varga, *Mater. Sci. Forum* **185–188**, 759 (1995).
59. J. H. Potgieter, W. O. Barnard, G. Myburg, K. Varga, P. Baradlai, and L. Tomcsányi, *J. Appl. Electrochem.* **26**, 1103 (1996).
60. K. Varga, P. Baradlai, D. Hanzel, W. Meisel, and A. Vértes, *Electrochim. Acta* **42**, 1157 (1997).

61. A. Kolics, K. Varga, and E. Maleczki, *J. Radioanal. Nucl. Chem. Letters* **175**, 339 (1993).
62. B. E. Conway, *Prog. Surf. Sci.* **49**, 331 (1995).
63. G. Horányi, E. M. Rizmayer, and P. Joó, *J. Electroanal. Chem.* **152**, 211 (1983).
64. P. Zelenay, L. M. Rice-Jackson, and A. Wieckowski, *J. Electroanal. Chem.* **283**, 389 (1990).
65. A. Kolics, A. E. Thomas, and A. Wieckowski, *J. Chem. Soc., Faraday Trans.* **92**, 3727 (1996).
66. T. Wilke, X. Gao, C. G. Takoudis, and M. J. Weaver, *J. Catal.* **130**, 62 (1991).
67. M. Szklarczyk, A. Czerwinski, and J. Sobkowski, *J. Electroanal. Chem.* **132**, 263 (1982).
68. P. Mrozek, Y. Sung, M. Han, M. Gamboa-Aldeco, A. Wieckowski, C. Chen, and A. A. Gewirth, *Electrochim. Acta* **40**, 17 (1995).
69. D. M. Kolb, in *Advances in Electrochemistry and Electrochemical Engineering*, H. Gerischer and C. W. Tobias (eds.), Vol. 11, John Wiley and Sons, New York, 1978, p. 127.
70. a. S. Swathirajan and S. Bruckenstein, *J. Electroanal. Chem.* **146**, 137 (1983). b. S. Swathirajan and S. Bruckenstein, *Electrochim. Acta* **28**, 865 (1983). c. S. Swathirajan and S. Bruckenstein, *J. Electrochem. Soc.* **129**, 1202 (1982). d. S. Swathirajan, H. Mizota, and S. Bruckenstein, *J. Phys. Chem.* **86**, 2480 (1982). e. T. M. Riedhammer, L. S. Melnicki, and S. Bruckenstein, *Z. Phys. Chem.* **NF111**, 177 (1978). f. W. J. Lorenz, H. D. Hermann, N. Wüthrich, and F. Hilbert, *J. Electrochem. Soc.* **121**, 1167 (1974).
71. C. H. Chen, S. M. Vesecky, and A. A. Gewirth, *J. Am. Chem. Soc.* **114**, 451 (1992).
72. G. Myburg, K. Varga, W. O. Barnard, P. Baradlai, L. Tomcsányi, J. H. Potgieter, C.W. Louw, and M.J. van Staden, *Appl. Surf. Sci.* **136**, 29 (1998).

Author Index

Subject Index